Elements of Matrix Modeling and Computing with MATLAB®

Elements of Matrix Modeling and Computing with MATLAB®

Robert E. White

Chapman & Hall/CRC
Taylor & Francis Group
Boca Raton   London   New York

Chapman & Hall/CRC is an imprint of the
Taylor & Francis Group, an informa business

Chapman & Hall/CRC
Taylor & Francis Group
6000 Broken Sound Parkway NW, Suite 300
Boca Raton, FL 33487-2742

International Standard Book Number-10: 1-58488-627-7 (Hardcover)
International Standard Book Number-13: 978-1-58488-627-3 (Hardcover)

Library of Congress Cataloging-in-Publication Data

White, R. E. (Robert E.)
   Elements of matrix modeling and computing with MATLAB / Robert E. White.
      p. cm.
   Includes bibliographical references and index.
   ISBN-13: 978-1-58488-627-3 (acid-free paper)
   ISBN-10: 1-58488-627-7 (acid-free paper)
   1. Matrices--Data processing. 2. MATLAB. I. Title.

QA188.W53 2007
512.9'4340285--dc22   2006024089

Visit the Taylor & Francis Web site at
http://www.taylorandfrancis.com

and the CRC Press Web site at
http://www.crcpress.com

Contents

List of Figures

List of Tables

Preface

An important objective of this book is to provide "math-on-time" for second year students of science and engineering. The student should have had one semester of calculus. The student most likely would take this matrix course concurrently with the second semester of calculus or would use this text for independent study of these important topics. This text fills in often missed topics in the first year of calculus including complex numbers and functions, matrices, algebraic systems, curve fitting, elements of linear differential equations, transform methods and some computation tools.

Chapters one and two have introductory material on complex numbers, 2D and 3D vectors and their products, which are often covered in the beginning of multivariable calculus. Here a connection is established between the geometric and algebraic approaches to these topics. This is continued into chapters three, four and five where higher order algebraic systems are solved via row operations, inverse matrices and LU factorizations. Linearly independent vectors and subspaces are used to solve over and under determined systems. Chapters six and seven describe first and second order linear differential equations and introduce eigenvalues and eigenvectors for the solution of linear systems of initial value problems. The last two chapters use transform methods to filter distorted images or signals. The discrete Fourier transform is introduced via the continuous versions of the Laplace and Fourier transforms. The discrete Fourier transform properties are derived from the Fourier matrix representation and are used to do image filtering in the frequency domain.

The first five chapters can be used as a two-credit course (28 50-minute classes). Among the nine chapters there is more than enough material for a three-credit course. This three-credit matrix course when coupled with a nine- or ten-credit calculus sequence can serve as a more "diverse" alternative to the traditional twelve-credit calculus sequence. The twelve-credit calculus sequence can be adapted to this alternative by reducing the precalculus, moving some of 2D and 3D vectors and differential equations into the matrix course, and using computing tools to do the complicated computations and graphing.

Most sections have some applications, which should indicate the utility of the mathematics being studied. Seven basic applications are developed in various sections of the text and include circuits, trusses, mixing tanks, heat conduction, data modeling, motion of a mass and image filters. The applications are developed from very simple models to more complex models. The reader can locate sections pretaining to a particular application by using the index.

MATLAB$^{\circledR}$ is used to do some of the more complicated computations. Although the primary focus is to develop by-hand calculation skills, most sections at the end have some MATLAB calculations. The MATLAB m-files used in the text are listed in the index and are included in the book's Web site: http://www4.ncsu.edu/~white. The approach to using computing tools includes: first, learn the math and by-hand calculations; second, use a computing tool to confirm the by-hand calculations; third, use the computing tool to do more complicated calculations and applications.

I hope this book will precipitate discussions concerning the core mathematical course work that scientists and engineers are required to study. Discrete models and computing have become more common, and this has increased the need for additional study of matrix computation, and numerical and linear algebra. The precise topics, skills, theory and appropriate times to teach these are certainly open for discussion. The matrix algebra topics in this book are a small subset of most upper level linear algebra courses, which should be enhanced and taken by a number of students. This book attempts to make a bridge from two- and three-variable problems to more realistic problems with more variables, but it emphasizes skills more than theory.

I thank my colleagues who have contributed to many discussions about the content of this text. And, many thanks go to my personal friends and Liz White who have listened to me emote during the last year.

Bob White

MATLAB is a registered trademark of The MathWorks, Inc. For product information, please contact:

The MathWorks, Inc.
3 Apple Hill Drive
Natick, MA 01760-2098 USA
Tel: 508-647-7000
Fax: 508-647-7001
E-mail: info@mathworks.com
Web: www.mathworks.com <http://www.mathworks.com/>.

Introduction

One can view an $m \times n$ matrix as a table of objects with m rows and n columns. The objects are usually real or complex numbers, but they could be characters or records of information. A simple example is data for the last 12 months of car sales where there are $m = 12$ rows and $n = 2$ columns. The first column will have the month's number and the second column will have the number of cars sold in the corresponding month. By examining the data one would like to make a prediction about futures sales. This is where the modeling enters. If the graph of the sales versus months "looks" like a straight line, then the data may be modeled by a linear function of time $y = \widehat{m}t + c$. The slope $\widehat{m}$ and intercept c must be chosen so that the computed sales are "close" to the car sales data. This is done by appropriate manipulations of the two column vectors and computing a solution of the resulting system of algebraic equations. Once $\widehat{m}$ and c have been found, the predicted sales for t larger than 12 can easily be calculated by evaluating the linear function. The modeling process is complicated by incorrect sales data, changing prices and other models such as a parabolic function of time.

This text examines a variety of applications, which have matrix models and often have algebraic systems that must be solved either by-hand calculations or using a computing tool. Applications to projectiles, circuits, mixing tanks, trusses, heat conduction, motion of a mass, curve fitting and image enhancement will be initially modeled in very simple ways and then revisited so as to make the model more accurate. This is typical of the modeling process where there is an application, a model, mathematical method, computations and assessment of the results. Then this cycle is repeated so as to enhance the application's model.

The first two chapters deal with problems in two- and three-dimensional space where the matrices have no more than three rows or columns. Here geometric insight can be used to understand the models. In Section 2.5 the extension to higher dimensions is indicated for vectors and matrices, solution to larger algebraic systems, more complicated curve fitting, time dependent problems with systems of differential equations and image modeling. Chapters three, four and five have the basic matrix methods that are required to solve systems in higher dimensions. Chapters six and seven contain time dependent models and introduce linear systems of differential equations. The last two chapters are an introduction to image and signal processing.

Most sections have some by-hand matrix calculations in the numbered examples, some applications and some MATLAB computations, see [4] and [6]. The focus is on the by-hand calculations, and one should carefully study the numbered examples. Each numbered example usually has two exercises associated with it. There are also additional exercises, which may fill in some parts of the text, be related to applications or use MATLAB. This text is not intended to be a tutorial on MATLAB, but there are a number of short codes that may help you understand the topics being discussed. The by-hand calculations should be done, and MATLAB should be used to confirm these calculations. This will give you confidence in both your understanding of the by-hand matrix computation and the use of MATLAB. Larger dimensional problems can easily be done using MATLAB or other computer software.

The following matrices are used in Chapters 3, 4, 5 and 9, and they can be generalized to larger matrices enabling one to cross the bridge from models with few variables to many variables.

$$Z = \begin{bmatrix} 0 & 0 & 0 \\ 0 & 0 & 0 \\ 0 & 0 & 0 \end{bmatrix} \qquad I = \begin{bmatrix} 1 & 0 & 0 \\ 0 & 1 & 0 \\ 0 & 0 & 1 \end{bmatrix}$$

$$E_{32}(-3) = \begin{bmatrix} 1 & 0 & 0 \\ 0 & 1 & 0 \\ 0 & -3 & 1 \end{bmatrix} \qquad U = \begin{bmatrix} 1 & 7 & -10 \\ 0 & 2 & 4 \\ 0 & 0 & 3 \end{bmatrix}$$

$$[A\ d] = \begin{bmatrix} 2 & -1 & 0 & 200 \\ -1 & 2 & -1 & 0 \\ 0 & -1 & 2 & 70 \end{bmatrix} \qquad [U\ \widehat{d}] = \begin{bmatrix} 1 & -1 & 0 & 200 \\ 0 & 3/2 & -1 & 100 \\ 0 & 0 & 4/3 & 410/3 \end{bmatrix}$$

$$A = \begin{bmatrix} 2 & -1 & 0 & 0 \\ -1 & 2 & -1 & 0 \\ 0 & -1 & 2 & -1 \\ 0 & 0 & -1 & 2 \end{bmatrix} \qquad A^{-1} = (1/10) \begin{bmatrix} 8 & 6 & 4 & 2 \\ 6 & 12 & 8 & 4 \\ 4 & 8 & 12 & 6 \\ 2 & 4 & 6 & 8 \end{bmatrix}$$

$$LS = \begin{bmatrix} 1 & 1 \\ 2 & 1 \\ 3 & 1 \\ 4 & 1 \end{bmatrix} \qquad REF = \begin{bmatrix} 1 & 2 & 3 & 4 & 5 \\ 0 & 0 & 1 & 2 & 1 \\ 0 & 0 & 0 & 0 & 0 \end{bmatrix}$$

$$\mathbb{F}_4 = \begin{bmatrix} 1 & 1 & 1 & 1 \\ 1 & z & z^2 & z^3 \\ 1 & z^2 & 1 & z^2 \\ 1 & z^3 & z^2 & z \end{bmatrix}$$

Chapter 1

Vectors in the Plane

This chapter contains geometric and algebraic descriptions of objects in two dimensional space, $\mathbb{R}^2$, and in the complex plane, $\mathbb{C}$. The objects include vectors, lines, complex valued functions and some curves. Fundamental operations include vector addition and dot product. The basic properties of complex numbers and complex valued functions are introduced. Applications to navigation, work, torque, areas and signal representation via phasors are given.

1.1 Floating Point and Complex Numbers

In this section we first discuss the integers and rational numbers. The floating point numbers, which are used in computers, are a finite subset of the rational numbers. The real and complex numbers are natural extensions of these. The complex numbers also can be represented by directed line segments or vectors in the plane. Although initially complex numbers may appear to be of questionable value, they will be used extensively in the chapters on differential equations and image processing.

1.1.1 Rational Numbers

The *integers* are the set of whole numbers and include both positive, negative and zero

$$\mathbb{Z} \equiv \{\cdots - 2, -1, 0, 1, 2, \cdots\}.$$

The addition and product of two integers are also integers. Any integer can be uniquely factored into a product of *prime numbers* (an integer that is only divisible by itself and one). For example, $90 = 5^1 3^2 2^1$.

The *rational numbers* are fractions of integers m/n where n is not zero and m and n are integers

$$\mathbb{Q} \equiv \{m/n : m, n \in \mathbb{Z}, n \neq 0\}.$$

1

The set of rational numbers has a countable but infinite number of elements. Also, the addition and product of two rational numbers are rational numbers.

1.1.2   Real Numbers

Any *real number* is approximated by a sequence of rational numbers. Traditionally, one uses a base ten decimal expansion with $x_i, e \in \mathbb{Z}$ and $0 \leq x_i < 10$

$$
\begin{aligned}
x &= \pm(.x_1 \cdots x_d \cdots)10^e \\
&\equiv \pm(x_1/10 + \cdots + x_d/10^d + \cdots)10^e.
\end{aligned}
$$

This expansion is either a bounded increasing or decreasing sequence of rational numbers and, therefore, by the completeness axiom for the real numbers it must converge.

$$
\mathbb{R} \equiv \{\pm(x_1/10 + \cdots + x_d/10^d + \cdots)10^e : x_i, e \in \mathbb{Z}, \ 0 \leq x_i < 10\}.
$$

Real numbers contain the rational numbers, but not all real numbers are rational. For example, consider $x = 3^{1/2}$ where 3 is a prime number. If x were a rational number, then $x = m/n$ giving $3 = m^2/n^2$ and $3^1 n^2 = m^2$. The left side has an odd number of prime factors 3, and the right side has an even number of prime factors 3. This contradicts the unique factorization property and, hence, x cannot be a rational number.

1.1.3   Floating Point Numbers

Computers use a finite subset of the rational numbers to approximate any real number. This set of numbers may depend on the computer being used. However, they do have the same general form and are called floating point numbers. Any real number x can be represented by an infinite decimal expansion $x = \pm(.x_1 \cdots x_d \cdots)10^e$, and by truncating this we can define the chopped floating point numbers.

Let x be any real number and denote a *floating point number* by

$$
\begin{aligned}
fl(x) &= \pm.x_1 \cdots x_d 10^e \\
&\equiv \pm(x_1/10 + \cdots + x_d/10^d)10^e.
\end{aligned}
$$

This is a floating point number with base equal to 10 where x_1 is not equal to zero, x_i are integers between 0 and 9, the exponent e is an integer between given integers $-f$ and g and d is a positive integer called the precision of the floating point system

$$
\mathbb{F} \equiv \{\pm(x_1/10 + \cdots + x_d/10^d)10^e : x_1 \neq 0, \ 0 \leq x_i < 10, \ 0 < d\}.
$$

Associated with each real number, x, and its floating point approximate number, $fl(x)$, is the *floating point error*, $fl(x) - x$. This error decreases as the *precision*, d, increases. Each computer calculation has some floating point error. Moreover, as additional floating point calculations are done, there may be an accumulation of these floating point errors.

Example 1.1.1. Let $x = -1.5378$ and $fl(x) = -0.154\ 10^1$ where $d = 3$. The floating point error is

$$fl(x) - x = -.0022.$$

The error will accumulate with any further operations containing $fl(x)$, for example, $fl(x)^2 = .237\ 10^1$ and

$$fl(x)^2 - x^2 = 2.37 - 2.36482884 = .00517116.$$

Repeated calculations using floating point numbers can accumulate to significant errors.

1.1.4 Complex Numbers

Any real number times itself must be positive unless it is zero. The square root of any negative number cannot be positive and, therefore, not a real number! The set of complex numbers is defined to be real numbers plus another real number times $i \equiv (-1)^{1/2}$

$$\mathbb{C} \equiv \{a + bi : a, b \in \mathbb{R} \text{ and } i = (-1)^{1/2}\}.$$

If $z \in \mathbb{C}$, then $\text{real}(z) = a$ is called the *real* part and $\text{imag}(z) = b$ is called the *imaginary* part. Two *complex numbers are equal* if and only if both the real and imaginary parts are equal.

Complex numbers are often represented graphically where the real part is located on the horizontal axis and the imaginary part is located on the vertical axis. The complex number is then either viewed as point in the complex plane or as *directed line* segment or *arrow* from the origin $0 + i0$ to the complex point $z = a + bi$. Figure 1.1.1 illustrates this for $z = 2 + 1i$. One point of confusion concerns the starting point of the arrow, which is the origin, but often the arrow is moved parallel to its original position, as is illustrated by the top dashed arrow in Figure 1.1.1 that is parallel to z. This is done to depict addition of two arrows by placing the beginning of the arrow at the end of another arrow.

There are several important operations: conjugate of z, modulus of z, addition, subtraction, product and division. The conjugate of z is another complex number formed by the reflection of z about the real axis. The modulus of z is the length of the vector. The algebraic operations are defined in the natural way by using $i^2 = -1$.

Definitions 1.1.1. Attributes of complex numbers $z = a + bi$ and $w = c + di$.

$$
\begin{aligned}
z &\equiv w \text{ if and only if } a = c \text{ and } b = d, \\
\bar{z} &\equiv a - bi \text{ or } conj(z) \equiv a - bi, \\
|z| &\equiv (a^2 + b^2)^{1/2} \text{ or } norm \text{ or } modulus \text{ of } z \text{ and} \\
\arg(z) &\equiv \arctan(b/a) \text{ or } argument \text{ or } angle \text{ of } z.
\end{aligned}
$$

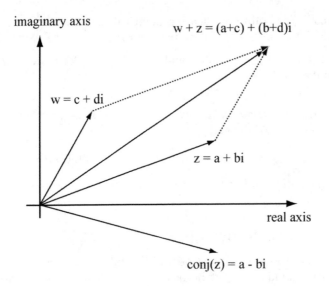

imaginary axis

$w + z = (a+c) + (b+d)i$

$w = c + di$

$z = a + bi$

real axis

$conj(z) = a - bi$

Figure 1.1.1: Complex Numbers as Arrows

Definition 1.1.2. The *polar representation* of a complex number with $\theta = \arg(z)$ is
$$z = |z| \left(\cos(\theta) + i \sin(\theta) \right).$$

Example 1.1.2. Let $z = \sqrt{3} + i$. Then
$$\overline{z} = \sqrt{3} - i, \ 3z = 3\sqrt{3} + 3i \text{ and } |z| = 4^{1/2} = 2.$$
Also, $arg(z) = \arctan(1/\sqrt{3}) = \pi/6$ and
$$z = 2(\cos(\pi/6) + i \sin(\pi/6)).$$

Solving equations with unknown complex numbers $z = x + iy$ requires one to find two real numbers x and y. So, there must be two real equations, which can be found by setting equal the real and imaginary parts of a given complex equation. For example, consider the linear complex equation

$$z + 7 + 2i = 3 + 4i.$$

This is equivalent to equating the real and imaginary parts

$$\text{real}(z + 7 + 2i) = 3 \text{ and}$$
$$\text{imag} (z + 7 + 2i) = 4.$$

Addition of complex numbers is defined by adding the real and imaginary parts, $z + 7 + 2i = (x + iy) + 7 + 2i = (x + 7) + (y + 2)i$ and, then, the above two real equations are

$$x + 7 = 3 \text{ and}$$
$$y + 2 = 4.$$

Thus, the complex solution of $z + 7 + 2i = 3 + i4$ is $z = -4 + 2i$.

Definitions 1.1.3. Operations with complex numbers $z = a+bi$ and $w = c+di$.

$$z \pm w \equiv (a \pm c) + (b \pm d)i$$
$$zw \equiv (a + bi)(c + di) = (ac - bd) + (bc + ad)i \text{ and}$$
$$z/w \equiv \frac{a + bi}{c + di} = \frac{a + bi}{c + di}\frac{c - di}{c - di} = \frac{ac + bd}{c^2 + d^2} + \frac{bc - ad}{c^2 + d^2}i.$$

Example 1.1.3. Let $z = 2 + i$ and $w = 1 + 3i$. Then
$zw = -1 + 7i$, $z/w = 5/10^{1/2} - 5/10^{1/2} i$ and
$2z - 5w = (4 + 2i) - (5 + 15i) = -1 - 13i$.

Example 1.1.4. Let $z = \cos(2\pi/3) + i \sin(2\pi/3)$. Note the modulus is one so that the angle between the z vector and the real axis is $2\pi/3$. Compute z^3
$$z^3 = (-1/2 + \sqrt{3}/2 \ i)^3$$
$$= (-1/2 + \sqrt{3}/2 \ i)^2(-1/2 + \sqrt{3}/2 \ i)$$
$$= (-1/2 - \sqrt{3}/2 \ i)(-1/2 + \sqrt{3}/2 \ i)$$
$$= 1.$$

Example 1.1.5. The modulus can be computed as the square root of the product of the complex number and the conjugate of the complex number

$$|z|^2 = z\bar{z} = (a + bi)(a - bi) = a^2 + b^2.$$

This is useful in expressing ratios of complex number in terms of real and imaginary parts

$$\frac{1+i}{3+2i} = \frac{1+i}{3+2i}\frac{3-2i}{3-2i}$$
$$= \frac{(3+2) + (3-2)i}{3^2 + 2^2}$$
$$= \frac{5}{14} + \frac{1}{14}i.$$

The zero complex number has a zero real and imaginary parts, and we simply write $0 = 0 + i0$. The unit complex number has real part equal to one, imaginary part equal to zero, and we write $1 = 1 + i0$. The multiplicative inverse of a non-zero complex number $z = a + bi$ is another complex number

$$z^{-1} = \frac{1}{a + bi}\frac{a - bi}{a - bi}$$
$$= \frac{a}{a^2 + b^2} + \frac{-b}{a^2 + b^2}i.$$

The following eleven algebraic properties are easily established. The first five properties are the additive group properties, the next five are the multiplicative group properties for non-zero complex numbers and the last is the distributive property.

Theorem 1.1.1 *(Field Properties of Complex Numbers)* Let $z, w, w_1, w_2 \in \mathbb{C}$. Then

$$z + w \in \mathbb{C}, \ z + 0 = z, \ z + (-z) = 0, \ z + w = w + z,$$
$$z + (w_1 + w_2) = (z + w_1) + w_2;$$
$$zw \in \mathbb{C}, \ z1 = z, \ z(z^{-1}) = 1 \text{ when } z \neq 0, \ zw = wz,$$
$$z(w_1 w_2) = (zw_1)w_2;$$
$$z(w_1 + w_2) = zw_1 + zw_2.$$

Example 1.1.6. This example illustrates how one can solve *complex algebraic equations* where we must solve for both the real and imaginary parts of an unknown complex number $z = x + iy$. Here we use the above field properties

$$\frac{1 + 4i}{z + 1} = 2 + i$$
$$1 + 4i = (2 + i)(z + 1)$$
$$= (2 + i)z + (2 + i).$$

Now solve for z

$$z = \frac{(1 + 4i) - (2 + i)}{(2 + i)}$$
$$= \frac{-1 + 3i}{2 + i} \frac{2 - i}{2 - i}$$
$$= \frac{1}{5} + i\frac{7}{5}.$$

So, $x = 1/5$, $y = 7/5$, which is easy to verify

$$\frac{1 + 4i}{(1/5 + (7/5)i) + 1} = \frac{5 + 20i}{6 + 7i}$$
$$= \frac{5 + 20i}{6 + 7i} \frac{6 - 7i}{6 - 7i}$$
$$= \frac{170 + 85i}{85}$$
$$= 2 + i.$$

The complex numbers are extensions of the real numbers, and many of the algebraic and absolute value properties extend to the complex numbers. For example, if a and c are real numbers and $|a|$ is the absolute value, then $|ac| = |a| |c|$. When a and c are extended to complex numbers and absolute value is extended to the modulus of a complex number, then one can show

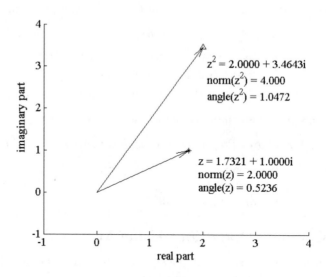

Figure 1.1.2: Norm(z^2) and Angle(z^2)

$|zw| = |z| \, |w|$. In order to see this is true, let $z = a + bi$ and $w = c + di$ and show $|zw|^2 = |z|^2 \, |w|^2$:

$$
\begin{aligned}
zw &= (a + bi)(c + di) = (ac - bd) + (bc + ad)i \\
|zw|^2 &= (ac - bd)^2 + (bc + ad)^2 \\
&= (ac)^2 - 2(ac)(bd) + (bd)^2 + \\
&\quad (bc)^2 + 2(bc)(ad) + (ad)^2 \\
&= a^2(c^2 + d^2) + b^2(c^2 + d^2) \\
&= (a^2 + b^2)(c^2 + d^2) \\
&= |z|^2 \, |w|^2.
\end{aligned}
$$

If $z = w$, then we could write this as $|z^2| = |z|^2$

Another property of the product zw is $\arg(zw) = \arg(z) + \arg(w)$, and this with $z = w$ is illustrated in Figure 1.1.2. Figure 1.1.2 was generated by the MATLAB code complex_prod.m and command quiver.m to generate the vectors. The proof follows from the trigonometric identities

$$
\begin{aligned}
\cos(\theta + \varphi) &= \cos(\theta)\cos(\varphi) - \sin(\theta)\sin(\varphi) \text{ and} \\
\sin(\theta + \varphi) &= \sin(\theta)\cos(\varphi) + \cos(\theta)\sin(\varphi).
\end{aligned}
$$

Let $z = |z|\,(\cos(\theta) + i\sin(\theta))$ and $w = |w|\,(\cos(\varphi) + i\sin(\varphi))$ so that

$$
\begin{aligned}
zw &= |z|\,(\cos(\theta) + i\sin(\theta))\,|w|\,(\cos(\varphi) + i\sin(\varphi)) \\
   &= |z|\,|w|\,(\cos(\theta)\cos(\varphi) - \sin(\theta)\sin(\varphi)) + \\
   &\quad i(\sin(\theta)\cos(\varphi) + \cos(\theta)\sin(\varphi)) \\
   &= |zw|\,(\cos(\theta + \varphi) + i\sin(\theta + \varphi)).
\end{aligned}
$$

Additional properties are summarized in the following theorem.

Theorem 1.1.2 *(Norm Properties of Complex Numbers) Let $z, w \in \mathbb{C}$. Then*

$|zw| = |z|\,|w|\,, \arg(zw) = \arg(z) + \arg(w),$
$|z + w| \le |z| + |w|$ and
$|z| = 0$ if and only if $z = 0 + i0$.

1.1.5   Complex Operations Using MATLAB

MATLAB also uses i to represent the square root of minus one. The following are some complex numbers and operations:

```
>> z = 1 + i*2
   z =
         1.0000 + 2.0000i
>> w = 2*z +1
   w =
         3.0000 + 4.0000i
>> z^2
   ans =
         -3.0000 + 4.0000i
>> w*z
   ans =
         -5.0000 +10.0000i
>> z/w
   ans =
         0.4400 + 0.0800i.
```

The MATLAB command quiver(x,y,dx,dy,s) produces a plot of a vector that begins at the point (x,y) and ends at the point (x+dx,y+dy), see Figure 1.1.2. The following graphs a vector representation of a complex number:

```
   z =
         1.0000 + 2.0000i
>> plot(z,'*')
>> hold on
>> quiver(0,0,real(z),imag(z),1)
```

This produces a graph with the complex number marked by a *, and then inserts an arrow from the origin to the complex number. The MATLAB code complex_prod.m also illustrates this.

1.1.6 Exercises

1.   Consider the real number $1/3$.
    (a).   Find $fl(1/3)$ with base 10 and three digits, $d = 3$.
    (b).   Compute $(fl(1/3))^2$ and the error $= (fl(1/3))^2 - (1/3)^2$.
2.   Let p be prime number. Show $p^{1/2}$ is not a rational number.
3.   Let $z = 7 - i$ and $w = 2 + 3i$.
    (a).   Graph z, w, $\overline{w}$ and $2w + 3z$.
    (b).   Compute zw and z/w.
4.   Let $z = -2 + 3i$ and $w = 2 - 3i$.
    (a).   Graph z, w, $\overline{w}$ and $2w + 3z$.
    (b).   Compute zw and z/w.
5.   For $z = \cos(2\pi/3) + i \sin(2\pi/3)$ compute $1 + z + z^2$.
6.   For $z = \cos(2\pi/4) + i \sin(2\pi/4)$ compute $1 + z + z^2 + z^3$.
7.   Use the properties in Theorem 1.1.1 to solve for the complex number $z = x + yi$

$$z + 7 + 2i = 3 + 4i.$$

8.   Prove the remaining properties in Theorem 1.1.1.
9.   Find the real and imaginary parts of

$$\frac{2 + i}{3 - 7i}.$$

10.   Find the real and imaginary parts of

$$\frac{2 + i}{10 + 7i} + 5.$$

11.   Solve for the complex number $z = x + yi$

$$\frac{2 + 4i}{z + i} = 7 + i.$$

12.   Solve for the complex number $z = x + yi$

$$\frac{2 + 4i}{2z - i} = -1 + i.$$

13.   Prove the remaining properties in Theorem 1.1.2.
14.   Does the quadratic formula hold when the coefficients a, b and c are complex numbers? You may wish to consider some special cases such as $a = 1, b = 0$ and $c = i$. One needs to be able to define the square root of a complex number!
15.   Use MATLAB and enter $z = -2 + 3i$ and $w = 2 - 3i$.
    (a).   Graph z and w.
    (b).   Compute zw and z/w.
    (c).   Use quiver() to plot the vectors for z and w.
16.   Consider the calculations in Figure 1.1.2 given by the MATLAB code complex_prod.m. Let $z = \sqrt{3} + 1i$.
    (a).   Use by-hand calculations to find z^2, z^3, their norms and arguments.
    (b).   Modify complex_prod.m to confirm your by-hand calculations.

1.2   Complex Valued Functions

Mappings from the complex numbers into the complex numbers can be viewed as functions from the plane into the plane. Several examples such as $az + b$ or z^2 were introduced in the previous section and here additional functions will be described. An important objective is to develop the *Euler formula* for the exponential function of an imaginary number

$$e^{iy} = \cos(y) + i\sin(y).$$

This formula is extremely useful and has applications to trigonometric identities, the solution of algebraic and differential equations, circuits and signal processing.

1.2.1   Linear and Power Functions

The *linear complex function* is

$$f(z) = az + b.$$

The constants a and b are given complex numbers, and $z = x+iy$ is any complex number where x and y are real numbers. This means

$$f(z) = u(x,y) + iv(x,y) \text{ where}$$

the functions $u(x,y)$ and $v(x,y)$ have real values. The solution of $f(z) = d = d_1 + id_2$ is equivalent to equating the real and imaginary parts and reduces to solving two real equations for x and y

$$u(x,y) = d_1 \text{ and } v(x,y) = d_2. \tag{1.2.1}$$

Example 1.2.1. Consider solving $f(z) = (1+i)z + 3 - i = 7i$. Let $z = x + iy$ so that

$$
\begin{aligned}
f(z) &= (1+i)(x+iy) + 3 - i \\
&= (x - y + 3) + i(x + y - 1).
\end{aligned}
$$

Thus, $u(x,y) = x - y + 3$ and $v(x,y) = x + y - 1$. In order to solve $f(z) = 7i = 0 + 7i$, equate the real and imaginary parts as in equation (1.2.1)

$$x - y + 3 = 0 \text{ and } x + y - 1 = 7.$$

This algebraic system can be solved by adding the two equations $2x + 2 = 7$ so that $x = 5/2$ and $y = 11/2$. Thus, the solution of $f(z) = (1+i)z + 3 - i = 7i$ is $z = 5/2 + (11/2)i$. The solution can also be found by using the properties in Theorem 1.1.1

$$z = \frac{d-b}{a} = \frac{7i - (3-i)}{1+i} = \frac{-3+8i}{1+i} = \frac{-3+8i}{1+i}\frac{1-i}{1-i} = \frac{5+11i}{2}.$$

The *linear rational complex function* is

$$f(z) = \frac{az + b}{cz + d}$$

where constants a, b, c and d are complex numbers.

Example 1.2.2. Let $a = 1, b = i, c = 1$ and $d = -i$ giving

$$f(z) = \frac{z + i}{z - i}.$$

This function can be represented by two real valued functions $u(x, y)$ and $v(x, y)$

$$
\begin{aligned}
f(z) &= \frac{z + i}{z - i} \\
&= \frac{(x + iy) + i}{(x + iy) - i} \\
&= \frac{x + i(y + 1)}{x + i(y - 1)} \frac{x - i(y - 1)}{x - i(y - 1)} \\
&= \frac{x^2 + y^2 - 1}{x^2 + (y - 1)^2} + i\frac{2x}{x^2 + (y - 1)^2} \\
&= u(x, y) + iv(x, y).
\end{aligned}
$$

The solution of $f(z) = 2 + i$ can be found by equating the real and imaginary parts $u(x, y) = 2$, $v(x, y) = 1$ and solving for x and y. Or, one can use the properties in Theorem 1.1.1 to find z as follows:

$$
\begin{aligned}
\frac{z + i}{z - i} &= 2 + i \\
z + i &= (2 + i)(z - i) = (2 + i)z - (2 + i)i \\
z(1 - (2 + i)) &= -(2 + i)i - i \\
z(-1 - i) &= -3i + 1 \\
z &= \frac{-1 + 3i}{1 + i}\frac{1 - i}{1 - i} = \frac{2 + 4i}{2} = 1 + 2i.
\end{aligned}
$$

A *quadratic complex function* is

$$f(z) = az^2 + bz + c.$$

The coefficients a, b and c are given complex numbers, and one would like to solve $f(z) = az^2 + bz + c = 0$. In order to generalize the quadratic formula to the complex case, the square root of a complex number needs to be defined. Since the square root function is the inverse of the square function, let us focus on

$$f(z) = z^2.$$

In Section 1.1 we used the polar representation $z = |z|\,(\cos(\theta) + i\sin(\theta))$ and Theorem 1.1.2 with $z = w$ to obtain

$$z^2 = |z|^2\,(\cos(2\theta) + i\sin(2\theta)).$$

This suggests that the square root of complex number should be

$$z^{1/2} \equiv |z|^{1/2}\,(\cos(\theta/2) + i\sin(\theta/2)). \tag{1.2.2}$$

A *complex square root* function $f(z) = \sqrt{z} = u + iv$ must have complex values so that its square is z

$$
\begin{aligned}
(\sqrt{z})^2 &= (u+iv)^2 \\
x + iy &= (u^2 - v^2) + i2uv.
\end{aligned}
$$

Thus, one must choose u and v such that $x = u^2 - v^2$ and $y = 2uv$. The use of trigonometric identities confirms

$$u = |z|^{1/2}\cos(\theta/2) \text{ and } v = |z|^{1/2}\sin(\theta/2)$$

are the solution of these two equations and, hence, equation (1.2.2) is correct for the definition of square root of z.

Example 1.2.3. Use equation (1.2.2) to find the square root of $z = 1 + \sqrt{3}i$. The modulus of z is 2 and the angle is $\pi/3$ or 60 degrees so that $z = 2(\cos(\pi/3) + i\sin(\pi/3))$. Equation (1.2.2) gives

$$
\begin{aligned}
(1 + \sqrt{3}i)^{1/2} &= 2^{1/2}(\cos(\pi/6) + i\sin(\pi/6)) \\
&= \sqrt{2}(\sqrt{3}/2 + i(1/2)) \\
&= \sqrt{3/2} + i(1/\sqrt{2}).
\end{aligned}
$$

Note the solution of the complex quadratic equation $z^2 = 1 + \sqrt{3}i$ is plus or minus this square root.

Figure 1.2.1 illustrates the linear, square and square root functions of a complex variable; this was generated by the MATLAB code complex_power.m. The complex numbers in the domain correspond to the domain semicircle with radius equal to three. The linear function has values in the complex plane where the domain semicircle has been doubled in radius and shifted to the right. The square function has values in the complex plane where the domain semicircle has been expanded in both the radial direction and in an angular direction to one complete circle of radius equal to three squared. The smallest curve depicts the square root function. Now, the domain semicircle has been contracted in both the radial and angular components to one half of the semicircle and with radius equal to the square root of three.

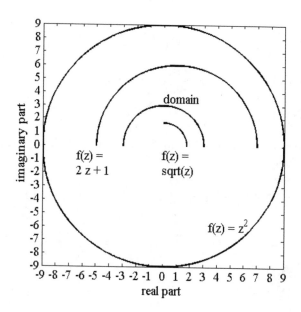

Figure 1.2.1: Affine, Square and Square Root of z

1.2.2   Exponential Function e^z

The exponential function of a complex variable $z = x + iy$ at first look is
$e^z = e^{x+iy} = e^x e^{iy}$. But, what is the meaning of e^{iy}? Find functions $u(y)$ and
$v(y)$ such that

$$e^{iy} = u(y) + iv(y)$$

has properties similar to the exponential function of a real variable. Two properties that one would like to hold are

$$e^{i0} \;=\; 1 \text{ and} \tag{1.2.3}$$

$$\frac{d}{dy}e^{iy} \;=\; ie^{iy}. \tag{1.2.4}$$

By equating the real and imaginary parts in equations (1.2.3) and (1.2.4) we
have

$$u(0) \;=\; 1 \text{ and } v(0) = 0 \text{ and} \tag{1.2.5}$$

$$\frac{du}{dy} \;=\; -v \text{ and } \frac{dv}{dy} = u. \tag{1.2.6}$$

Note the $u(y) = \cos(y)$ and $v(y) = \sin(y)$ satisfy these equations.

  In order to justify using the exponential function to represent this complex
valued function, use the power series expansions, which are studied in the second

semester of calculus,

$$e^x = 1 + x + \frac{x^2}{2!} + \frac{x^3}{3!} + \frac{x^4}{4!} + \frac{x5}{5!} + \cdots,$$

$$\cos(x) = 1 - \frac{x^2}{2!} + \frac{x^4}{4!} - \cdots \text{ and}$$

$$\sin(x) = x - \frac{x^3}{3!} + \frac{x^5}{5!} - \cdots.$$

Extend the exponential function's domain from the real numbers to the complex numbers by replacing x by iy

$$
\begin{aligned}
e^{iy} &= 1 + iy + \frac{(iy)^2}{2!} + \frac{(iy)^3}{3!} + \frac{(iy)^4}{4!} + \frac{(iy)5}{5!} + \cdots \\
&= 1 + iy - \frac{y^2}{2!} - i\frac{y^3}{3!} + \frac{y^4}{4!} + i\frac{y^5}{5!} + \cdots \\
&= [1 - \frac{y^2}{2!} + \frac{y^4}{4!} - \cdots] + i[y - \frac{y^3}{3!} + \frac{y^5}{5!} - \cdots] \\
&= \cos(y) + i\sin(y).
\end{aligned}
$$

Definition 1.2.1. *Euler's Formula* is

$$e^{iy} = \cos(y) + i\sin(y).$$

The general *complex exponential function* is

$$f(z) = e^z = e^{x+iy} = e^x e^{iy} = e^x(\cos(y) + i\sin(y)).$$

Also, any complex variable $z = x + iy$ can be represented by using $e^{i\theta} = \cos(\theta) + i\sin(\theta)$ where $\theta = \arctan(y/x)$. This gives the *exponential representation of a complex variable*

$$
\begin{aligned}
z &= x + iy \\
&= |z|(\cos(\theta) + i\sin(\theta)) \\
&= |z|e^{i\theta}.
\end{aligned}
$$

1.2.3   Application to Trigonometric Identities

Four consequences of Euler's formula are presented. First, since the cosine is an even function and the sine function is an odd function,

$$
\begin{aligned}
e^{-ix} &= \cos(-x) + i\sin(-x) \\
&= \cos(x) - i\sin(x) \\
&= \overline{e^{ix}}.
\end{aligned}
$$

Second, this leads to the representation of the cosine and sine functions

$$\cos(x) \;=\; (e^{ix} + e^{-ix})/2 \text{ and} \tag{1.2.7}$$
$$\sin(x) \;=\; (e^{ix} - e^{-ix})/(2i). \tag{1.2.8}$$

Third, consider the product of e^{ix} and e^{iy} and use the trigonometric identities for the sum of angles

$$
\begin{aligned}
e^{ix}e^{iy} \;&=\; (\cos(x) + i\sin(x))(\cos(y) + i\sin(y)) \\
&=\; [\cos(x)\cos(y) - \sin(x)\sin(y)] \\
&\quad +i[\sin(x)\cos(y) + \cos(x)\sin(y)] \\
&=\; \cos(x+y) + i\sin(x+y) \\
&=\; e^{i(x+y)}.
\end{aligned}
$$

Fourth, another identity, which follows from the repeated use of $e^{ix}e^{iy} = e^{i(x+y)}$ with $y = x$, is *de Moivre's formula*

$$
\begin{aligned}
(\cos(x) + i\sin(x))^n \;&=\; (e^{ix})^n \\
&=\; e^{ixn} \\
&=\; \cos(nx) + i\sin(nx). \tag{1.2.9}
\end{aligned}
$$

This formula for different choice of n leads to a variety of trigonometric identities. Equation (1.2.9) can be used to reduce computations with large exponents.

Example 1.2.4. Use de Moivre's formula to evaluate $(3 + 4i)^{10}$. First, write $3 + 4i$ in exponential form $3 + 4i = 5e^{i\theta}$ where $\theta = \tan^{-1}(4/3)$. Second, use equation (1.2.9) with $n = 10$ on the second factor

$$
\begin{aligned}
(3 + 4i)^{10} \;&=\; (5e^{i\theta})^{10} \\
&=\; 5^{10}(e^{i\theta})^{10} \\
&=\; 5^{10}e^{i\theta 10} \\
&=\; 5^{10}(\cos(10\theta) + i\sin(10\theta)) \\
&=\; 5^{10}(-0.9885 + 0.1512i).
\end{aligned}
$$

1.2.4 Application to Solution of $z^n = d$

Consider the case $n = 2$ where the solution of $z^2 = d$ is plus or minus the square root of d. Let $d = |d|\, e^{i\theta}$ and

$$(|d|\, e^{i\theta})^{1/2} = |d|^{1/2}\, e^{i\theta/2}.$$

The second solution is $-|d|^{1/2}\, e^{i\theta/2} = |d|^{1/2}\, e^{i\theta/2}e^{i2\pi/2}$.

There should be n solutions of $z^n = 1$. In order to discover these, represent one by $1 = e^{i2\pi}$ and, hence,

$$1^{1/n} = (e^{i2\pi})^{1/n} = e^{i2\pi/n}.$$

This leads to all n solutions $z = (e^{i2\pi/n})^j = e^{i(2\pi/n)j}$ where $j = 0, 1, ..., n-1$. They are solutions because

$$
\begin{aligned}
z^n &= (e^{i(2\pi/n)j})^n \\
&= e^{i(2\pi)j} \\
&= \cos((2\pi)j) + i\sin((2\pi)j) \\
&= 1.
\end{aligned}
$$

Definition 1.2.2. Let $j = 0, 1, \cdots, n-1$. Then $e^{i(2\pi/n)j}$ are solutions to $z^n = 1$ and are called the nth *roots of unity.*

Example 1.2.5. Let $n = 3$ and find the three solutions of $z^3 = 1$. Here $2\pi/n = 2\pi/3$ or 120 degrees. The three solutions are

$$
\begin{aligned}
e^{i(2\pi/3)0} &= \cos((2\pi/3)0) + i\sin((2\pi/3)0) = 1 \\
e^{i(2\pi/3)1} &= \cos((2\pi/3)1) + i\sin((2\pi/3)1) = -0.5000 + 0.8660i \\
e^{i(2\pi/3)2} &= \cos((2\pi/3)2) + i\sin((2\pi/3)2) = -0.5000 - 0.8660i.
\end{aligned}
$$

Another special case with $n = 12$ is illustrated in Figure 1.2.2 where the 12 vectors with radius equal to one are the solutions. The other 12 vectors have increasing moduli and have staggered angles

$$(1.05e^{(i2\pi(1+.05)/n)})^j.$$

The figure was generated by the MATLAB code euler_form.m.

The solution of the more general problem $z^n = d$ is straightforward once the complex number d is written in polar form

$$
\begin{aligned}
d &= |d|\,e^{i\theta} = |d|\,e^{i\theta}e^{i2\pi} \\
z &= d^{1/n} = |d|^{1/n}(e^{i\theta}e^{i2\pi})^{1/n} \\
&= |d|^{1/n}e^{i\theta/n}e^{i2\pi/n}. \qquad (1.2.10)
\end{aligned}
$$

Define $w \equiv e^{i2\pi/n}$ and use the fact $(w^j)^n = 1$ to obtain all n solutions $z = d^{1/n}w^j$ where $j = 0, 1, ..., n-1$. A summary of the above is given in the following theorem.

Theorem 1.2.1 (*Solutions of $z^n = d$*). *Let $w \equiv e^{i2\pi/n}$ and $d = |d|\,e^{i\theta}$. Then w satisfies $\overline{w} = w^{-1}, w^n = 1$ and $1 + w + \cdots + w^{n-1} = 0$. Moreover, the solutions of $z^n = d$ are given by $z = |d|^{1/n}e^{i\theta/n}w^j = d^{1/n}w^j$ where $j = 0, 1, \cdots, n-1$.*

Example 1.2.6. Let $n = 3$ and find the three solutions of $z^3 = 1 + i$. Here $2\pi/n = 2\pi/3$ and $d = 1 + i = 2^{1/2}e^{i\pi/4}$. The three solutions are

$$
\begin{aligned}
2^{1/6}e^{i\pi/12}e^{i(2\pi/3)0} &= 2^{1/6}(\cos(\pi/12 + (2\pi/3)0) + i\sin(\pi/12 + (2\pi/3)0)) \\
&= 1.0842 + 0.2905i \\
2^{1/6}e^{i\pi/12}e^{i(2\pi/3)1} &= 2^{1/6}(\cos(\pi/12 + (2\pi/3)1) + i\sin(\pi/12 + (2\pi/3)1)) \\
&= -0.7937 + 0.7937i \\
2^{1/6}e^{i\pi/12}e^{i(2\pi/3)2} &= 2^{1/6}(\cos(\pi/12 + (2\pi/3)2) + i\sin(\pi/12 + (2\pi/3)2)) \\
&= -0.2905 - 1.0842i.
\end{aligned}
$$

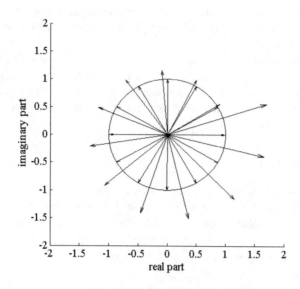

Figure 1.2.2: Solutions of $z^{12} = 1$

1.2.5 Complex Functions Using MATLAB

Complex functions are evaluated using MATLAB as are real functions. The computations in Example 1.2.6 are easy to do. The second solution is

```
>> a = exp(i*2*pi/3)
   a =
       -0.5000 + 0.8660i
>> b = 2^(1/6)*exp(i*pi/12)
   b =
       1.0842 + 0.2905i
>> a*b
   ans =
       -0.7937 + 0.7937i
>> z = a*b
   z =
       -0.7937 + 0.7937i
>> z^3
   ans =
       1.0000 + 1.0000i.
```

The MATLAB code complex_power.m, see Figure 1.2.1, uses an implied loop to generate a row vector of values for the parameter t. Then a row vector of complex values is computed to form the semicircle, which is to be the domain of three functions. Array operations given by $z.^p$ are used to compute the square and square root functions. An array operation on a row vector produces another row vector whose components are given by the operation. For example, if $a = [2\ 3]$, then $a.\hat{}2 = [2^2\ 3^2]$.

The MATLAB code euler_form.m, see Figure 1.2.2, uses a for-loop to generate a sequence of vectors associated with the complex exponential function raised to a variety of powers. The vectors are created by the MATLAB command quiver(x,y,dx,dy,1), which graphs a vector beginning at the point (x, y) and ending at $(x + dx, y + dy)$.

1.2.6   Exercises

1.   Let $f(z) = (2 + i)z + 7 + i$ where $z = x + iy$
   (a).   Find $u(x, y)$ and $v(x, y)$ such that $f(z) = u + iv$.
   (b).   Find the solution of $f(z) = 10 - 2i$.
2.   Let $f(z) = (-3 + i)z + 1 + 2i$ where $z = x + iy$
   (a).   Find $u(x, y)$ and $v(x, y)$ such that $f(z) = u + iv$.
   (b).   Find the solution of $f(z) = 1 - 2i$.
3.   Let $f(z) = (z + 2i)/(z - i)$ where $z = x + iy$
   (a).   Find $u(x, y)$ and $v(x, y)$ such that $f(z) = u + iv$.
   (b).   Find the solution of $f(z) = 1 - 2i$.
4.   Let $f(z) = (2z + i)/(z + i)$ where $z = x + iy$
   (a).   Find $u(x, y)$ and $v(x, y)$ such that $f(z) = u + iv$.
   (b).   Find the solution of $f(z) = 1 + 2i$.
5.   Use equation (1.2.2) to find the square root of $1 + i$.
6.   Use equation (1.2.2) to find the square root of $-1 + \sqrt{3}i$.
7.   Use the complex quadratic formula to solve $az^2 + bz + c = 0$.
   (a).   Solve the quadratic equation $z^2 + 2z - i = 0$.
   (b).   Check your answer.
8.   Consider the completing square method of developing the quadratic formula.
   (a).   Prove: if z and w are complex numbers and $zw = 0$, then either $z = 0$ or $w = 0$.
   (b).   Use this fact to derive the quadratic formula for complex numbers.
9.   Use de Moivre's formula in equation (1.2.9) to compute $(1 + i)^{10}$.
10.   Use de Moivre's formula in equation (1.2.9) to compute $(3 + 2i)^7$.
11.   Use Theorem 1.2.1 to find all three solutions to $z^3 = 3 + 4i$.
12.   Use Theorem 1.2.1 to find all four solutions to $z^4 = -1 + \sqrt{3}i$.
13.   Let $w \equiv e^{i2\pi/n}$. Show w satisfies $\overline{w} = w^{-1}, w^n = 1$ and $1 + w + \cdots + w^{n-1} = 0$.
14.   Use MATLAB to verify the third equation in Example 1.2.6.
15.   Use MATLAB to experiment with different size semicircles in the code complex_power.m.
16.   Use MATLAB to experiment with powers $n = 3, 6$ and 12 in the code euler_form.m.

1.3 Vectors in $\mathbb{R}^2$

Complex numbers are represented either by ordered pairs of real numbers or as directed line segments. The first number in the pair is the real part and the second number is the imaginary part. The directed line segment or arrow starts at the origin and ends at the complex point. There are a number of important similar examples such as displacement in the plane of a mass, force acting on a mass as well as the velocity and acceleration of a mass. As in the complex numbers, these quantities can be added, subtracted and scaled, and have length and angles.

1.3.1 Vector Definition and Notations

There are a number of notations of vectors. One is the $\overrightarrow{i}$, $\overrightarrow{j}$ unit vectors often used in physics textbooks. Other notations list the components in either a row or a column. There is also some confusion about the difference between a point in a plane and a vector in a plane; often the notation is the same! Part of the problem is the need to have vectors with more than two or three components, which will be considered after the first two chapters.

Definition 1.3.1. A *vector in* $\mathbb{R}^2$ is an ordered list of two real numbers $\overrightarrow{a} = [a_1 \ a_2]$. One can visualize this by forming the directed line segment from the origin point $(0,0)$ to the point (a_1, a_2).

Notation. Points in the plane $\mathbb{R}^2$ will be denoted by (a_1, a_2), and vectors will be considered as either row or column vectors:

$$\overrightarrow{a} = [a_1 \ a_2] \text{ denotes a } row \ vector,$$

$$\mathbf{a} = \begin{bmatrix} a_1 \\ a_2 \end{bmatrix} \text{ denotes a } column \ vector.$$

$\mathbf{a}^T = [a_1 \ a_2]$ is called the *transpose* of the column vector $\mathbf{a}$ so that $\mathbf{a}^T = \overrightarrow{a}$.

Example 1.3.1. The standard *unit basis vectors* are

$$\overrightarrow{i} = [1 \ 0] \text{ or } \mathbf{e}_1 = \begin{bmatrix} 1 \\ 0 \end{bmatrix} \text{ and}$$

$$\overrightarrow{j} = [0 \ 1] \text{ or } \mathbf{e}_2 = \begin{bmatrix} 0 \\ 1 \end{bmatrix}.$$

Any vector can be represented as a linear combination of unit vectors. For example, if $\overrightarrow{a} = [2 \ 3]$, then either by row vectors

$$\overrightarrow{a} = 2\overrightarrow{i} + 3\overrightarrow{j} \text{ or by column vectors}$$

$$\mathbf{a} = 2\mathbf{e}_1 + 3\mathbf{e}_2.$$

These vectors are illustrated in Figure 1.3.1.

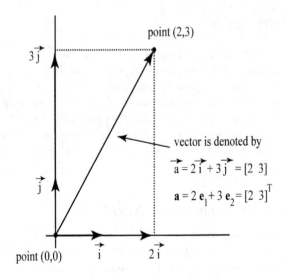

Figure 1.3.1: A Vector in the Plane

Movement of a mass in a plane involves displacement, velocity and force vectors. If one moves a mass from the origin to a point (d_1, d_2), then there is displacement in both the horizontal and the vertical directions. This requires two numbers, and we agree to list the horizontal term first so that

$$\vec{d} = [d_1 \quad d_2] = \text{displacement vector.}$$

In order to move an object, a force must be exerted on the mass, and this will also have horizontal and vertical components. By adjusting the size of the components, the direction and magnitude of the force can be determined to obtain the desired motion. Let f_1 be the force in the horizontal direction, f_2 be the vertical force and denote the force vector as

$$\vec{f} = [f_1 \quad f_2] = \text{force vector.}$$

The resulting motion will have speeds in the horizontal and vertical directions. The velocity vector lists these speeds so that the velocity vector also has a direction and magnitude. The speed in the horizontal direction is v_1 and when this is coupled with the vertical speed v_2 we get a velocity vector

$$\vec{v} = [v_1 \quad v_2] = \text{velocity vector.}$$

Newton's law of motion can be applied in both the horizontal and vertical directions

$$f_1 = m\frac{dv_1}{dt} \text{ and } f_2 = m\frac{dv_2}{dt}.$$

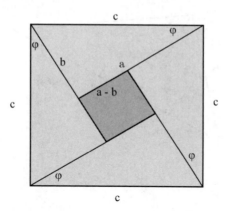

Figure 1.3.2: $c^2 = a^2 + b^2$

The vector notation for Newton's law of motion is

$$
\begin{aligned}
\overrightarrow{f} &= [f_1 \ \ f_2] \\
&= [m\frac{dv_1}{dt} \ \ m\frac{dv_2}{dt}] \\
&= m\overrightarrow{a} \ \text{where} \\
\overrightarrow{a} &= [\frac{dv_1}{dt} \ \ \frac{dv_2}{dt}] = \text{acceleration vector.}
\end{aligned}
$$

As in complex vectors there are operations with vectors such as addition, length and angles. Two very useful theorems are related to the length of a side of a triangle, which is opposite one of the angles. When the angle is 90 degrees, the Pythagorean theorem gives a very useful relationship between the three sides. There are number of proofs for the Pythagorean theorem, but the following is a very nifty proof that is based on geometry. Consider Figure 1.3.2 where the right triangle has sides of length a, b and c with c being opposite the 90 degree angle. The figure is drawn with a larger than b. The area of the big square is c squared, and it is the union of four identical triangles and a small square

$$
\begin{aligned}
c^2 &= 4(ab/2) + (a-b)^2 \\
&= 2ab + a^2 - 2ab + b^2 \\
&= a^2 + b^2.
\end{aligned}
$$

Theorem 1.3.1 *(Pythagorean) Consider a right triangle with sides a, b and c. If c is opposite the right angle, then*

$$c^2 = a^2 + b^2.$$

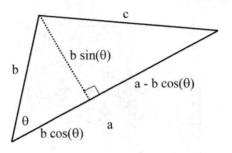

Figure 1.3.3: $c^2 = b^2 + a^2 - 2ab\cos(\theta)$

An important consequence of the above, where $\sin(\theta) = b/c$ and $\cos(\theta) = a/c$, is the trigonometric identity

$$1 = \sin^2(\theta) + \cos^2(\theta).$$

Another result pertains to the case when the angle is not 90 degrees as is illustrated in the triangle in Figure 1.3.3 where the triangle with sides a, b and c is a union of two right triangles. The left right triangle has base equal to $b\cos(\theta)$ and height equal to $b\sin(\theta)$. The other right triangle has the same height, but the base is $a - b\cos(\theta)$ (assuming $a > b$). Therefore, applying the Pythagorean theorem to the right triangle on the right side gives

$$
\begin{aligned}
c^2 &= (b\sin(\theta))^2 + (a - b\cos(\theta))^2 \\
&= b^2\sin^2(\theta) + a^2 - 2ab\cos(\theta) + b^2\cos^2(\theta) \\
&= b^2(\sin^2(\theta) + \cos^2(\theta)) + a^2 - 2ab\cos(\theta) \\
&= b^2 + a^2 - 2ab\cos(\theta).
\end{aligned}
$$

Theorem 1.3.2 *(Cosine Law) Consider a triangle with sides a, b and c. If c is opposite the angle θ, then*

$$c^2 = b^2 + a^2 - 2ab\cos(\theta).$$

1.3.2   Vector Operations

Vectors can be scaled, added and subtracted by either viewing the vectors geometrically or as symbolic objects. Figure 1.3.4 illustrates vector scaling by either elongating or shrinking the vector $\vec{b}$. Two vectors $\vec{a}$ and $\vec{b}$ can be added by moving vector $\vec{b}$ parallel to its original position so that its beginning point is the end point of vector $\vec{a}$. The addition corresponds to the diagonal of the parallelogram formed by vectors $\vec{a}$ and $\vec{b}$. If $-\vec{a}$ is added to $\vec{b}$, then the subtraction is the other diagonal from the end of $\vec{a}$ to the end of $\vec{b}$.

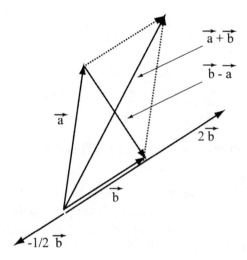

Figure 1.3.4: $\vec{a} + \vec{b}, \vec{a} - \vec{b}$ and $s\vec{b}$

Definitions 1.3.2. Let $\vec{a} = [a_1\ a_2]$, $\vec{b} = [b_1\ b_2]$ and $\vec{0} = [0\ 0]$ be the zero vector. $\vec{a}$ *equals* $\vec{b}$ means both $a_1 = b_1$ and $a_2 = b_2$. The *addition of two vectors* is another vector whose components are the sum of the components of the two vectors

$$\vec{a} + \vec{b} \equiv [\ a_1 + b_1 \quad a_2 + b_2\ ].$$

Let s be a real number. The *scalar product of a vector* is another vector whose components are the scalar times the components of the vector

$$s\vec{a} \equiv [\ sa_1 \quad sa_2\ ].$$

These operations are similar to those for complex numbers. However, it is important to note that vector products, which could be similar to complex products, are not defined. There are special vector products called "dot", "cross", "box" and "array" products, which will be introduced in the subsequent sections. The following theorem lists some basic algebraic properties of the above operations.

Theorem 1.3.3 *(Vector Space Properties) Let* $\vec{a}, \vec{b}$ *and* $\vec{c}$ *be vectors and let* s *and* t *be real numbers. Then the following rules hold*

$$\begin{aligned}
\vec{a} + \vec{b} &= \vec{b} + \vec{a},\ (\vec{a} + \vec{b}) + \vec{c} = \vec{a} + (\vec{b} + \vec{c}), \\
\vec{a} + \vec{0} &= \vec{a}\ and\ \vec{a} + (-\vec{a}) = \vec{0}; \\
(s+t)\vec{a} &= s\vec{a} + t\vec{a},\ s(t\vec{a}) = (st)\vec{a},\ 1\vec{a} = \vec{a}\ and \\
s(\vec{a} + \vec{b}) &= s\vec{a} + s\vec{b}.
\end{aligned}$$

The proofs are straightforward. The proof of $\vec{a} + \vec{b} = \vec{b} + \vec{a}$ uses the commutative property for addition of real numbers

$$
\begin{aligned}
\vec{a} + \vec{b} &= [\ a_1 + b_1 \quad a_2 + b_2\ ] \\
&= [\ b_1 + a_1 \quad b_2 + a_2\ ] \\
&= \vec{b} + \vec{a}.
\end{aligned}
$$

Example 1.3.2. Let $\vec{a} = [2\ \ 4]$, $\vec{b} = [-1\ \ 2]$ and $s = 3$. Then $\vec{a} + \vec{b} = [1\ \ 6]$, $\vec{a} - \vec{b} = [3\ \ 2]$ and $s\vec{b} = [-3\ \ 6]$.

Example 1.3.3. One can use the above basic properties to solve vector equations for an unknown vector $\vec{x} = [x_1\ \ x_2]$. Suppose vectors $\vec{a}$ and $\vec{b}$ are given and $\vec{x}$ must satisfy the vector equation $\vec{a} + \vec{x} = \vec{b}$. Solve this by adding the vector $-\vec{a}$ to both sides and use the algebraic rules

$$
\begin{aligned}
(\vec{a} + \vec{x}) - \vec{a} &= \vec{b} - \vec{a} \\
(\vec{x} + \vec{a}) - \vec{a} &= \vec{b} - \vec{a} \\
\vec{x} + (\vec{a} - \vec{a}) &= \vec{b} - \vec{a} \\
\vec{x} &= \vec{b} - \vec{a}.
\end{aligned}
$$

Definitions 1.3.3. Let $\vec{a} = [a_1\ a_2]$. The length or *Euclidean norm of the vector* $\vec{a}$ is given by the Pythagorean theorem

$$
\|\vec{a}\| \equiv \sqrt{a_1^2 + a_2^2}.
$$

The *angle of the vector* is

$$
\theta = \arctan(a_2/a_1).
$$

The *unit vector in the direction of* $\vec{a}$ is

$$
\frac{\vec{a}}{\|\vec{a}\|} = \frac{1}{\|\vec{a}\|}\vec{a}.
$$

The cosine law may be rewritten in terms of the norms of the vectors $\vec{a}, \vec{b}, \vec{c} = \vec{a} - \vec{b}$ and the angle θ between the vectors $\vec{a}$ and $\vec{b}$

$$
\left\|\vec{a} - \vec{b}\right\|^2 = \|\vec{a}\|^2 + \left\|\vec{b}\right\|^2 - 2\|\vec{a}\|\left\|\vec{b}\right\|\cos(\theta).
$$

This allows one to compute the angle between two given vectors. In fact, this can be simplified to easily compute $\cos(\theta)$, $\sin(\theta)$ and the area of the parallelogram formed by two vectors. In Section 1.4 this will be applied to the computation of work and torque.

Example 1.3.4. If $\overrightarrow{a} = [2\ \ 4]$ and $\overrightarrow{b} = [-1\ \ 2]$, then $\overrightarrow{a} - \overrightarrow{b} = [3\ \ 2]$, $\left\|\overrightarrow{a} - \overrightarrow{b}\right\|^2 = 13$, $\|\overrightarrow{a}\|^2 = 20$, $\left\|\overrightarrow{b}\right\|^2 = 5$ so that

$$
\begin{aligned}
13 &= 20 + 5 - 2\sqrt{20}\sqrt{5}\cos(\theta) \\
&= 20 + 5 - 20\cos(\theta) \\
\cos(\theta) &= 12/20 = 3/5 \\
\theta &= \cos^{-1}(3/5) \approx 53.13 \text{ degrees.}
\end{aligned}
$$

The unit vector associated with $\overrightarrow{a}$ is

$$
\frac{\overrightarrow{a}}{\|\overrightarrow{a}\|} = \frac{[2\ \ 4]}{2\sqrt{5}} = [1/\sqrt{5}\ \ 2/\sqrt{5}].
$$

1.3.3 Application to Navigation of an Aircraft

A pilot wishes to fly northeast with a speed of 400 miles per hour. The means the desired velocity vector has length equal to 400 and the speed in the east and north directions are equal. Let this desired velocity be the vector $\overrightarrow{v} = [v_1\ \ v_2]$ where $\|\overrightarrow{v}\| = 400$ and $v_1 = v_2 > 0$

$$
v_1^2 + v_1^2 = 400^2.
$$

Thus, $v_1 = v_2 = 400/\sqrt{2}$. An added difficulty is the wind velocity, which is coming from the northwest and has a speed equal to 100 miles per hour. Let $\overrightarrow{w} = [w_1\ \ w_2]$ where $\|\overrightarrow{w}\| = 100$ and $w_1 = -w_2 > 0$

$$
w_1^2 + w_1^2 = 100^2.
$$

Thus, $w_1 = -w_2 = 100/\sqrt{2}$. What should the heading velocity $\overrightarrow{h} = [h_1\ \ h_2]$ be so that the desired velocity is attained? The aircraft must be pointed north of northeast and have a speed larger than 400. The desired velocity $\overrightarrow{v}$ is the resultant of the wind velocity $\overrightarrow{w}$ and the heading velocity $\overrightarrow{h}$, which means

$$
\overrightarrow{v} = \overrightarrow{w} + \overrightarrow{h}.
$$

By the basic rules for vector operations

$$
\begin{aligned}
\overrightarrow{h} &= \overrightarrow{v} - \overrightarrow{w} \\
[h_1\ h_2] &= [400/\sqrt{2}\ \ 400/\sqrt{2}] - [100/\sqrt{2}\ \ -100/\sqrt{2}] \\
&= [300/\sqrt{2}\ \ 500/\sqrt{2}].
\end{aligned}
$$

The speed of the aircraft should be

$$
\left\|\overrightarrow{h}\right\| = \sqrt{(300/\sqrt{2})^2 + (500/\sqrt{2})^2} = 100\sqrt{18} \approx 424.26,
$$

and the angle from the east direction should be

$$
\arctan((500/\sqrt{2})/(300/\sqrt{2})) = \arctan(5/3) \approx 59.04 \text{ degrees.}
$$

1.3.4   Vectors Using MATLAB

The vector notation in MATLAB is very similar to what has been used in this text and uses square brackets to enclose both row and column vectors. The overhead arrow and bold notations are not used to distinguish row and column vectors. A row vector such as $a = [2\ 5]$ is converted to a column vector by using a prime, $[2\ 5]'$.

```
>> a = [2  5]
   a =
        2 5
>> e1 =[1  0]'
   e1 =
        1
        0
>> e2 = [0  1]'
   e2 =
        0
        1
>> aa = a(1)*e1 + a(2)*e2
   aa =
        2
        5
>> b = [-1  5]
   b =
        -1 5
>> c = 2*a + 3*b
   c =
        1 25
```

The MATLAB command quiver(x,y,dx,dy,s) can be used to graph vectors starting at the point (x, y) and ending at the point $(x + dx, y + dy)$ and using $s = 1$. The following generates a graph of the two vectors $\overrightarrow{a} = [2\ 5]$ and $\overrightarrow{b} = [-1\ 3]$.

```
>> quiver(0,0,2,5,1)
>> hold on
>> quiver(0,0,-1,3,1)
```

1.3.5   Exercises

1.   Consider the vector $\overrightarrow{a} = [2\ 5]$. Write it in $\overrightarrow{i}\ \overrightarrow{j}$ and column forms.
2.   Consider the vector $\overrightarrow{a} = [-1\ 3]$. Write it in $\overrightarrow{i}\ \overrightarrow{j}$ and column forms.
3.   Let $\overrightarrow{a} = [2\ 5]$ and $\overrightarrow{b} = [-1\ 3]$. Graph $\overrightarrow{a},\ \overrightarrow{b},\ -\overrightarrow{b},\ \overrightarrow{a} + \overrightarrow{b}$ and $\overrightarrow{a} - \overrightarrow{b}$.
4.   Let $\overrightarrow{a} = [2\ \ -3]$ and $\overrightarrow{b} = [-1\ \ 5]$. Graph $2\overrightarrow{a},\ \overrightarrow{b},\ -\overrightarrow{b},\ \overrightarrow{a} + \overrightarrow{b}$ and $\overrightarrow{a} - 2\overrightarrow{b}$.

5.   Let $\vec{a} = [2\ 5]$ and $\vec{b} = [-1\ 3]$. Solve $\vec{a} + \vec{x} = \vec{b}$.

6.   Let $\vec{a} = [1\ 7]$ and $\vec{b} = [-1\ 3]$. Solve $2\vec{a} = \vec{b} + \vec{x}$.

7.   Consider Theorem 1.3.3. Prove $(\vec{a} + \vec{b}) + \vec{c} = \vec{a} + (\vec{b} + \vec{c})$.

8.   Consider Theorem 1.3.3. Prove $s(\vec{a} + \vec{b}) = s\vec{a} + s\vec{b}$.

9.   Let $\vec{a} = [1\ 7]$. Compute $\|\vec{a}\|$, the angle and unit vector of $\vec{a}$.

10.   Let $\vec{a} = [1\ -3]$. Compute $\|\vec{a}\|$, the angle and unit vector of $\vec{a}$.

11.   Let $\vec{a} = [2\ 5]$ and $\vec{b} = [-1\ 3]$. Graph these vectors and use the cosine law to find the angle between them.

12.   Let $\vec{a} = [1\ 2]$ and $\vec{b} = [-1\ -3]$. Graph these vectors and use the cosine law to find the angle between them.

13.   Consider the aircraft heading problem. If the desired velocity is to the northeast with speed 500 and the wind velocity has speed 120 and from the north, find the heading velocity.

14.   Consider the aircraft heading problem. If the desired velocity is to the northwest with speed 500 and the wind velocity has speed 120 and from the east, find the heading velocity.

15.   Use MATLAB to compute $2\vec{a}$, $\vec{b}$, $-\vec{b}$, $\vec{a} + \vec{b}$ and $\vec{a} - 2\vec{b}$ where $\vec{a} = [2\ -3]$ and $\vec{b} = [-1\ 5]$.

16.   Use MATLAB to graph $\vec{a}$ and $\vec{b}$ where $\vec{a} = [2\ -3]$ and $\vec{b} = [-1\ 5]$.

1.4   Dot Product and Work

In this section we continue the study of the parallelogram that is formed by two vectors $\vec{a}$ and $\vec{b}$ in the plane. The vector version of the cosine law allows one to compute $\cos(\theta)$ where θ is the angle between the vectors $\vec{a}$ and $\vec{b}$. This calculation is important because it leads to a number of applications such as an easy derivation of trigonometric identities, area of the parallelogram and calculation of work and torque in 2D and 3D space.

1.4.1   Cosine Identity and Dot Product

Let $\vec{a} = [a_1\ a_2]$ and $\vec{b} = [b_1\ b_2]$ have the angle θ between the vectors. The vector version of the cosine law is

$$\left\|\vec{a} - \vec{b}\right\|^2 = \|\vec{a}\|^2 + \left\|\vec{b}\right\|^2 - 2\|\vec{a}\|\left\|\vec{b}\right\|\cos(\theta). \tag{1.4.1}$$

By definition of vector subtraction and the norm

$$\begin{aligned}
\left\|\vec{a} - \vec{b}\right\|^2 &= (a_1 - b_1)^2 + (a_2 - b_2)^2 \\
&= a_1^2 - 2a_1 b_1 + b_1^2 + a_2^2 - 2a_2 b_2 + b_2^2 \\
&= \|\vec{a}\|^2 + \left\|\vec{b}\right\|^2 - 2(a_1 b_1 + a_2 b_2).
\end{aligned}$$

Place this into the left side of equation (1.4.1) and cancel the norms

$$\|\vec{a}\|^2 + \left\|\vec{b}\right\|^2 - 2(a_1b_1 + a_2b_2) = \|\vec{a}\|^2 + \left\|\vec{b}\right\|^2 - 2\|\vec{a}\|\left\|\vec{b}\right\|\cos(\theta)$$

$$-2(a_1b_1 + a_2b_2) = -2\|\vec{a}\|\left\|\vec{b}\right\|\cos(\theta)$$

$$a_1b_1 + a_2b_2 = \|\vec{a}\|\left\|\vec{b}\right\|\cos(\theta). \tag{1.4.2}$$

Equation (1.4.2) is of fundamental importance. The left side is an easy calculation, and the right side makes a connection with the geometry. At the end of this section an application will be given to work being done in a constant force field, which is independent of three possible paths.

Theorem 1.4.1 *(Cosine Identity) Let* $\vec{a} = [a_1 \ a_2]$ *and* $\vec{b} = [b_1 \ b_2]$ *be non-zero vectors and let* θ *be the angle between the two vectors. Then the* $\cos(\theta)$ *can be computed by*

$$\cos(\theta) = \frac{a_1b_1 + a_2b_2}{\|\vec{a}\|\left\|\vec{b}\right\|}.$$

Definition 1.4.1. The *dot product of two vectors* $\vec{a} = [a_1 \ a_2]$ and $\vec{b} = [b_1 \ b_2]$ is given by the real number on either side of equation (1.4.2). The following are common notations for the dot product of row or column vectors, respectively,

$$\vec{a} \bullet \vec{b} \equiv a_1b_1 + a_2b_2 = \|\vec{a}\|\left\|\vec{b}\right\|\cos(\theta) \text{ or}$$

$$\mathbf{a}^T\mathbf{b} \equiv a_1b_1 + a_2b_2 = \|\mathbf{a}\|\|\mathbf{b}\|\cos(\theta).$$

The cosine identity can be restated as either

$$\cos(\theta) = \frac{\vec{a} \bullet \vec{b}}{\|\vec{a}\|\left\|\vec{b}\right\|} \text{ or}$$

$$\cos(\theta) = \frac{\mathbf{a}^T\mathbf{b}}{\|\mathbf{a}\|\|\mathbf{b}\|}.$$

Also, the norm of a vector can be written as either

$$\|\vec{a}\|^2 = \vec{a} \bullet \vec{a} \text{ or}$$

$$\|\mathbf{a}\|^2 = \mathbf{a}^T\mathbf{a}.$$

Since $-1 \leq \cos(\theta) \leq 1$, the inequality, which is called the *Cauchy-Schwarz* inequality, is

$$\left|\vec{a} \bullet \vec{b}\right| \leq \|\vec{a}\|\left\|\vec{b}\right\| \text{ or}$$

$$\left|\mathbf{a}^T\mathbf{b}\right| \leq \|\mathbf{a}\|\|\mathbf{b}\|.$$

Definitions 1.4.2. Let $\vec{a}$ and $\vec{b}$ be non-zero vectors. Two *vectors are parallel* if and only if they are multiples of each other, that is, there is a real number s such that $\vec{a} = s\,\vec{b}$. Two *vectors are perpendicular (also called orthogonal)* if and only if the $\cos(\theta)$ is zero, that is, $\vec{a} \bullet \vec{b} = \mathbf{a}^T\mathbf{b} = 0$.

Example 1.4.1. Let $\vec{a} = [2\ \ 3]$ and $\vec{b} = [-1\ \ 2]$. Then $\left\| \vec{a} \right\|^2 = 2^2 + 3^2 = 13$, $\left\| \vec{b} \right\|^2 = (-1)^2 + 2^2 = 5$, and $\vec{a} \bullet \vec{b} = 2(-1) + 3(2) = 4$.

$$\cos(\theta) = \frac{\vec{a} \bullet \vec{b}}{\left\| \vec{a} \right\| \left\| \vec{b} \right\|} = \frac{4}{\sqrt{13}\sqrt{5}} = \frac{4}{\sqrt{65}} \text{ and}$$

$$\theta = \cos^{-1}\left(\frac{4}{\sqrt{65}}\right) \approx 60.26 \text{ degrees.}$$

Example 1.4.2. Consider the following three vectors written as column vectors

$$\mathbf{a} = \begin{bmatrix} 2 \\ 3 \end{bmatrix}, \mathbf{b} = \begin{bmatrix} 22 \\ 33 \end{bmatrix} \text{ and } \mathbf{c} = \begin{bmatrix} -6 \\ 4 \end{bmatrix}.$$

Vectors $\mathbf{a}$ and $\mathbf{b}$ are parallel because $\mathbf{a} = (1/11)\mathbf{b}$. Vector $\mathbf{a}$ is perpendicular to $\mathbf{c}$ because the dot product $\mathbf{a}^T\mathbf{c} = 2(-6) + 3(4) = 0$. Also, $\mathbf{b}$ and $\mathbf{c}$ must be perpendicular. Consider a vector $\mathbf{d}$ whose first component is 2, but the second component is unknown. Choose the second component so that the vector $\mathbf{d}$ is perpendicular to vector $\mathbf{a}$. Let the vector $\mathbf{d}$ be represented by its transpose $\mathbf{d}^T = [2\ \ x]$. The vector $\mathbf{d}$ being perpendicular to $\mathbf{a}$ requires their dot product to be zero

$$\begin{aligned} \mathbf{a}^T\mathbf{d} &= [\,2\ \ 3\,]\begin{bmatrix} 2 \\ x \end{bmatrix} \\ &= 2(2) + 3x = 0. \end{aligned}$$

Thus, $x = -4/3$ and $\mathbf{d}^T = [2\ \ -4/3]$.

Definition 1.4.3. The *projection of* $\vec{b}$ *onto* $\vec{a}$ is a vector with direction the same as $\vec{a}$ and length equal to $\left\| \vec{b} \right\| \cos(\theta)$

$$\frac{\vec{a}}{\left\| \vec{a} \right\|} \left\| \vec{b} \right\| \cos(\theta) = \frac{\vec{a}}{\left\| \vec{a} \right\|^2} \vec{a} \bullet \vec{b}.$$

The dot product has a number of interesting algebraic properties. One is the order of the vectors does not change the value of the dot product.

$$\begin{aligned} \vec{a} \bullet \vec{b} &= a_1 b_1 + a_2 b_2 \\ &= b_1 a_1 + b_2 a_2 \\ &= \vec{b} \bullet \vec{a}. \end{aligned}$$

Additional dot product properties listed in the following theorem are also easy to prove.

Theorem 1.4.2 *(Properties of Dot Product) Let $\vec{a}, \vec{b}, \vec{c}$ be given vectors and let s be a real number. The dot product properties are*

$$\vec{a} \bullet \vec{b} = \vec{b} \bullet \vec{a}, \; s(\vec{a} \bullet \vec{b}) = (s\vec{a}) \bullet \vec{b},$$
$$\vec{a} \bullet (\vec{b} + \vec{c}) = \vec{a} \bullet \vec{b} + \vec{a} \bullet \vec{c} \; and$$
$$\left| \vec{a} \bullet \vec{b} \right| \leq \| \vec{a} \| \| \vec{b} \| \; (Cauchy\text{-}Schwarz\ inequality).$$

Since the norm of a vector can be written in terms of the dot product

$$\| \vec{a} \|^2 = \vec{a} \bullet \vec{a},$$

several important properties of a norm can be derived from the above dot product properties. The following norm properties are similar to the absolute value of a real number.

Theorem 1.4.3 *(Properties of Norm) Let $\vec{a}$ and $\vec{b}$ be given vectors and let s be a real number. The norm properties are*

$$\| \vec{a} \| \geq 0; \; \| \vec{a} \| \geq 0 \; if\ and\ only\ if\ \vec{a} = \vec{0},$$
$$\| s\vec{a} \| = |s| \| \vec{a} \| \; and$$
$$\left\| \vec{a} + \vec{b} \right\| \leq \| \vec{a} \| + \left\| \vec{b} \right\| \; (triangle\ inequality).$$

The triangle inequality follows from the dot product properties

$$\left\| \vec{a} + \vec{b} \right\|^2 = (\vec{a} + \vec{b}) \bullet (\vec{a} + \vec{b})$$
$$= \vec{a} \bullet (\vec{a} + \vec{b}) + \vec{b} \bullet (\vec{a} + \vec{b})$$
$$= \vec{a} \bullet \vec{a} + \vec{a} \bullet \vec{b} + \vec{b} \bullet \vec{a} + \vec{b} \bullet \vec{b}$$
$$= \| \vec{a} \|^2 + 2\vec{a} \bullet \vec{b} + \left\| \vec{b} \right\|^2$$
$$\leq \| \vec{a} \|^2 + 2 \left| \vec{a} \bullet \vec{b} \right| + \left\| \vec{b} \right\|^2.$$

Next apply the *Cauchy-Schwarz* inequality to get

$$\left\| \vec{a} + \vec{b} \right\|^2 \leq \| \vec{a} \|^2 + 2 \| \vec{a} \| \left\| \vec{b} \right\| + \left\| \vec{b} \right\|^2$$
$$= (\| \vec{a} \| + \left\| \vec{b} \right\|)^2.$$

1.4.2   Applications to Angles and Areas

The dot product version of the cosine identity can be used to derive the formula for the cosine of the sum of two angles

$$\cos(\theta + \phi) = \cos(\theta)\cos(\phi) - \sin(\theta)\sin(\phi).$$

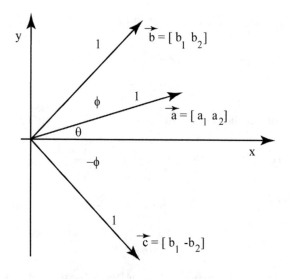

Figure 1.4.1: Trigonometric Identity and Dot Product

Choose two vectors as illustrated in Figure 1.4.1 where both vectors have unit length. Then the first and second components are the cosine and sine of their angles. Let the vectors $\vec{a}$ and $\vec{b}$ have angles equal to θ and ϕ, respectively, and let the vector $\vec{c}$ be the reflection vector $\vec{a}$ about the horizontal axis so that

$$\begin{aligned}
\vec{a} &= [\cos(\theta) \quad \sin(\theta)] \\
\vec{b} &= [\cos(\phi) \quad \sin(\phi)] \text{ and} \\
\vec{c} &= [\cos(\phi) \quad -\sin(\phi)].
\end{aligned}$$

The angle between vectors $\vec{a}$ and $\vec{c}$ is $\theta - (-\phi) = \theta + \phi$ and, therefore, by application of the cosine identity to the vectors $\vec{a}$ and $\vec{c}$

$$\begin{aligned}
\cos(\theta + \phi) &= \frac{\vec{a} \bullet \vec{c}}{\|\vec{a}\| \|\vec{c}\|} \\
&= \frac{\cos(\theta)\cos(\phi) + \sin(\theta)(-\sin(\phi))}{1(1)}.
\end{aligned}$$

Another application of the dot product is to the computation of the area in a parallelogram formed by two vectors, which is depicted in Figure 1.4.2. Here the base of the parallelogram is the length of the vector $\vec{a}$, $\|\vec{a}\|$, and the height of the parallelogram is $\left\| \vec{b} \right\| \sin(\theta)$ and, thus,

$$
\begin{aligned}
(area)^2 &= (\|\vec{a}\| \|\vec{b}\| \sin(\theta))^2 \\
&= (\|\vec{a}\| \|\vec{b}\|)^2 (1 - \cos^2(\theta)).
\end{aligned}
$$

Use the cosine identity for $\cos(\theta)$ to obtain

$$
\begin{aligned}
(area)^2 &= (\|\vec{a}\| \|\vec{b}\|)^2 (1 - (\frac{\vec{a} \bullet \vec{b}}{\|\vec{a}\| \|\vec{b}\|})^2) \\
&= \|\vec{a}\|^2 \|\vec{b}\|^2 - (\vec{a} \bullet \vec{b})^2 \\
&= (a_1^2 + a_2^2)(b_1^2 + b_2^2) - (a_1 b_1 + a_2 b_2)^2 \\
&= (a_1 b_2 - b_1 a_2)^2. \tag{1.4.3}
\end{aligned}
$$

This formula may also be expressed using the *determinant* of a 2×2 matrix formed by the column vectors that define the parallelogram

$$
\det([\mathbf{a}\ \ \mathbf{b}]) = \det(\begin{bmatrix} a_1 & b_1 \\ a_2 & b_2 \end{bmatrix}) \equiv a_1 b_2 - b_1 a_2.
$$

Then equation (1.4.3) is

$$
(area)^2 = (\det([\mathbf{a}\ \ \mathbf{b}]))^2.
$$

Theorem 1.4.4 *(Area Equivalence) Let two non-zero vectors be represented as column vectors* $\mathbf{a}$ *and* $\mathbf{b}$*. The area of the parallelogram formed by these vectors being non-zero is equivalent to the vectors not being parallel. In symbolic terms, this means* $\det([\mathbf{a}\ \ \mathbf{b}]) \neq 0$ *is equivalent to* $\mathbf{a} \neq c\mathbf{b}$ *for all non-zero real numbers* c*. Moreover, area is equal to the absolute value of* $\det([\mathbf{a}\ \ \mathbf{b}])$*.*

Another point of view generalizes to higher dimensional vectors. The matrix product may be written as a linear combination of the column vectors $\mathbf{a}$ and $\mathbf{b}$

$$
\begin{bmatrix} a_1 & b_1 \\ a_2 & b_2 \end{bmatrix} \begin{bmatrix} r \\ s \end{bmatrix} = \begin{bmatrix} 0 \\ 0 \end{bmatrix}
$$

$$
\begin{bmatrix} a_1 r + b_1 s \\ a_2 r + b_2 s \end{bmatrix} = \begin{bmatrix} 0 \\ 0 \end{bmatrix}
$$

$$
r \begin{bmatrix} a_1 \\ a_2 \end{bmatrix} + s \begin{bmatrix} b_1 \\ b_2 \end{bmatrix} = \begin{bmatrix} 0 \\ 0 \end{bmatrix}
$$

$$
r\mathbf{a} + s\mathbf{b} = \mathbf{0}.
$$

Thus, the above are also equivalent to $r\mathbf{a} + s\mathbf{b} = \mathbf{0}$ implies both r and s must be zero. Such vectors are called *linearly independent,* which will be studied in

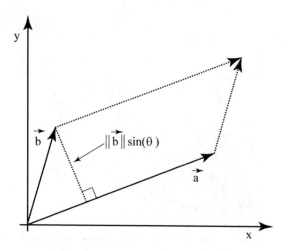

Figure 1.4.2: Area and Dot Product

more detail in Section 5.1. In Figure 1.4.3 the area of the larger parallelogram, formed by the scaled vectors $r\mathbf{a}$ and $s\mathbf{b}$, is rs times the area of the smaller parallelogram. This follows from determinants

$$
\begin{aligned}
\det([r\mathbf{a} \quad s\mathbf{b}]) \quad &= \quad \det\left(\begin{bmatrix} a_1 r & b_1 s \\ a_2 r & b_2 s \end{bmatrix}\right) \\
&= \quad a_1 r(b_2 s) - b_1 s(a_2 r) \\
&= \quad rs \det([\mathbf{a} \quad \mathbf{b}]).
\end{aligned}
$$

1.4.3 Applications to Work and Torque

This application is concerned with moving a mass from one point in a plane to an other point in the plane. One can do this by using a ramp or using a wheel. In the case of ramp the displacement vector is along the top of the ramp and the force vector may not be parallel to the ramp, see Figure 1.4.4. For the wheel one must be able to apply a non-zero force tangent to the perimeter of the wheel, see Figure 1.4.5.

The movement of an object up a ramp requires a force with both non-zero horizontal and vertical components. In order to reduce the computation of work to the scalar formula, work equals distance times the force, we must find the component of the force vector in the direction of the displacement vector. In Figure 1.4.4 this is $\left\| \overrightarrow{f} \right\| \cos(\theta)$ so that the work is $\left\| \overrightarrow{d} \right\| \left\| \overrightarrow{f} \right\| \cos(\theta)$. If the given vectors are $\overrightarrow{d}$ and $\overrightarrow{f}$, then the work is computed by the cosine identity and

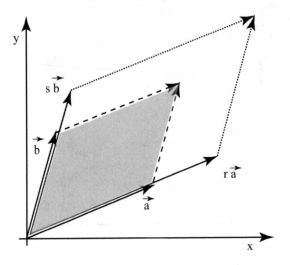

Figure 1.4.3: Linearly Independent Vectors

the dot product

$$work = \overrightarrow{d} \bullet \overrightarrow{f}.$$

For example, if the bottom edge of the ramp is 10 and the vertical part is 5, then $\overrightarrow{d} = [10 \ 5]$. Suppose the force vector is 100 in the horizontal direction and 10 in the vertical direction so that $\overrightarrow{f} = [100 \ 10]$. Then the work is $\overrightarrow{d} \bullet \overrightarrow{f} = 10(100) + 5(10) = 1050$.

For the rotation of a wheel the force vector may not be tangent to the wheel and so one must compute the component of the force acting in this direction. This computation, in contrast to the above work problem, uses the sine of the angle and not the cosine. Figure 1.4.5 reveals why this is the case. The scalar value for torque is the radius times the tangent component of the force $\|\overrightarrow{r}\| \|\overrightarrow{f}\| \sin(\theta)$. This value is related to plus or minus the area of the parallelogram formed by the radius vector and the force vector.

1.4.4   Application to Work and Constant Force Vector

Consider an object that is to be moved from one point in a plane to another point. Three possible paths are the direct path (path one in Figure 1.4.6), the horizontal and then vertical (path two), and vertical and then horizontal (path three). If the force vector is a constant vector, then we claim the work done will be independent of the three paths. Let the direct path be the displacement vector $\overrightarrow{a} = \overrightarrow{d} = [d_1 \ d_2]$ and the constant force vector be $\overrightarrow{b} = \overrightarrow{f} = [f_1 \ f_2]$. Apply the cosine identity in equation (1.4.2)

$$d_1 f_1 + d_2 f_2 = \|\overrightarrow{d}\| \|\overrightarrow{f}\| \cos(\theta).$$

Work is $\| \vec{f} \| \, \| \vec{d} \| \cos(\theta)$

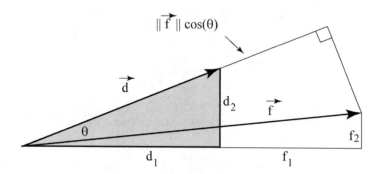

Figure 1.4.4: Work and a Ramp

Magnitude of torque is $\| \vec{r} \| \, \| \vec{f} \| \sin(\theta)$

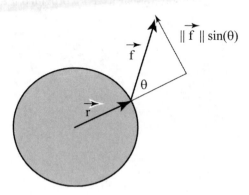

Figure 1.4.5: Torque on a Wheel

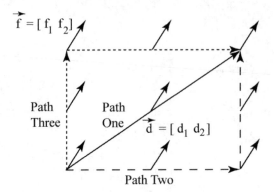

Figure 1.4.6: Work with Independent Paths

The work along the direct path is $\left\|\vec{d}\right\| \left\|\vec{f}\right\| \cos(\theta)$. The work along the second path is $d_1 f_1 + d_2 f_2$, and the work along the third path is $d_2 f_2 + d_1 f_1$, which is the same as path two because the addition of real numbers commutes. Thus, the work done is independent of the three paths taken for constant displacement and force vectors.

1.4.5   Dot Product Using MATLAB

The dot product of two vectors can be computed in two ways. The MATLAB command dot(a,b) will do this if the vectors are both row or column vectors. If the vectors are both column vectors, then one can use the transpose version $\mathbf{a}^T \mathbf{b}$, and in MATLAB this would appear as $a' * b$ provided a and b were column vectors.

```
>> a
    a =
        2  1
>> b
    b =
        1 -1
>> dot(a,b)
    ans =
        1
>> a'*b
    ans =
        2 -2
        1 -1
```

In the above a and b were initially row vectors so that $a' * b$ will be a column vector times a row vector, which results in a 2×2 matrix. By transposing them to column vectors one can compute the dot product as a row vector times a column vector.

```
>> cola = a'
   cola =
      2
      1
>> colb = b'
   colb =
      1
      -1
>> colb'*cola
   ans =
      1
```

1.4.6 Exercises

1.   Verify the cosine identity in equation (1.4.2) for

$$\overrightarrow{a} = [\sqrt{3}\ \ 1] \text{ and } \overrightarrow{b} = [1\ \ \sqrt{3}].$$

2.   Verify the cosine identity in equation (1.4.2) for the vectors in column form

$$\mathbf{a} = \begin{bmatrix} \sqrt{3} \\ 1 \end{bmatrix} \text{ and } \mathbf{b} = \begin{bmatrix} 1 \\ 1 \end{bmatrix}.$$

3.   Use the cosine identity in equation (1.4.2) to find the angle between the vectors in column form

$$\mathbf{a} = \begin{bmatrix} 1 \\ 2 \end{bmatrix} \text{ and } \mathbf{b} = \begin{bmatrix} -3 \\ 1 \end{bmatrix}.$$

4.   Use the cosine identity in equation (1.4.2) to find the angle between the vectors $\overrightarrow{a} = [2\ \ 1]$ and $\overrightarrow{b} = [1\ \ -1]$.

5.   Find x such that $\overrightarrow{a} = [3\ \ 1]$ and $\overrightarrow{b} = [x\ \ -1]$ are perpendicular.

6.   Find x such that the following column vectors are perpendicular

$$\mathbf{a} = \begin{bmatrix} 1 \\ 2 \end{bmatrix} \text{ and } \mathbf{b} = \begin{bmatrix} 3 \\ x \end{bmatrix}.$$

7.   Consider Theorem 1.4.2.

   (a).   Prove $s(\overrightarrow{a} \bullet \overrightarrow{b}) = (s\overrightarrow{a}) \bullet \overrightarrow{b} = \overrightarrow{a} \bullet (s\overrightarrow{b})$.

   (b).   Verify $\overrightarrow{a} \bullet (\overrightarrow{b} + \overrightarrow{c}) = \overrightarrow{a} \bullet \overrightarrow{b} + \overrightarrow{a} \bullet \overrightarrow{c}$ for the vectors $\overrightarrow{a} = [2\ \ 1]$, $\overrightarrow{b} = [3\ \ 4]$ and $\overrightarrow{c} = [1\ \ -1]$.

   (c).   What is the difference between "prove" and "verify"?

8.   Consider Theorem 1.4.3. Prove the first three norm properties.

9.   Use the formula in equation (1.4.3) to compute the area of the parallelogram given by the column vectors

$$\mathbf{a} = \begin{bmatrix} 1 \\ 2 \end{bmatrix} \text{ and } \mathbf{b} = \begin{bmatrix} -3 \\ 1 \end{bmatrix}.$$

10.   Use the formula in equation (1.4.3) to compute the area of the parallelogram given by the vectors $\overrightarrow{a} = [2 \ 1]$ and $\overrightarrow{b} = [1 \ -1]$.

11.   Let $\overrightarrow{a} = [2 \ 1]$ and $\overrightarrow{b} = [1 \ -1]$. Verify the area of the parallelogram
given by the vectors $6\overrightarrow{a}$ and $3\overrightarrow{b}$ is 18 times the area of the parallelogram given
by the vectors $\overrightarrow{a}$ and $\overrightarrow{b}$.

12.   Consider the work in moving an object up a ramp whose horizontal edge
is 25 and vertical edge is 5. If the force vector is $\overrightarrow{f} = [10 \ 1]$, find the work.

13.   Consider the work in moving an object up a ramp that requires a force
vector with force of 3 in the horizontal direction and 1 in the vertical direction.
If the displacement vector is $\overrightarrow{d} = [30 \ 6]$, find the work.

14.   Use MATLAB to compute the dot product of $\overrightarrow{a} = [2 \ 1]$ and $\overrightarrow{b} = [3 \ 4]$.

15.   Use MATLAB to compute the area of the parallelogram between $\overrightarrow{a} = [1 \ 2]$ and $\overrightarrow{b} = [-3 \ 1]$.

1.5   Lines and Curves in $\mathbb{R}^2$ and $\mathbb{C}$

In this section lines in the plane will be represented by three variations on
algebraic equations and three variations on vector equations. We will use these
descriptions to find the point on a line that is closest to a given point not on
the line. One method uses calculus to solve the problem, and the other method
uses geometry to generate the same solution. Both these approaches can be
used to solve more general applied problems. Finally, curves in the plane will
be described by parametric equations. An important class of curves related to
signal processing can be generated by Euler's formula and complex variables.

1.5.1   Lines in the Plane

The three most common algebraic representations of a line are the *slope-intercept*,
the *single equation* and the *parametric equations*, respectively:

$$y = mx + c \text{ or } \frac{y - y_0}{x - x_0} = m, \tag{1.5.1}$$

$$ax + by + \hat{c} = 0 \text{ or } a(x - x_0) + b(y - y_0) = 0, \tag{1.5.2}$$

$$x = d_1t + x_0, \ y = d_2t + y_0 \text{ or } \frac{x - x_0}{d_1} = \frac{y - y_0}{d_2} = t. \tag{1.5.3}$$

Example 1.5.1. Suppose a line contains the point $(1, 2)$ and satisfies the
equation $3x + 4y - 11 = 0$. Since $x = 1$ and $y = 2$ satisfies this equation, the

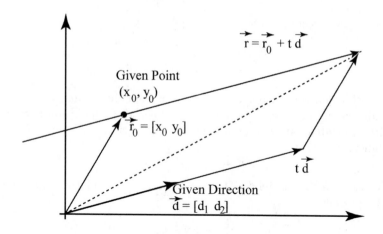

Figure 1.5.1: Line Given a Point and Direction

point is in fact on the line. Solve the given equation for y to get

$$y = \frac{-3}{4}x + \frac{11}{4} \quad \text{or} \quad \frac{y-2}{x-1} = \frac{-3}{4} = m.$$

In order to find the parametric equation (1.5.3), use the second form for the slope

$$\frac{y-2}{x-1} = \frac{-3}{4} \quad \text{or} \quad \frac{y-2}{-3} = \frac{x-1}{4} = t.$$

This means $y = -3t + 2$ and $x = 4t + 1$. Note there are other possible choices for the parameter such as $t = 2\hat{t}$.

Three vector representations are *one point and a direction vector* $\overrightarrow{d}$, *two distinct points*, and *one point and a normal vector* $\overrightarrow{n}$. The first representation is given by the vector version of the parametric equations

$$\begin{aligned} [x \; y] &= t[d_1 \; d_2] + [x_0 \; y_0] \\ \overrightarrow{r} &= t\overrightarrow{d} + \overrightarrow{r}_0. \end{aligned} \quad (1.5.4)$$

This is illustrated in Figure 1.5.1 where the direction vector and the point are given; the given point is (x_0, y_0) and the vector $\overrightarrow{r}_0$ starts at the origin and ends this given point. The vector $t\overrightarrow{d}$ in the figure depicts an extension of the given direction vector. Then any point on the line, represented by the vector $\overrightarrow{r}$ beginning at the origin and ending at the point on the line, is a vector sum of $t\overrightarrow{d}$ and $\overrightarrow{r}_0$.

If two points are given on the line by (x_0, y_0) and (x_1, y_1), then consider the two vectors starting at the origin and ending at these points, $\overrightarrow{r}_0 = [x_0 \; y_0]$ and $\overrightarrow{r}_1 = [x_1 \; y_1]$. Then $\overrightarrow{r}_1 - \overrightarrow{r}_0$ is a vector that is parallel to the line and can be used as the direction vector

$$\begin{aligned}
\vec{r} &= t\vec{d} + \vec{r}_0 \\
&= t(\vec{r}_1 - \vec{r}_0) + \vec{r}_0.
\end{aligned} \qquad (1.5.5)$$

If a point and a normal vector are given, then the normal vector must be perpendicular to any vector parallel to the line. If $\vec{r}$ and $\vec{r}_0$ are two vectors that are associated with two points on the line, then $\vec{r} - \vec{r}_0$ and $\vec{n}$ must be perpendicular so that

$$\vec{n} \bullet (\vec{r} - \vec{r}_0) = 0. \qquad (1.5.6)$$

This vector equation is related to the second version of the single equation (1.5.2) representation

$$\begin{aligned}
a(x - x_0) + b(y - y_0) &= 0 \\
[a \ b] \bullet [x - x_0 \ \ y - y_0] &= 0 \\
\vec{n} \bullet (\vec{r} - \vec{r}_0) &= 0.
\end{aligned}$$

Example 1.5.2. Find the vector equation (1.5.4) for the line with point $(2, 1)$ and parallel to the vector $[1 \ 4]$.

$$\begin{aligned}
\vec{r} &= t\vec{d} + \vec{r}_0 \\
[x \ y] &= t[1 \ 4] + [2 \ 1] \text{ or} \\
x &= 1t + 2 \text{ and } y = 4t + 1.
\end{aligned}$$

One choice for a normal vector is $\vec{n} = [4 \ -1]$ because $\vec{n} \bullet [1 \ 4] = 0$.

$$\begin{aligned}
\vec{n} \bullet (\vec{r} - \vec{r}_0) &= 0 \\
[4 \ -1] \bullet [x - 2 \ \ y - 1] &= 0 \\
4(x - 2) + (-1)(y - 1) &= 0 \\
4x - y - 7 &= 0.
\end{aligned}$$

Although the equations look different, it is easy to verify that the parametric equations do satisfy this last single equation representation.

1.5.2   Minimum Distance from a Point to a Line

Let a line be represented by a direction vector $\vec{d}$ and point (x_0, y_0). Let a point (p_1, p_2) not be on this line. Let $\vec{r}_0 = [x_0 \ y_0]$ and $\vec{p} = [p_1 \ p_2]$ be the vectors associated with these two points. Any point on the line will then be a function of the parameter t

$$\vec{r}(t) = t\vec{d} + \vec{r}_0.$$

The distance squared from the point not on the line and any point on the line is a function of t

$$f(t) \equiv (\vec{r}(t) - \vec{p}) \bullet (\vec{r}(t) - \vec{p}). \qquad (1.5.7)$$

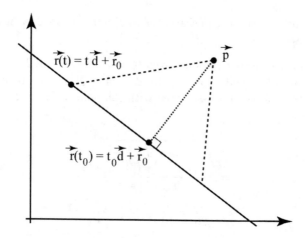

Figure 1.5.2: Minimum Distance of Point to a Line

The objective is to find the value of $t = t_0$ so that

$$f(t_0) = \min_t f(t).$$

This is illustrated in Figure 1.5.2.

The first method for solving this minimization problem uses calculus. We will find $t = t_0$ so that

$$\frac{df}{dt}(t_0) = 0 \text{ and } \frac{d^2 f}{dt^2}(t_0) > 0.$$

In order to compute the derivatives, use the basic properties of the dot product

$$
\begin{aligned}
f(t) &= (\overrightarrow{r}(t) - \overrightarrow{p}) \bullet (\overrightarrow{r}(t) - \overrightarrow{p}) \\
&= (t\overrightarrow{d} + \overrightarrow{r}_0 - \overrightarrow{p}) \bullet (t\overrightarrow{d} + \overrightarrow{r}_0 - \overrightarrow{p}) \\
&= (t\overrightarrow{d} + \overrightarrow{r}_0 - \overrightarrow{p}) \bullet t\overrightarrow{d} + (t\overrightarrow{d} + \overrightarrow{r}_0 - \overrightarrow{p}) \bullet (\overrightarrow{r}_0 - \overrightarrow{p}) \\
&= (t\overrightarrow{d}) \bullet (t\overrightarrow{d}) + 2(t\overrightarrow{d}) \bullet (\overrightarrow{r}_0 - \overrightarrow{p}) + (\overrightarrow{r}_0 - \overrightarrow{p}) \bullet (\overrightarrow{r}_0 - \overrightarrow{p}) \\
&= t^2 \overrightarrow{d} \bullet \overrightarrow{d} + 2t\overrightarrow{d} \bullet (\overrightarrow{r}_0 - \overrightarrow{p}) + (\overrightarrow{r}_0 - \overrightarrow{p}) \bullet (\overrightarrow{r}_0 - \overrightarrow{p}).
\end{aligned}
$$

This is a quadratic function of t whose derivatives are easily computed

$$\frac{df}{dt} = 2t\overrightarrow{d} \bullet \overrightarrow{d} + 2\overrightarrow{d} \bullet (\overrightarrow{r}_0 - \overrightarrow{p}) \text{ and } \frac{d^2 f}{dt^2} = 2\overrightarrow{d} \bullet \overrightarrow{d} > 0.$$

Thus, the t_0 where the first derivative is zero is

$$t_0 = \frac{-\overrightarrow{d} \bullet (\overrightarrow{r}_0 - \overrightarrow{p})}{\overrightarrow{d} \bullet \overrightarrow{d}}.$$

Since the second derivative is positive, the minimum distance is the square root of $f(t)$ evaluated at t_0.

Example 1.5.3. Consider the line given by $x + 2y = 4$. The line contains the points $(0, 2)$ and $(4, 0)$ and is parallel to $\vec{d} = [4 \ \ -2]$. If the point not on this line is $(2, 5)$, then the vector associated with the point is $\vec{p} = [2 \ \ 5]$.

$$
\begin{aligned}
t_0 &= \frac{-\vec{d} \bullet (\vec{r}_0 - \vec{p})}{\vec{d} \bullet \vec{d}} \\
&= \frac{-[4 \ \ -2] \bullet ([0 \ \ 2] - [2 \ \ 5])}{[4 \ \ -2] \bullet [4 \ \ -2]} \\
&= \frac{2}{20} = \frac{1}{10}.
\end{aligned}
$$

The desired point on the line is $\vec{r}(\frac{1}{10}) = \frac{1}{10}\vec{d} + \vec{r}_0 = \frac{1}{10}[4 \ \ -2] + [0 \ \ 2] = [0 \ \ 2] = [\frac{4}{10} \ \ \frac{18}{10}]$.

For the geometric approach to this problem observe in Figure 1.5.2 that vector from $\vec{p}$ to $\vec{r}(t_0)$ must be perpendicular to the given line. This means the direction vector of the line $\vec{d}$ must be perpendicular to $\vec{r}(t_0) - \vec{p}$. Now use the dot product to obtain

$$
\begin{aligned}
\vec{d} \bullet (\vec{r}(t_0) - \vec{p}) &= 0 \\
\vec{d} \bullet (t_0 \vec{d} + \vec{r}_0 - \vec{p}) &= 0 \\
t_0 \vec{d} \bullet \vec{d} + \vec{d} \bullet (\vec{r}_0 - \vec{p}) &= 0
\end{aligned}
$$

and, hence, we get the same t_0.

Theorem 1.5.1 (*Minimum Distance to a Line*) *Let a given point, given by position vector $\vec{p}$, not be on a line $\vec{r}(t) = t\vec{d} + \vec{r}_0$. The distance squared $f(t)$ is defined by equation (1.5.7).*

$$
f(t_0) = \min_t f(t) \ \text{has solution } t_0 = \frac{-\vec{d} \bullet (\vec{r}_0 - \vec{p})}{\vec{d} \bullet \vec{d}}.
$$

Variations on this important result will be considered in Section 2.3 for lines in space, and in Section 4.2 for data fitting using "least squares." If $\vec{r}_0$ is the zero vector and the other vectors are written as column vectors, then the minimization problem is solved by the solution of

$$
\mathbf{d}^T\mathbf{d}\, t_0 = \mathbf{d}^T\mathbf{p}.
$$

Here **d** is a column vector with two rows, and t_0 is called a *least squares* solution to $\mathbf{d}\,t = \mathbf{p}$, which has two equations and one unknown.

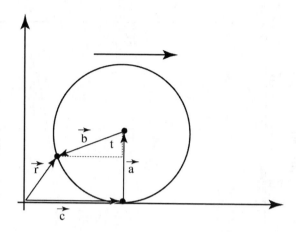

Figure 1.5.3: Cycloid and Wheel

1.5.3 Curves in the Plane and Phasors

The parametric representation of a line has components that are linear polynomials of the parameter, $at + b$. If one or both of the component functions are not linear polynomials, then the set of points will not be a line. For example, if $x = t$ and $y = at^2$, then we have a parabola. In this case the position vectors for points on the parabola are $\overrightarrow{r} = [t \ \ at^2]$. Another example is a circle with radius a where one representation is $x = a\cos(t)$ and $y = a\sin(t)$. The position vector is $\overrightarrow{r} = [a\cos(t) \ \ a\sin(t)]$. Complex variables and Euler's formula can also describe this circle by $z = ae^{it} = a\cos(t) + ia\sin(t)$.

A more complicated curve is the cycloid, which is the curve given by a point on a wheel of radius a as the wheel rolls in a straight line. As is illustrated in Figure 1.5.3, the position vector is given by

$$
\begin{aligned}
\overrightarrow{r} &= \overrightarrow{c} + \overrightarrow{a} + \overrightarrow{b} \\
&= [ta \ \ 0] + [0 \ \ a] + [-a\sin(t) \ \ -a\cos(t)] \\
&= [a(t - \sin(t)) \ \ a(1 - \cos(t)].
\end{aligned}
$$

The angle t is the parameter, and it is given in radians so that the arclength opposite the angle is at. Figure 1.5.4 depicts this curve with $a = 1$ and was generated by the MATLAB code cycloid.m.

Euler's formula can be used to generate sine and cosine curves with different amplitudes, a, frequencies, f, and phases, θ. By adding the associated complex numbers one can easily generate very complicated signals such as a person's voice. Consider just one tone and one sine function of time t

$$
a\sin(2\pi ft + \theta).
$$

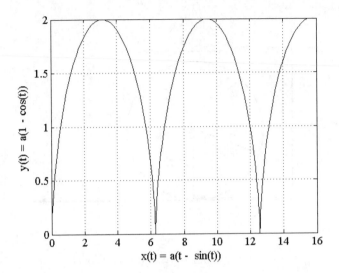

Figure 1.5.4:　Cycloid

The sine function is the imaginary part of

$$ae^{i(2\pi ft+\theta)} = ae^{i2\pi ft}e^{i\theta}.$$

At time $t = 0$, this is $ae^{i\theta}$, which is called a *phasor* where θ is the phase angle. In order to have more than one tone, additional frequencies must be used. Consider two possible tones

$$a_1 \sin(2\pi f_1 t + \theta_1) \text{ and } a_2 \sin(2\pi f_2 t + \theta_2).$$

The signal associated with these two is simply the addition of the two sine functions. In terms of complex variables this is the imaginary part of

$$z = a_1 e^{i(2\pi f_1 t+\theta_1)} + a_2 e^{i(2\pi f_2 t+\theta_2)}.$$

By letting time vary the complex variable will move in the complex plane, and one can plot its imaginary part as a function of time. This is illustrated in Figure 1.5.5, which was generated by the MATLAB code phasor.m.

1.5.4　Curves Using MATLAB

We will concentrate on graphing a curve by using parametric equations. If we have values for the points on the curve and they are stored in two row vectors of equal length, then the MATLAB command plot() can be used to generate a graph. For example, in the following two row vectors each with 4 components are created.

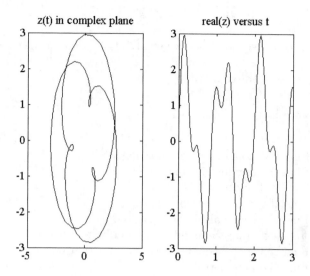

Figure 1.5.5: Two-tone Signal

```
>> x = [1 2 3 4];
>> y = [1 4 9 16];
>> plot(x,y)
```

The command plot(x,y) will generate a graph connecting the four points $(1, 1)$, $(2, 4)$, $(3, 9)$ and $(4, 16)$.

If one needs to have more data points, then this can be done by for-loops or by implied loops called vector operations. The for-loops is more flexible, but it is slower. The following for-loop creates three row vectors that have 10 components.

```
>> for j = 1:10
        t(j) = j*.5;
        x(j) = t(j)+1;
        y(j) = t(j)^2;
    end
>> plot(x,y)
```

The following uses vector operations to generate the same graph. The MATLAB command y = t.^2 is an array operation, which produces another row vector with the same number of components as the row vector t and whose components are the squares of those in t. For example, if t = [1 2 3], then t.^2 = [1 4 9].

```
>> t = .5:.5:5;
>> x = t + 1;
>> y = t.^2;
```

>> plot(x,y)

The MATLAB code cycloid.m, see Figure 1.5.4, uses vector operations, and the code phasor.m, see Figure 1.5.5, uses for-loops with a MATLAB command pause so that one can step through the loop.

1.5.5   Exercises

1.   Let $x - 3y = 9$.
  (a).   Find the slope-intercept equation.
  (b).   Find the parametric equations.
2.   Let $x = 2t + 1$ and $y = 3t - 2$.
  (a).   Find the slope-intercept equation.
  (b).   Find the single algebraic equation.
3.   Let the line contain the point $(2, 7)$ and be parallel to $[1 \ \ -1]$.
  (a).   Find the vector equation using a direction vector.
  (b).   Find the parametric equations.
4.   Let the line contain the point $(2, 7)$ and be perpendicular to $[1 \ \ -1]$.
  (a).   Find the vector equation using a normal vector.
  (b).   Find the single algebraic equation.
5.   Find the point on the line $\overrightarrow{r} = t[1 \ \ -2] + [0 \ \ 1]$ that is closest to the point $(5, 5)$.
6.   Find the point on the line given by $x = 2t + 2$ and $y = -t + 1$ that is closest to the point $(5, 5)$.
7.   Use MATLAB to create a graph of the line given by $x = t + 2$ and $y = 2t$ with $0 \le t \le 5$.
8.   Use MATLAB to create a graph of the curve given by $x = t + 2$ and $y = 2t^2$ with $0 \le t \le 3$.
9.   Use the MATLAB code cycloid.m to experiment with different wheels, $a = 0.5, 1$ and 2.
10.   Use the MATLAB code phasor.m to experiment with a variety of the inputs:
  (a).   Vary the frequencies.
  (b).   Vary the amplitudes.
  (c).   Vary the phase angles.

Chapter 2

Vectors in Space

Vectors in space are introduced, and the dot, cross and box products are studied. Lines and planes are carefully described as well as extensions to higher dimensional space. Applications to work, torque, inventories and visualizations are included.

2.1   Vectors and Dot Product

A point in space can be located in a number of ways, but here the Cartesian coordinate system will be used. You may wish to visualize this from the interior of a room looking down into the corner. The corner is the origin; the x-axis is the intersection of the left wall and floor; the y-axis is the intersection of the right wall and the floor; the intersection of the left and right walls is the z-axis. This is illustrated in Figure 2.1.1. The point (a, b, c) is located by moving a units in the x-axis, then moving b units parallel to the y-axis, and moving c units parallel to the z-axis. The distance from the origin to the point is given by two applications of the Pythagorean theorem to the right triangles in Figure 2.1.1. Associated with the point (a, b, c) is the *position vector* from the origin to this point.

Definition 2.1.1. A *vector in* $\mathbb{R}^3$ is an ordered list of three real numbers $\overrightarrow{a} = [a_1 \ a_2 \ a_3]$. One can visualize this by forming the directed line segment from the origin point $(0, 0, 0)$ to the point (a_1, a_2, a_3).

Notation. Points in $\mathbb{R}^3$ will be denoted by (a_1, a_2, a_3) and vectors will be represented by either row or column vectors:

$$\overrightarrow{a} = [a_1 \ a_2 \ a_3] \text{ indicates a } \textit{row vector},$$

$$\mathbf{a} = \begin{bmatrix} a_1 \\ a_2 \\ a_3 \end{bmatrix} \text{ indicates a } \textit{column vector}.$$

$\mathbf{a}^T = [a_1 \ a_2 \ a_3]$ is called the *transpose* of the column vector $\mathbf{a}$ so that $\mathbf{a}^T = \overrightarrow{a}$.

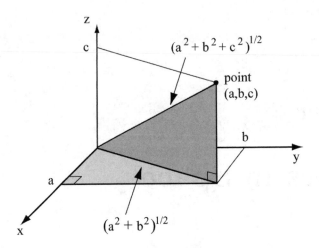

Figure 2.1.1:   Point in Space

Example 2.1.1. The standard *unit basis vectors* are

$$\overrightarrow{i} \;=\; [1 \; 0 \; 0] \text{ or } \mathbf{e}_1 = \begin{bmatrix} 1 \\ 0 \\ 0 \end{bmatrix},$$

$$\overrightarrow{j} \;=\; [0 \; 1 \; 0] \text{ or } \mathbf{e}_2 = \begin{bmatrix} 0 \\ 1 \\ 0 \end{bmatrix} \text{ and}$$

$$\overrightarrow{k} \;=\; [0 \; 0 \; 1] \text{ or } \mathbf{e}_3 = \begin{bmatrix} 0 \\ 0 \\ 1 \end{bmatrix}.$$

Any vector can be represented as a linear combination of unit basis vectors

$$\overrightarrow{a} \;=\; [a_1 \; a_2 \; a_3] = a_1 \overrightarrow{i} + a_2 \overrightarrow{j} + a_3 \overrightarrow{k} \text{ or}$$

$$\mathbf{a} \;=\; \begin{bmatrix} a_1 \\ a_2 \\ a_3 \end{bmatrix} = a_1 \mathbf{e}_1 + a_2 \mathbf{e}_2 + a_3 \mathbf{e}_3.$$

For example, if $\overrightarrow{a} = [2 \; 3 \; 4]$, then either by row vectors

$$\overrightarrow{a} \;=\; 2 \overrightarrow{i} + 3 \overrightarrow{j} + 4 \overrightarrow{k} \text{ or by column vectors}$$

$$\mathbf{a} \;=\; 2\mathbf{e}_1 + 3\mathbf{e}_2 + 4\mathbf{e}_3.$$

These vectors are depicted in Figure 2.1.2.

The movement of a mass in space is described by vectors with three components such as displacement, force, velocity and acceleration vectors. *Newton's*

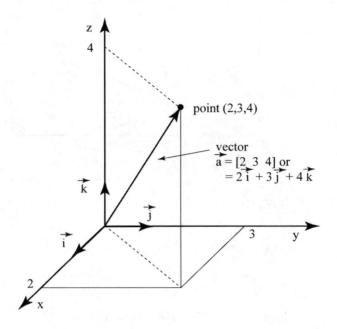

Figure 2.1.2: Vector in Space

law of motion in space can be derived by applying it to each of the three directions

$$f_1 = m\frac{dv_1}{dt}, \quad f_2 = m\frac{dv_2}{dt} \text{ and } f_3 = m\frac{dv_3}{dt}$$

where m is the mass, $\overrightarrow{f} = [f_1 \ f_2 \ f_3]$ is the force vector and $\overrightarrow{v} = [v_1 \ v_2 \ v_3]$ is the velocity vector. In vector notation this means

$$
\begin{aligned}
\overrightarrow{f} &= [f_1 \ f_2 \ f_3] \\
&= [m\frac{dv_1}{dt} \quad m\frac{dv_2}{dt} \quad m\frac{dv_3}{dt}] \\
&= m\overrightarrow{a} \text{ where} \\
\overrightarrow{a} &= [\frac{dv_1}{dt} \quad \frac{dv_2}{dt} \quad \frac{dv_3}{dt}] \text{ is the acceleration vector.}
\end{aligned}
$$

2.1.1 Vector Operations

Vectors can be scaled, added and subtracted by either viewing the vectors geometrically or as symbolic objects. Two non-parallel vectors in space form a plane in space so that the geometric approach to these operations is similar to vectors with two components. Figure 2.1.3 illustrates vector addition and scaling by either elongating or shrinking the vector. Two vectors $\overrightarrow{a}$ and $\overrightarrow{b}$ can be added by moving vector $\overrightarrow{b}$ parallel to its original position so that its

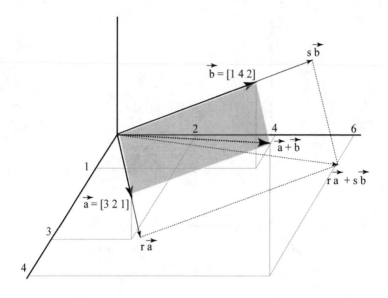

Figure 2.1.3:   Vector Addition in Space

beginning point is the end point of vector $\overrightarrow{a}$. The addition $\overrightarrow{a} + \overrightarrow{b}$ corresponds to the diagonal of the parallelogram formed by vectors $\overrightarrow{a}$ and $\overrightarrow{b}$. If $-\overrightarrow{b}$ is added to $\overrightarrow{a}$, then the subtraction $\overrightarrow{a} - \overrightarrow{b}$ is the other diagonal from the end of $\overrightarrow{b}$ to the end of $\overrightarrow{a}$.

Definitions 2.1.2. Let $\overrightarrow{a} = [a_1 \ a_2 \ a_3]$, $\overrightarrow{b} = [b_1 \ b_2 \ b_3]$ and $\overrightarrow{0} = [0 \ 0 \ 0]$ be the zero vector. $\overrightarrow{a}$ *equals* $\overrightarrow{b}$ means $a_1 = b_1$, $a_2 = b_2$ and $a_3 = b_3$. The *addition of two vectors* is another vector whose components are the sum of the components of the two vectors

$$\overrightarrow{a} + \overrightarrow{b} \equiv [\ a_1 + b_1 \quad a_2 + b_2 \quad a_3 + b_3 \ ].$$

Let s be a real number. The *scalar product of a vector* is another vector whose components are the scalar times the components of the vector

$$s\overrightarrow{a} \equiv [\ sa_1 \quad sa_2 \quad sa_3 \ ].$$

The following theorem lists some basic algebraic properties of the above operations for vectors in space $\mathbb{R}^3$, which are analogous to those in the plane $\mathbb{R}^2$.

Theorem 2.1.1 *(Vector Space Properties)* *Let* $\vec{a}, \vec{b}$ *and* $\vec{c}$ *be vectors in* $\mathbb{R}^3$ *and let* s *and* t *be real numbers. Then the following rules hold*

$$
\begin{aligned}
\vec{a} + \vec{b} &= \vec{b} + \vec{a}, \ (\vec{a} + \vec{b}) + \vec{c} = \vec{a} + (\vec{b} + \vec{c}), \\
\vec{a} + \vec{0} &= \vec{a} \text{ and } \vec{a} + (-\vec{a}) = \vec{0}; \\
(s + t)\vec{a} &= s\vec{a} + t\vec{a}, \ s(t\vec{a}) = (st)\vec{a}, \ 1\vec{a} = \vec{a} \text{ and} \\
s(\vec{a} + \vec{b}) &= s\vec{a} + s\vec{b}.
\end{aligned}
$$

The proofs are straightforward. The proof of $s(\vec{a} + \vec{b}) = s\vec{a} + s\vec{b}$ uses the distributive property of addition

$$
\begin{aligned}
s(\vec{a} + \vec{b}) &= [\ s(a_1 + b_1) \quad s(a_2 + b_2) \quad s(a_3 + b_3) \ ] \\
&= [\ sa_1 + sb_1 \quad sa_2 + sb_2 \quad sa_3 + sb_3 \ ] \\
&= [\ sa_1 \quad sa_2 \quad sa_3 \ ] + [\ sb_1 \quad sb_2 \quad sb_3 \ ] \\
&= s\vec{a} + s\vec{b}.
\end{aligned}
$$

Example 2.1.2. Let $\vec{a} = [2 \ 4 \ 1]$, $\vec{b} = [-1 \ 2 \ 4]$ and $s = 3$. Then

$$
\begin{aligned}
\vec{a} + \vec{b} &= [1 \ 6 \ 5], \\
\vec{a} - \vec{b} &= [3 \ 2 \ -3] \text{ and} \\
s\vec{b} &= [-3 \ 6 \ 12].
\end{aligned}
$$

Definition 2.1.3. Let $\vec{a} = [a_1 \ a_2 \ a_3]$. The length or *Euclidean norm of the vector* $\vec{a}$ is given by the Pythagorean theorem

$$
\|\vec{a}\| \equiv \sqrt{a_1^2 + a_2^2 + a_3^2}.
$$

As illustrated in Figure 2.1.3 two vectors in space form a plane and so the cosine law may be rewritten in terms of the norms of the vectors $\vec{a}, \vec{b}, \vec{c} = \vec{a} - \vec{b}$ and the angle θ between the vectors $\vec{a}$ and $\vec{b}$

$$
\left\| \vec{a} - \vec{b} \right\|^2 = \|\vec{a}\|^2 + \left\| \vec{b} \right\|^2 - 2\|\vec{a}\| \left\| \vec{b} \right\| \cos(\theta). \tag{2.1.1}
$$

This allows one to compute the angle between two given vectors.

Example 2.1.3. Consider the vectors in Figure 2.1.3 $\vec{a} = [3 \ 2 \ 1]$ and $\vec{b} = [1 \ 4 \ 2]$. The vector starting at the point $(1 \ 4 \ 2)$ and ending at $(3 \ 2 \ 1)$ is represented by $\vec{a} - \vec{b}$, and $\left\| \vec{a} - \vec{b} \right\|$ is the distance between the end points of the vectors $\vec{a}$ and $\vec{b}$. Then $\vec{a} - \vec{b} = [2 \ -2 \ -1]$, $\left\| \vec{a} - \vec{b} \right\|^2 = 9$, $\|\vec{a}\|^2 = 14$, $\left\| \vec{b} \right\|^2 = 21$, and equation (2.1.1) becomes

$$
\begin{aligned}
9 &= 14 + 21 - 2\sqrt{14}\sqrt{21} \cos(\theta) \\
&= 14 + 21 - 14\sqrt{6} \cos(\theta).
\end{aligned}
$$

Thus,

$$\cos(\theta) = 13/(7\sqrt{6}) \text{ and}$$
$$\theta = \cos^{-1}(13/(7\sqrt{6})) \approx 40.70 \text{ degrees.}$$

2.1.2   Cosine Identity and Dot Product

Let $\vec{a} = [a_1 \ \ a_2 \ \ a_3]$ and $\vec{b} = [b_1 \ \ b_2 \ \ b_3]$ form a plane in space and have the angle θ between the vectors. The vector version of the cosine law is equation (2.1.1). By definition of vector subtraction and the norm

$$\left\| \vec{a} - \vec{b} \right\|^2 = (a_1 - b_1)^2 + (a_2 - b_2)^2 + (a_3 - b_3)^2$$
$$= a_1^2 - 2a_1 b_1 + b_1^2 + a_2^2 - 2a_2 b_2 + b_2^2$$
$$+ a_3^2 - 2a_3 b_3 + b_3^2$$
$$= \left\| \vec{a} \right\|^2 + \left\| \vec{b} \right\|^2 - 2(a_1 b_1 + a_2 b_2 + a_3 b_3).$$

Place this into the left side of equation (2.1.1) and cancel the norms to get

$$a_1 b_1 + a_2 b_2 + a_3 b_3 = \left\| \vec{a} \right\| \left\| \vec{b} \right\| \cos(\theta). \tag{2.1.2}$$

As in the case for vectors with two components, equation (2.1.2) is of fundamental importance because the left side is an easy calculation, and the right side makes a connection with the geometry.

Theorem 2.1.2 *(Cosine Identity)* Let $\vec{a} = [a_1 \ \ a_2 \ \ a_3]$ and $\vec{b} = [b_1 \ \ b_2 \ \ b_3]$ *be non-zero vectors and let θ be the angle between the two vectors. Then $\cos(\theta)$ can be computed by*

$$\cos(\theta) = \frac{a_1 b_1 + a_2 b_2 + a_3 b_3}{\left\| \vec{a} \right\| \left\| \vec{b} \right\|}.$$

Definition 2.1.4. The *dot product of two vectors* $\vec{a} = [a_1 \ \ a_2 \ \ a_3]$ *and* $\vec{b} = [b_1 \ \ b_2 \ \ b_3]$ is given by the real number on either side of equation (2.1.2). The following are common notations for the dot product of row or column vectors, respectively,

$$\vec{a} \bullet \vec{b} \equiv a_1 b_1 + a_2 b_2 + a_3 b_3 = \left\| \vec{a} \right\| \left\| \vec{b} \right\| \cos(\theta) \text{ or}$$
$$\mathbf{a}^T \mathbf{b} \equiv a_1 b_1 + a_2 b_2 + a_3 b_3 = \| \mathbf{a} \| \, \| \mathbf{b} \| \cos(\theta).$$

The cosine identity can be restated as either

$$\cos(\theta) = \frac{\vec{a} \bullet \vec{b}}{\left\| \vec{a} \right\| \left\| \vec{b} \right\|} \text{ or}$$

$$\cos(\theta) = \frac{\mathbf{a}^T \mathbf{b}}{\| \mathbf{a} \| \, \| \mathbf{b} \|}.$$

Also, the norm of a vector can be written as either

$$\|\vec{a}\|^2 = \vec{a} \bullet \vec{a} \text{ or}$$
$$\|\mathbf{a}\|^2 = \mathbf{a}^T\mathbf{a}.$$

Definitions 2.1.5. Let $\vec{a}$ and $\vec{b}$ be non-zero vectors. Two *vectors are parallel* if and only if they are multiples of each other, that is, there is a real number s such that $\vec{a} = s\,\vec{b}$. Two *vectors are perpendicular or orthogonal* if and only if the $\cos(\theta)$ is zero, that is, $\vec{a} \bullet \vec{b} = \mathbf{a}^T\mathbf{b} = 0$.

Example 2.1.4. Consider the vectors in Figure 2.1.3 $\vec{a} = [3\ 2\ 1]$ and $\vec{b} = [1\ 4\ 2]$. Then $\|\vec{a}\|^2 = 14$, $\left\|\vec{b}\right\|^2 = 21$ and $\vec{a} \bullet \vec{b} = 3(1) + 2(4) + 1(2) = 13$ so that

$$\cos(\theta) = \frac{\vec{a} \bullet \vec{b}}{\|\vec{a}\|\left\|\vec{b}\right\|} = \frac{13}{\sqrt{14}\sqrt{21}} = \frac{13}{7\sqrt{6}} \text{ and}$$

$$\theta = \cos^{-1}\left(\frac{13}{7\sqrt{6}}\right) \approx 40.70 \text{ degrees.}$$

The dot product has a number of interesting algebraic properties. One is it is associative with respect to a scalar multiple

$$s(\vec{a} \bullet \vec{b}) = s(a_1b_1 + a_2b_2 + a_3b_3)$$
$$= (sa_1)b_1 + (sa_2)b_2 + (sa_3)b_3$$
$$= (s\vec{a}) \bullet \vec{b}.$$

Additional properties are listed in the following theorems, which are analogous to Theorems 1.4.2 and 1.4.3, and are also easy to prove. The *Cauchy-Schwarz inequality* in (2.1.3) follows directly from the cosine identity in (2.1.2). Real valued functions that satisfy equations (2.1.4)-(2.1.6) are called *norms*, and they are analogous to the absolute value of a single real number. An interpretation of the triangle inequality is the minimum distance between two points is a straight line. These concepts generalize to higher dimensional vectors, see Sections 2.5 and 3.1.

Theorem 2.1.3 (*Properties of Dot Product*) *Let $\vec{a}, \vec{b}, \vec{c}$ be given vectors in space $\mathbb{R}^3$ and let s be a real number. Then the following hold*

$$\vec{a} \bullet \vec{b} = \vec{b} \bullet \vec{a}, \ s(\vec{a} \bullet \vec{b}) = (s\vec{a}) \bullet \vec{b},$$
$$\vec{a} \bullet (\vec{b} + \vec{c}) = \vec{a} \bullet \vec{b} + \vec{a} \bullet \vec{c} \text{ and}$$
$$\left|\vec{a} \bullet \vec{b}\right| \leq \|\vec{a}\|\left\|\vec{b}\right\| \text{ (Cauchy-Schwarz inequality);} \quad (2.1.3)$$

Theorem 2.1.4 *(Properties of Norm) Let $\vec{a}$, $\vec{b}$ be given vectors in space $\mathbb{R}^3$ and let s be a real number. Then the following hold*

$$\|\vec{a}\| \geq 0;\ \vec{a} = \vec{0} \text{ if and only if } \|\vec{a}\| = 0, \tag{2.1.4}$$
$$\|s\vec{a}\| = |s|\,\|\vec{a}\| \text{ and} \tag{2.1.5}$$
$$\left\|\vec{a} + \vec{b}\right\| \leq \|\vec{a}\| + \left\|\vec{b}\right\| \text{ (triangle inequality).} \tag{2.1.6}$$

2.1.3 Applications to Work

Consider moving a mass from one point in space to an other point where the component changes are recorded in a displacement vector $\vec{d} = [d_1\ d_2\ d_3]$. A force in each direction is required, which can be recorded in a force vector $\vec{f} = [f_1\ f_2\ f_3]$. For constant displacement and force vectors, the work is the product of displacement distance and the length of the force vector projected on the displacement vector, $\left\|\vec{f}\right\|\cos(\theta)$. Then the work is $\left\|\vec{d}\right\|\left\|\vec{f}\right\|\cos(\theta)$. The work is given by the cosine identity and the dot product

$$work = \vec{d} \bullet \vec{f} = d_1 f_1 + d_2 f_2 + d_3 f_3.$$

For example, if the mass is moved from the point $(1,1,1)$ to the point $(2,3,4)$, then $\vec{d} = [1\ 2\ 3]$. Suppose the force vector's components are 100 in the x-direction, 10 in the y-direction and 5 in the z-direction so that $\vec{f} = [100\ 10\ 5]$. Then the work is $\vec{d} \bullet \vec{f} = 1(100) + 2(10) + 3(5) = 135$.

2.1.4 Dot Product Using MATLAB

The MATLAB command dot(a,b) computes the dot product if the vectors are both row or column vectors. If the vectors are both column vectors, then one can use the transpose version $\mathbf{a}^T\mathbf{b}$. In MATLAB this would appear as $a' * b$ provided a and b were column vectors.

```
>> a = [1 -1 3]
    a =
         1 -1 3
>> b = [4 1 7]
    b =
         4 1 7
>> dot(a,b)
    ans =
         24
>> a_col = a'
>> b_col = b'
>> a_col'*b_col
    ans = 24
```

The MATLAB command norm(a) computes $\|\vec{a}\|$, which is the same as $(\vec{a} \bullet \vec{a})^{1/2}$. This is called the *Euclidean* norm and is one of many important norms (see the MATLAB command help norm).

>> norm(a)
  ans =
     3.3166
>> (dot(a,a))^.5
  ans =
     3.3166

The next calculations illustrate the MATLAB command quiver3(x, y, z, dx, dy, dz, 1), which will generate a vector in space starting at the point (x, y, z) and ending at the point $(x + dx, y + dy, z + dz)$.

>> quiver3(0,0,0,2,3,-1,1)
>> hold on
>> quiver3(0,0,0,3,1,2,1)

The input parameters may be arrays with the same size. Use arrays as inputs (see MATLAB command help quiver3 for an interesting example).

>> x = [0 0];y = [0 0]; z = [0 0];
>> dx = [2 3];dy = [3 1];dz = [-1 2];
>> hold off
>> quiver3(x,y,z,dx,dy,dz,1)

2.1.5 Exercises

1.   Plot the points $(3, 1, 2)$ and $(1, 3, 1)$. Indicate the corresponding vectors starting at the origin.

2.   Plot the points $(3, 1, -1)$ and $(1, 1, 3)$. Indicate the corresponding vectors starting at the origin.

3.   Find the unit basis vector and column vector representations of

$$\vec{a} = [3 \ \ 1 \ \ 2] \text{ and } \vec{b} = [1 \ \ 3 \ \ 1].$$

4.   Find the unit basis vector and row vector representations of

$$\mathbf{a} = \begin{bmatrix} 3 \\ 1 \\ -1 \end{bmatrix} \text{ and } \mathbf{b} = \begin{bmatrix} 1 \\ 1 \\ 3 \end{bmatrix}.$$

5.   Let $\vec{a} = [3 \ \ 1 \ \ 2]$ and $\vec{b} = [1 \ \ 3 \ \ 1]$.
   (a).   Compute $-2\vec{b}$ and $2\vec{a} + \vec{b}$.
   (b).   Find the vector $\vec{x} = [x_1 \ \ x_2 \ \ x_3]$ so that $2\vec{a} + \vec{x} = \vec{b}$.

6.   Let $\vec{a} = 2\vec{i} + 4\vec{j} - \vec{k}$ and $\vec{b} = -\vec{i} + 2\vec{j} + 3\vec{k}$
   (a).   Compute $2\vec{b}$ and $2\vec{a} + 3\vec{b}$.

(b).   Find the vector $\vec{x} = x_1 \vec{i} + x_2 \vec{j} + x_3 \vec{k}$ so that $3\vec{a} + \vec{x} = \vec{b}$.

7.   Let $\vec{a} = \vec{i} + 4\vec{j} - 2\vec{k}$ and $\vec{b} = -\vec{i} + \vec{j} + 3\vec{k}$

(a).   Compute $\|\vec{a}\|$ and $\|\vec{b}\|$.

(b).   Find $\vec{a} \bullet \vec{b}$.

8.   Let $\vec{a} = [0 \ 1 \ 2]$ and $\vec{b} = [1 \ 4 \ -1]$.

(a).   Compute $\|\vec{a}\|$ and $\|\vec{b}\|$.

(b).   Find $\vec{a} \bullet \vec{b}$.

9.   Use the cosine identity in equation (2.1.2) to find the angle between the vectors in column form

$$\mathbf{a} = \begin{bmatrix} 1 \\ 2 \\ -1 \end{bmatrix} \text{ and } \mathbf{b} = \begin{bmatrix} 1 \\ 3 \\ -1 \end{bmatrix}.$$

10.   Use the cosine identity in equation (2.1.2) to find the angle between the vectors $\vec{a} = [3 \ 1 \ -1]$ and $\vec{b} = [1 \ -1 \ 1]$.

11.   Consider Theorem 2.1.3.

(a).   Prove $s(\vec{a} \bullet \vec{b}) = (s\vec{a}) \bullet \vec{b} = \vec{a} \bullet (s\vec{b})$.

(b).   Verify $\vec{a} \bullet (\vec{b} + \vec{c}) = \vec{a} \bullet \vec{b} + \vec{a} \bullet \vec{c}$ for the vectors $\vec{a} = [2 \ 1 \ 3]$, $\vec{b} = [3 \ 4 \ 1]$ and $\vec{c} = [1 \ -1 \ 4]$.

(c).   What is the difference between "prove" and "verify"?

12.   Consider Theorem 2.1.4. Prove the results of lines (2.1.4) and (2.1.5).

13.   Consider the work in moving an object along an incline in space whose x-direction is 25, y-direction is 20 and z-direction is 5. If the force vector is $\vec{f} = [10 \ 1 \ 2]$, find the work.

14.   Consider the work in moving an object along an incline in space whose displacement vector is $\vec{d} = 20\vec{i} + 15\vec{j} + 3\vec{k}$. If the force vector is $\vec{f} = 10\vec{i} + 2\vec{j} + 1\vec{k}$, find the work.

15.   Use MATLAB to compute the dot product of $\vec{a} = [2 \ 1 \ -1]$ and $\vec{b} = [3 \ 4 \ 2]$ and their norms.

16.   Use MATLAB command quiver3(x,y,z,dx,dy,dz,1) to plot the vectors $\vec{a} = [2 \ 1 \ -1]$ and $\vec{b} = [3 \ 4 \ 2]$.

2.2   Cross and Box Products

In Section 1.4 the cosine identity was used to find the area of a parallelogram formed by two vectors in a plane. Then this was used to compute the torque for the rotation about a point in the plane. The cross product will allow us to generalize these to vectors in space and not just the xy-plane. The box product is a combination of the dot and cross product that will enable us to easily compute the volume of a parallelepiped formed by three vectors in space. These concepts will be related to determinants of 3×3 matrices.

2.2.1   Area and Torque in the xy-Plane

Any vector in the plane $\vec{a} = [a_1 \ a_2]$ can be viewed as space vector whose third component is zero, $\vec{a} = [a_1 \ a_2 \ 0]$. We use some of the results in Section 1.4 for vectors in space that are restricted to one of the coordinate planes. Let $\vec{a} = [a_1 \ a_2 \ 0]$ and $\vec{b} = [b_1 \ b_2 \ 0]$ not be parallel. Then the *area* is the area of the parallelogram formed by $\vec{a}$ and $\vec{b}$ is

$$
\begin{aligned}
(area)^2 &= (\|\vec{a}\| \, \|\vec{b}\| \sin(\theta))^2 \\
&= (a_1 b_2 - b_1 a_2)^2 \\
&= (\det(\begin{bmatrix} a_1 & b_1 \\ a_2 & b_2 \end{bmatrix}))^2.
\end{aligned}
$$

An application to torque, which involves rotation about the z-axis, is when $\vec{a} = \vec{r}$ gives the point to be rotated and $\vec{b} = \vec{f}$ is the force vector. Here $\|\vec{f}\| \sin(\theta)$ is the component of the force acting tangent to the rotation, and it can be directed in either a clockwise or counterclockwise motion. The convention is to use the clockwise motion as given by a right hand threaded screw where a clockwise rotation of a screw driver sinks the screw into a mass. If one views the screw as being below the xy-plane, and rotates the x-axis towards the y-axis, then the screw should move in a positive z-direction. Define the cross product of the unit vectors according to this convention:

$$
\begin{aligned}
\vec{i} \times \vec{j} &\equiv \vec{k} \\
\vec{j} \times \vec{k} &\equiv \vec{i} \text{ and} \\
\vec{k} \times \vec{i} &\equiv \vec{j}.
\end{aligned}
$$

The cross product of $\vec{a} = [a_1 \ a_2 \ 0]$ and $\vec{b} = [b_1 \ b_2 \ 0]$ is defined to be a vector in the z-direction with magnitude equal to $a_1 b_2 - b_1 a_2$

$$[a_1 \ a_2 \ 0] \times [b_1 \ b_2 \ 0] \equiv [0 \ 0 \ a_1 b_2 - b_1 a_2].$$

Note, if $\vec{a} = [1 \ 0 \ 0] = \vec{i}$ and $\vec{b} = [0 \ 1 \ 0] = \vec{j}$, then $\vec{i} \times \vec{j} = [0 \ 0 \ 1(1) - 0(0)] = \vec{k}$. This is illustrated in Figure 2.2.1 where the three possible planes are viewed from the back (negative) sides. In the case $\vec{a} = \vec{r}$ and $\vec{b} = \vec{f}$ the third component of $\vec{r} \times \vec{f}$ is the torque for rotation about the z-axis.

2.2.2   Cross Product and Torque in Space

In order to understand the cross product definition for any vectors in space, consider the area of the parallelogram formed by two vectors in space as illustrated in Figure 2.2.2. The area will be non-zero provided the two vectors are not parallel. The area that is projected into the xy-plane is the absolute value of

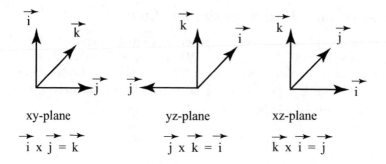

xy-plane   yz-plane   xz-plane

$\vec{i} \times \vec{j} = \vec{k}$ $\vec{j} \times \vec{k} = \vec{i}$ $\vec{k} \times \vec{i} = \vec{j}$

Figure 2.2.1:   Unit Vector Cross Products

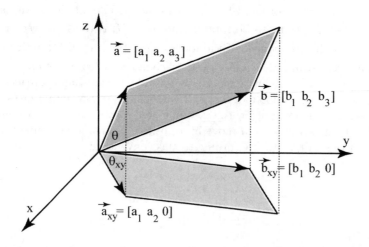

Figure 2.2.2:   Projected Area

the third component in the cross product of the two vectors projected into the xy-plane. One can also project into the yz-plane and the xz-plane with similar area formulae. The above analysis can be applied to the other two coordinate planes to yield

$$[a_1 \ 0 \ a_3] \times [b_1 \ 0 \ b_3] \ \equiv \ [0 \ -(a_1b_3 - b_1a_3) \ 0] \text{ and}$$
$$[0 \ a_2 \ a_3] \times [0 \ b_2 \ b_3] \ \equiv \ [a_2b_3 - b_2a_3 \ 0 \ 0].$$

The cosine identity eventually leads to a formula for the *AREA* of the parallelogram formed by two vectors in space $\vec{a} = [a_1 \ a_2 \ a_3]$ and $\vec{b} = [b_1 \ b_2$

b_3]

$$
\begin{aligned}
(AREA)^2 &= (\|\vec{a}\|\,\|\vec{b}\|\sin(\theta))^2 \\
&= (\|\vec{a}\|\,\|\vec{b}\|)^2(1 - \cos^2(\theta)) \\
&= (\|\vec{a}\|\,\|\vec{b}\|)^2\left(1 - \left(\frac{\vec{a}\bullet\vec{b}}{\|\vec{a}\|\,\|\vec{b}\|}\right)^2\right) \\
&= \|\vec{a}\|^2\,\|\vec{b}\|^2 - (\vec{a}\bullet\vec{b})^2.
\end{aligned}
$$

Next substitute the vector components into the above and endure a messy algebraic computation to obtain

$$
\begin{aligned}
(AREA)^2 &= (a_1^2 + a_2^2 + a_3^2)(b_1^2 + b_2^2 + b_3^2) - (a_1b_1 + a_2b_2 + a_3b_3)^2 \\
&\;\;\vdots \\
&= (a_2b_3 - b_2a_2)^2 + (b_1a_3 - a_1b_3)^2 + (a_1b_2 - b_1a_2)^2. \quad (2.2.1)
\end{aligned}
$$

The formula in equation (2.2.1) means the area in space squared is the sum of the three "projected areas squared."

Definition 2.2.1. The *cross product* of $\vec{a} = [a_1\;\; a_2\;\; a_3]$ and $\vec{b} = [b_1\;\; b_2\;\; b_3]$ is another vector $\vec{a}\times\vec{b}$ whose components are given by the projected areas

$$
\begin{aligned}
\vec{a}\times\vec{b} &\equiv [a_2b_3 - b_2a_3 \;\; -a_1b_3 + b_1a_3 \;\; a_1b_2 - b_1a_2] \quad (2.2.2) \\
&\equiv (a_2b_3 - b_2a_2)\,\vec{i} - (a_1b_3 - b_1a_3)\,\vec{j} + (a_1b_2 - b_1a_2)\,\vec{k}.
\end{aligned}
$$

As we show below $\vec{a}\times\vec{b}$ is perpendicular to both $\vec{a}$ and $\vec{b}$ and, thus, it is perpendicular to the plane formed by $\vec{a}$ and $\vec{b}$. Equation (2.2.1) implies the norm of the cross product is equal to the area of the parallelogram formed by the two vectors. Determinants are often used as an easy way to remember the definition of a cross product of two vectors.

Definition 2.2.2. Let a 3×3 matrix be given by three row vectors $\vec{c}, \vec{a}$ and $\vec{b}$

$$
A = \begin{bmatrix} c_1 & c_2 & c_3 \\ a_1 & a_2 & a_3 \\ b_1 & b_2 & b_3 \end{bmatrix}.
$$

The *determinant* of this matrix is a real number defined by expanding across the top row of the matrix

$$\begin{aligned}
\det(A) &\equiv c_1 \det(\begin{bmatrix} a_2 & a_3 \\ b_2 & b_3 \end{bmatrix}) \\
&\quad -c_2 \det(\begin{bmatrix} a_1 & a_3 \\ b_1 & b_3 \end{bmatrix}) \\
&\quad +c_3 \det(\begin{bmatrix} a_1 & a_2 \\ b_1 & b_2 \end{bmatrix}) \\
&= c_1 (a_2 b_3 - b_2 a_3) \\
&\quad -c_2 (a_1 b_3 - b_1 a_3) \\
&\quad +c_3 (a_1 b_2 - b_1 a_2).
\end{aligned} \tag{2.2.3}$$

The components of the row vectors are usually real or complex numbers. An abuse of notation is to let the top row be replaced by a row of unit basis vectors so that

$$\vec{a} \times \vec{b} \equiv \det(\begin{bmatrix} \vec{i} & \vec{j} & \vec{k} \\ a_1 & a_2 & a_3 \\ b_1 & b_2 & b_3 \end{bmatrix}).$$

Example 2.2.1. Let $\vec{a} = [1 \ 3 \ -2]$ and $\vec{b} = [4 \ 1 \ 5]$.

$$\begin{aligned}
\vec{a} \times \vec{b} &= \det(\begin{bmatrix} \vec{i} & \vec{j} & \vec{k} \\ 1 & 3 & -2 \\ 4 & 1 & 5 \end{bmatrix}) \\
&= (3(5) - 1(-2)) \vec{i} - (1(5) - 4(-2)) \vec{j} + (1(1) - 4(3)) \vec{k} \\
&= 17 \vec{i} - 13 \vec{j} - 11 \vec{k} \\
&= [17 \ -13 \ -11].
\end{aligned}$$

The cross product is perpendicular to both the vectors $\vec{a}$ and $\vec{b}$. In order to prove this, it suffices to show the dot product of one of the vectors with the cross product is zero

$$\begin{aligned}
(\vec{a} \times \vec{b}) \bullet \vec{a} &= [a_2 b_3 - b_2 a_3 \ -a_1 b_3 + b_1 a_3 \ a_1 b_2 - b_1 a_2] \bullet [a_1 \ a_2 \ a_3] \\
&= (a_2 b_3 - b_2 a_3) a_1 + (-a_1 b_3 + b_1 a_3) a_2 + (a_1 b_2 - b_1 a_2) a_3 \\
&= a_2 b_3 a_1 - b_2 a_3 a_1 - a_1 b_3 a_2 + b_1 a_3 a_2 + a_1 b_2 a_3 - b_1 a_2 a_3 \\
&= 0.
\end{aligned}$$

Theorem 2.2.1 *(Fundamental Cross Product Properties) The cross product* $\overrightarrow{a} \times \overrightarrow{b}$ *satisfies the following*

$$(\overrightarrow{a} \times \overrightarrow{b}) \bullet \overrightarrow{a} = 0,$$
$$(\overrightarrow{a} \times \overrightarrow{b}) \bullet \overrightarrow{b} = 0,$$
$$\left\|\overrightarrow{a} \times \overrightarrow{b}\right\|^2 = (\|\overrightarrow{a}\| \|\overrightarrow{b}\| \sin(\theta))^2 \text{ and}$$
$$\left\|\overrightarrow{a} \times \overrightarrow{b}\right\|^2 = \|\overrightarrow{a}\|^2 \|\overrightarrow{b}\|^2 - (\overrightarrow{a} \bullet \overrightarrow{b})^2$$

(Lagrange identity).

Example 2.2.2. Let $\overrightarrow{a} = [1 \ 3 \ -2]$ and $\overrightarrow{b} = [4 \ 1 \ -5]$ be as in the first example so that $\overrightarrow{a} \times \overrightarrow{b} = [17 \ -13 \ -11]$. The area of the parallelogram formed by these vectors is

$$\left\|\overrightarrow{a} \times \overrightarrow{b}\right\| = \sqrt{17^2 + (-13)^2 + (-11)^2} = \sqrt{579} \approx 24.06.$$

The cross product has a number of interesting properties. The distributive property is valid

$$\begin{aligned}
\overrightarrow{a} \times (\overrightarrow{b} + \overrightarrow{c}) &= [a_2(b_3 + c_3) - (b_2 + c_2)a_3 \\
&\quad -a_1(b_3 + c_3) + (b_1 + c_1)a_3 \\
&\quad a_1(b_2 + c_2) - (b_1 + c_1)a_2] \\
&= [a_2b_3 - b_2a_3 \ - a_1b_3 + b_1a_3 \ a_1b_2 - b_1a_2] + \\
&\quad [a_2c_3 - c_2a_3 \ - a_1c_3 + c_1a_3 \ a_1c_2 - c_1a_2] \\
&= \overrightarrow{a} \times \overrightarrow{b} + \overrightarrow{a} \times \overrightarrow{c}.
\end{aligned}$$

However, the cross product is not commutative and is not associative!

Theorem 2.2.2 *(Algebraic Properties of Cross Product) Consider the cross product of vectors in space. The following hold*

$$\overrightarrow{a} \times \overrightarrow{b} = -\overrightarrow{b} \times \overrightarrow{a}, \ (s\overrightarrow{a}) \times \overrightarrow{b} = s(\overrightarrow{a} \times \overrightarrow{b}),$$
$$\overrightarrow{a} \times (\overrightarrow{b} + \overrightarrow{c}) = \overrightarrow{a} \times \overrightarrow{b} + \overrightarrow{a} \times \overrightarrow{c},$$
$$\overrightarrow{a} \times (\overrightarrow{b} \times \overrightarrow{c}) = (\overrightarrow{a} \bullet \overrightarrow{c})\overrightarrow{b} - (\overrightarrow{a} \bullet \overrightarrow{b})\overrightarrow{c} \text{ and}$$
$$\overrightarrow{a} \bullet (\overrightarrow{b} \times \overrightarrow{c}) = (\overrightarrow{a} \times \overrightarrow{b}) \bullet \overrightarrow{c}.$$

Example 2.2.3. Consider the torque vector

$$torque = \overrightarrow{r} \times \overrightarrow{f}.$$

Let $\overrightarrow{r} = [1 \ 3 \ 1]$ and $\overrightarrow{f} = [4 \ 1 \ 2]$. You may think of the origin as being the location of screw or bolt. Let a wrench be attached to the bolt and have its handle end at the end of the position vector $\vec{r}$. Apply a force at the handle end given by the vector $\vec{f}$. Find the components of torque governing the rotation about x,y and z axes.

$$
\begin{aligned}
\vec{r} \times \vec{f} &= \det\left(\begin{bmatrix} \vec{i} & \vec{j} & \vec{k} \\ 1 & 3 & 1 \\ 4 & 1 & 2 \end{bmatrix}\right) \\
&= (3(2) - 1(1))\,\vec{i} - (1(2) - 4(1))\,\vec{j} + (1(1) - 4(3))\,\vec{k} \\
&= 5\,\vec{i} + 2\,\vec{j} - 11\,\vec{k}.
\end{aligned}
$$

2.2.3   Box Product and Volume

Let three vectors in space not be in the same plane. They form a parallelepiped as is illustrated in Figure 2.2.3. The volume is equal to the area in the base formed by the two vectors $\vec{a}$ and $\vec{b}$ times the height. The direction of the height is perpendicular to the base, that is, it is parallel to the cross product $\vec{a} \times \vec{b}$. The height is then equal to the projection of the third vector $\vec{c}$ onto $\vec{a} \times \vec{b}$. Let ϕ be the angle between the vectors $\vec{c}$ and $\vec{a} \times \vec{b}$ so that the height is $\|\vec{c}\| \cos(\phi)$. Thus, the volume is plus or minus

$$
\left\|\vec{a} \times \vec{b}\right\| (\|\vec{c}\| \cos(\phi)) = (\vec{a} \times \vec{b}) \bullet \vec{c}.
$$

Definition 2.2.3. The *box product* of $\vec{a}$, $\vec{b}$ and $\vec{c}$ is the real number

$$
(\vec{a} \times \vec{b}) \bullet \vec{c}.
$$

Example 2.2.4.   Compute the volume of the parallelepiped formed by the vectors $\vec{a} = [1\ \ 2\ \ 1]$, $\vec{b} = [2\ \ 0\ \ 1]$ and $\vec{c} = [0\ \ 1\ \ 4]$.

$$
\begin{aligned}
\vec{a} \times \vec{b} &= \det\left(\begin{bmatrix} \vec{i} & \vec{j} & \vec{k} \\ 1 & 2 & 1 \\ 2 & 0 & 1 \end{bmatrix}\right) \\
&= (2(1) - 1(0))\,\vec{i} - (1(1) - 1(2))\,\vec{j} + (1(0) - 2(2))\,\vec{k} \\
&= 2\,\vec{i} + 1\,\vec{j} - 4\,\vec{k} = [2\ \ 1\ \ -4].
\end{aligned}
$$

The box product is

$$
\begin{aligned}
(\vec{a} \times \vec{b}) \bullet \vec{c} &= [2\ \ 1\ \ -4] \bullet [0\ \ 1\ \ 4] \\
&= 2(0) + 1(1) + (-4)4 = -15.
\end{aligned}
$$

The volume of the parallelepiped is 15.

The box product can be computed by using determinants. Let A be the 3×3 matrix with first, second and third rows $\overrightarrow{c}$, $\overrightarrow{a}$ and $\overrightarrow{b}$, respectively. Then

$$
\begin{aligned}
(\overrightarrow{a} \times \overrightarrow{b}) \bullet \overrightarrow{c} &= \overrightarrow{c} \bullet (\overrightarrow{a} \times \overrightarrow{b}) \\
&= c_1(a_2 b_3 - b_2 a_3) \\
&\quad -c_2(a_1 b_3 - b_1 a_3) \\
&\quad +c_3(a_1 b_2 - b_1 a_2) \\
&= \det(A).
\end{aligned}
$$

Other representations of the determinant follow from rearranging the computations of $\det(A)$ in equation (2.2.3) and noting

$$
\begin{aligned}
\det(A) &= \det\left(\begin{bmatrix} c_1 & c_2 & c_3 \\ a_1 & a_2 & a_3 \\ b_1 & b_2 & b_3 \end{bmatrix} \right) \\
&= \det\left(\begin{bmatrix} a_1 & a_2 & a_3 \\ b_1 & b_2 & b_3 \\ c_1 & c_2 & c_3 \end{bmatrix} \right) \quad (2.2.4) \\
&= \det\left(\begin{bmatrix} a_1 & b_1 & c_1 \\ a_2 & b_2 & c_2 \\ a_3 & b_3 & c_3 \end{bmatrix} \right). \quad (2.2.5)
\end{aligned}
$$

An interpretation of equation (2.2.4) is that $\overrightarrow{a} \bullet (\overrightarrow{b} \times \overrightarrow{c})$ gives plus or minus the volume when the base of the parallelepiped is generated by the vectors $\overrightarrow{b}$ and $\overrightarrow{c}$. Equation (2.2.5) is formed by interchanging the rows and columns of the previous matrix; the new matrix (2.2.5) is called the *transpose* of the matrix in (2.2.4). If the column vectors in (2.2.5) are represented by $\mathbf{a}, \mathbf{b}$ and $\mathbf{c}$, then the matrix in (2.2.5) may be written as $[\mathbf{a} \ \mathbf{b} \ \mathbf{c}]$ and $\det([\mathbf{a} \ \mathbf{b} \ \mathbf{c}])$ is plus or minus the volume.

Theorem 2.2.3 *(Volume Equivalence) Let three non-zero vectors be represented as column vectors* $\mathbf{a}, \mathbf{b}$ *and* $\mathbf{c}$. *The volume of the parallelepiped formed by these vectors being non-zero is equivalent to no vector of the three being in the plane formed by the other two vectors. In symbolic terms, this means* $\det([\mathbf{a} \ \mathbf{b} \ \mathbf{c}]) \neq 0$ *is equivalent to* $\mathbf{a} \neq d\mathbf{b}+e\mathbf{c}$ *or* $\mathbf{b} \neq d\mathbf{c}+e\mathbf{a}$ *or* $\mathbf{c} \neq d\mathbf{a}+e\mathbf{b}$ *for all non-zero real numbers* d *and* e. *Moreover, volume is equal to the absolute value of* $\det([\mathbf{a} \ \mathbf{b} \ \mathbf{c}])$.

Another point of view generalizes to higher dimensional vectors. The matrix product may be written as a linear combination of the column vectors $\mathbf{a} \ \mathbf{b}$ and

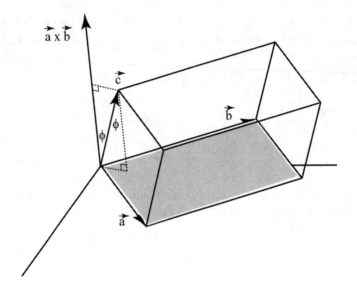

Figure 2.2.3: Box Product and Volume

c. In order to prove this, note

$$
\begin{bmatrix} a_1 & b_1 & c_1 \\ a_2 & b_2 & c_2 \\ a_3 & b_3 & c_3 \end{bmatrix} \begin{bmatrix} r \\ s \\ t \end{bmatrix} = \begin{bmatrix} 0 \\ 0 \\ 0 \end{bmatrix}
$$

$$
\begin{bmatrix} a_1 r + b_1 s + c_1 t \\ a_2 r + b_2 s + c_2 t \\ a_3 r + b_3 s + c_3 t \end{bmatrix} = \begin{bmatrix} 0 \\ 0 \\ 0 \end{bmatrix}.
$$

Next rewrite the left column vector as

$$
r \begin{bmatrix} a_1 \\ a_2 \\ a_3 \end{bmatrix} + s \begin{bmatrix} b_1 \\ b_2 \\ b_3 \end{bmatrix} + t \begin{bmatrix} c_1 \\ c_2 \\ c_3 \end{bmatrix} = \begin{bmatrix} 0 \\ 0 \\ 0 \end{bmatrix}
$$

$$
r\mathbf{a} + s\mathbf{b} + t\mathbf{c} = \mathbf{0}.
$$

Thus, the above are also equivalent to $r\mathbf{a} + s\mathbf{b} + t\mathbf{c} = \mathbf{0}$ implies all r, s and t must be zero. Such vectors are called *linearly independent* and will be further studied in Section 5.1. The vector equation $r\mathbf{a}+s\mathbf{b}+t\mathbf{c} = \mathbf{0}$ means the diagonal of the parallelepiped formed by the vectors $r\mathbf{a}$, $s\mathbf{b}$ and $t\mathbf{c}$ is the zero vector. See Figure 2.2.4 where the volume of the larger parallelepiped, given by the scaled vectors $r\mathbf{a}$, $s\mathbf{b}$ and $t\mathbf{c}$, is rst times the volume of the smaller parallelepiped. This

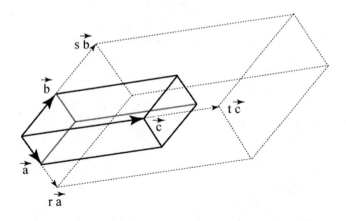

Figure 2.2.4: Determinant and Volume

follows from the determinant calculation

$$
\det([r\mathbf{a} \ \ s\mathbf{b} \ \ t\mathbf{c}]) = \det\left(\begin{bmatrix} a_1 r & b_1 s & c_1 t \\ a_2 r & b_2 s & c_2 t \\ a_3 r & b_3 s & c_3 t \end{bmatrix}\right)
$$
$$
= a_1 r((b_2 s)(c_3 t) - (c_2 t)(b_3 s)) + \cdots
$$
$$
= rst\,\det([\mathbf{a} \ \ \mathbf{b} \ \ \mathbf{c}]).
$$

Example 2.2.5. Compute the volume of the parallelepipeds given by the vectors $\mathbf{a}^T = [1\ 2\ 1]$, $\mathbf{b}^T = [2\ 0\ 1]$ and $\mathbf{c}^T = [0\ 1\ 4]$; $\widehat{\mathbf{a}}^T = [2\ 4\ 2]$, $\widehat{\mathbf{b}}^T = [6\ 0\ 3]$ and $\widehat{\mathbf{c}}^T = [0\ 4\ 16]$. Use determinants to find the first volume

$$
\text{(volume)} = abs(\det(\begin{bmatrix} 1 & 2 & 0 \\ 2 & 0 & 1 \\ 1 & 1 & 4 \end{bmatrix}))
$$
$$
= abs(1(0(4) - 1(1)) - 2(2(4) - 1(1)) + 0)
$$
$$
= abs(-15) = 15.
$$

The first volume is 15. The second volume is $2(3)(4)15 = 360$ because the new vectors are scalar multiples of the previous three with $r = 2$, $s = 3$ and $t = 4$.

2.2.4 Cross and Box Product Using MATLAB

The MATLAB commands det(A) and cross(a,b) compute the determinant of a square matrix A and the cross product of two vectors a and b. The following computations illustrate some of the above identities.

```
>> a = [1 2 1]; b = [2 0 1]; c = [0 1 4]
   c =
```

```
                   0 1 4
>> A = [a' b' c']          % convert rows to columns
        A =
                   1 2 0
                   2 0 1
                   1 1 4
>> det(A)
           ans =
                  -15
>> det(A')                 % det(A) = det(A')
           ans =
                  -15
>> newA = [c ; a; b]       % volume with base via a and b
        newA =
                   0 1 4
                   1 2 1
                   2 0 1
>> det(newA)
           ans =
                  -15
>> axb = cross(a,b)
        axb =
                   2 1 -4
>> dot(c,axb)              % box product
           ans =
                  -15
```

2.2.5   Exercises

1.   Let $\vec{a} = [2\ \ 3\ \ 1]$ and $\vec{b} = [3\ \ 0\ \ 1]$. Compute $\vec{a} \times \vec{b}$ and $\left\| \vec{a} \times \vec{b} \right\|$.

2.   Let $\vec{a} = 2\vec{i} + 4\vec{j} + \vec{k}$ and $\vec{b} = \vec{i} + 3\vec{j} - \vec{k}$. Compute $\vec{a} \times \vec{b}$ and $\left\| \vec{a} \times \vec{b} \right\|$.

3.   Let $\vec{a} = \vec{i} + 3\vec{j} + \vec{k}$ and $\vec{b} = 0\vec{i} + 3\vec{j} - \vec{k}$. Find the area of the parallelogram formed by these vectors.

4.   Let $\vec{a} = [2\ \ 0\ \ 1]$ and $\vec{b} = [3\ \ 1\ \ 1]$. Find the area of the parallelogram formed by these vectors.

5.   Show $\vec{a} \times \vec{b} = -\vec{b} \times \vec{a}$.

6.   Show $\vec{a} \times (\vec{b} \times \vec{c})$ may not be equal to $(\vec{a} \times \vec{b}) \times \vec{c}$. Do both computations for $\vec{a} = [2\ \ 3\ \ 1]$, $\vec{b} = [3\ \ 0\ \ 1]$ and $\vec{c} = [1\ \ 2\ \ 3]$.

7.   Consider the wrench torque in Example 2.2.3. If the ends of the wrench are given by $\vec{r} = [1\ \ 3\ \ 1]$ and the force vector has components in the x,y and z-directions equal to 2, 3 and 1, respectively, find the torque vector $\vec{r} \times \vec{f}$.

8.   Consider the wrench torque in Example 2.2.3. If the ends of the wrench are given by $\overrightarrow{r} = \overrightarrow{i} + 2\overrightarrow{j} + \overrightarrow{k}$ and the force vector is $\overrightarrow{f} = [2 \ 1 \ 4]$, compute the torque vector $\overrightarrow{r} \times \overrightarrow{f}$.

9.   A parallelepiped is given by three vectors $\overrightarrow{a} = 4\overrightarrow{i} + 1\overrightarrow{j} + \overrightarrow{k}$, $\overrightarrow{b} = \overrightarrow{i} + 5\overrightarrow{j} + \overrightarrow{k}$ and $\overrightarrow{c} = 2\overrightarrow{i} + \overrightarrow{j} + 5\overrightarrow{k}$.

    (a).   Use the box product to compute the volume.

    (b).   Use a determinant to find the volume.

10.   A parallelepiped is given by three vectors $\overrightarrow{a} = [1 \ 3 \ 1]$, $\overrightarrow{b} = [0 \ 3 \ 1]$ and $\overrightarrow{c} = [1 \ 2 \ 4]$.

    (a).   Use the box product to compute the volume.

    (b).   Use a determinant to find the volume.

11.   Prove the determinant identities in equations (2.2.4) and (2.2.5).

12.   Use the MATLAB command det(A) and cross(a,b) to do the computations in exercise nine.

13.   Use the MATLAB command det(A) and cross(a,b) to do the computations in exercise ten.

2.3   Lines and Curves in $\mathbb{R}^3$

This section extends the study of lines in the plane to lines in space. Vector and parametric representations of a line will be developed. Minimum distances from a point to a line and from line to a line will be analyzed. The cross product will be used to describe the minimum distance. More general curves including the helix and path of a projectile will be presented.

2.3.1   Lines in Space

Let a given point in space be represented by the position vector $\overrightarrow{r}_0 = [x_0 \ \ y_0$ $z_0]$ and let a given direction vector be $\overrightarrow{d} = [d_1 \ d_2 \ d_3]$. As for a line in a plane one can scale the direction vector and add it to the position vector whose end point is on the line to obtain the position vector for any point on the line

$$\overrightarrow{r}(t) = \overrightarrow{r}_0 + t\overrightarrow{d}.$$

This is called a *vector* equation for the line and is illustrated in Figure 2.3.1. The direction vector may be found from two distinct points on the line by subtracting the corresponding position vectors $\overrightarrow{d} = \overrightarrow{r}_1 - \overrightarrow{r}_0$. The *parametric* equations can be derived from the three scalar equations that correspond to the vector equation with $\overrightarrow{r}(t) = [x(t) \ y(t) \ z(t)]$

$$x(t) = x_0 + td_1, \ y(t) = y_0 + td_2 \text{ and } z(t) = z_0 + td_3.$$

Example 2.3.1. Find the vector and parametric representations of the line containing the two points $(1, 2, 4)$ and $(2, 5, 3)$. Let $\overrightarrow{r}_0 = [1 \ 2 \ 4]$ and $\overrightarrow{r}_1 = [2$

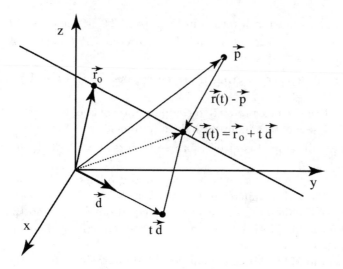

Figure 2.3.1: Vector Equation and Minimum Distance

$5\ 3]$ be the corresponding position vectors. A direction vector is computed by $\vec{d} = \vec{r}_1 - \vec{r}_0 = [2\ 5\ 3] - [1\ 2\ 4] = [1\ 3\ -1]$. Note, the choice for $\vec{r}_0$ and $\vec{r}_1$ will give a different direction vector, but it will be a multiple of the above. Also, one can choose either position vector to use in the vector representations

$$
\begin{aligned}
\vec{r}(t) &= \vec{r}_0 + t\vec{d} \\
[x(t)\ y(t)\ z(t)] &= [1\ 2\ 4] + t[1\ 3\ -1].
\end{aligned}
$$

The parametric representation is

$$x(t) = 1 + t,\ \ y(t) = 2 + 3t \text{ and } z(t) = 4 - t.$$

2.3.2   Minimum Distance to a Line

Consider a line given by $\vec{r}(t) = \vec{r}_0 + t\vec{d}$, and let a point (p_1, p_2, p_3) not be on this line and have position vector $\vec{p} = [p_1\ p_2\ p_3]$. The problem is to find the point on the line, given by some $t = t_0$, so that the distance from the point to the line is a minimum. The distance from the given end point of $\vec{p}$ to the end points of $\vec{r}(t)$ is the length of the vector $\vec{r}(t) - \vec{p} = (\vec{r}_0 - \vec{p}) + t\vec{d}$. Let $f(t) \equiv (\vec{r}(t) - \vec{p}) \bullet (\vec{r}(t) - \vec{p})$ so that one must solve the minimization problem

$$f(t_0) = \min_t f(t).$$

As in Section 1.5, this can be done by either calculus or by noting at $t = t_0$, the vector $\vec{r}(t) - \vec{p}$ must be perpendicular to the direction vector of the line (see Figure 2.3.1). This means the dot product of $\overrightarrow{r}(t) - \overrightarrow{p}$ with the direction vector $\overrightarrow{d}$ must be zero

$$
\begin{aligned}
\overrightarrow{d} \bullet (\overrightarrow{r}(t) - \overrightarrow{p}) &= 0 \\
\overrightarrow{d} \bullet ((\overrightarrow{r}_0 - \overrightarrow{p}) + t\overrightarrow{d}) &= 0 \\
\overrightarrow{d} \bullet (\overrightarrow{r}_0 - \overrightarrow{p}) + \overrightarrow{d} \bullet \overrightarrow{d} t &= 0.
\end{aligned}
$$

Thus, $t_0 = -\overrightarrow{d} \bullet (\overrightarrow{r}_0 - \overrightarrow{p})/(\overrightarrow{d} \bullet \overrightarrow{d})$. The minimum distance is the square root of $f(t_0)$.

The distance squared at $t = t_0$ simplifies to

$$
\begin{aligned}
f(t_0) &= (\overrightarrow{r}(t_0) - \overrightarrow{p}) \bullet (\overrightarrow{r}(t_0) - \overrightarrow{p}) \\
&= ((\overrightarrow{r}_0 - \overrightarrow{p}) + t_0 \overrightarrow{d}) \bullet (\overrightarrow{r}(t_0) - \overrightarrow{p}) \\
&= (\overrightarrow{r}_0 - \overrightarrow{p}) \bullet (\overrightarrow{r}(t_0) - \overrightarrow{p}) + 0 \\
&= (\overrightarrow{r}_0 - \overrightarrow{p}) \bullet ((\overrightarrow{r}_0 - \overrightarrow{p}) + t_0 \overrightarrow{d}).
\end{aligned}
$$

The computation of the minimum distance can be done by using the cross product and the Lagrange identity in Theorem 2.2.1

$$
\left\| \overrightarrow{a} \times \overrightarrow{b} \right\|^2 = \left\| \overrightarrow{a} \right\|^2 \left\| \overrightarrow{b} \right\|^2 - (\overrightarrow{a} \bullet \overrightarrow{b})^2. \tag{2.3.1}
$$

Next, insert $t_0 = -\overrightarrow{d} \bullet (\overrightarrow{r}_0 - \overrightarrow{p})/(\overrightarrow{d} \bullet \overrightarrow{d})$ into $f(t_0)$ and use the cross product identity with $\overrightarrow{a} = \overrightarrow{r}_0 - \overrightarrow{p}$ and $\overrightarrow{b} = \overrightarrow{d}$

$$
\begin{aligned}
f(t_0) &= (\overrightarrow{r}_0 - \overrightarrow{p}) \bullet (\overrightarrow{r}_0 - \overrightarrow{p}) + (\overrightarrow{r}_0 - \overrightarrow{p}) \bullet (t_0 \overrightarrow{d}) \\
&= \frac{(\overrightarrow{r}_0 - \overrightarrow{p}) \bullet (\overrightarrow{r}_0 - \overrightarrow{p}) \overrightarrow{d} \bullet \overrightarrow{d} - ((\overrightarrow{r}_0 - \overrightarrow{p}) \bullet \overrightarrow{d})^2}{\overrightarrow{d} \bullet \overrightarrow{d}} \\
&= \frac{\left\| (\overrightarrow{r}_0 - \overrightarrow{p}) \times \overrightarrow{d} \right\|^2}{\left\| \overrightarrow{d} \right\|^2}.
\end{aligned}
$$

Theorem 2.3.1 *(Minimum Distance from Point to Line) Let* $\overrightarrow{r}(t) = \overrightarrow{r}_0 + t\overrightarrow{d}$ *represent a line, and let a point* (p_1, p_2, p_3) *not be on this line and have position vector* $\overrightarrow{p} = [p_1 \ p_2 \ p_3]$. $\overrightarrow{r}(t_0) = \overrightarrow{r}_0 + t_0 \overrightarrow{d}$ *where* $t_0 = -\overrightarrow{d} \bullet (\overrightarrow{r}_0 - \overrightarrow{p})/\overrightarrow{d} \bullet \overrightarrow{d}$ *gives the point on the line that is closest to the point* (p_1, p_2, p_3). *Moreover, the minimum distance is*

$$
\frac{\left\| (\overrightarrow{r}_0 - \overrightarrow{p}) \times \overrightarrow{d} \right\|}{\left\| \overrightarrow{d} \right\|}.
$$

Example 2.3.2. Let a line be given by $\overrightarrow{r}_0 = [1 \ 1 \ 1]$ and $\overrightarrow{d} = [1 \ 2 \ 2]$, and find the minimum distance from the point $(4, 4, 1)$ to this line. According to the above theorem the point on the line that is closest is $\vec{r}(1) = \vec{r}_0 + 1\vec{d} = [1 \quad 1 \quad 1] + 1[1 \quad 2 \quad 2] = [2 \quad 3 \quad 3]$ where

$$
\begin{aligned}
t_0 &= -[1 \quad 2 \quad 2] \bullet ([1 \quad 1 \quad 1] - [4 \quad 4 \quad 1])/[1 \quad 2 \quad 2] \bullet [1 \quad 2 \quad 2] \\
&= -(-9)/9 = 1.
\end{aligned}
$$

One can compute the minimum distance directly by evaluating $\vec{r}(1)$ and computing the distance, or by using the above cross product formula, which only requires the two points and the direction vector. The direct calculation is

$$
\begin{aligned}
\vec{r}(1) - \vec{p} &= [2 \quad 3 \quad 3] - [4 \quad 4 \quad 1] = [-2 \quad -1 \quad 2] \\
\|\vec{r}(1) - \vec{p}\| &= 3.
\end{aligned}
$$

The cross product computation is

$$
\begin{aligned}
(\vec{r}_0 - \vec{p}) \times \vec{d} &= \det\left(\begin{bmatrix} \vec{i} & \vec{j} & \vec{k} \\ -3 & -3 & 0 \\ 1 & 2 & 2 \end{bmatrix} \right) \\
&= -6\,\vec{i} + 6\,\vec{j} - 3\,\vec{k} \\
&= [-6 \quad 6 \quad -3].
\end{aligned}
$$

This distance is the same as the direct approach

$$
\frac{\left\|(\vec{r}_0 - \vec{p}) \times \vec{d}\right\|}{\|\vec{d}\|} = \frac{9}{3} = 3.
$$

Another minimum distance problem is depicted in Figure 2.3.2, which involves two lines in space that are not parallel and do not intersect. We seek to find a point on one line and a point on the second line so that the distance between them is the smallest. Use the following vector representations for the two curves, where the parameter t locates a point on the first line and the parameter τ locates a point on the second line

$$
\begin{aligned}
\vec{r}(t) &= \vec{r}_0 + t\vec{d} \text{ and} \\
\vec{p}(\tau) &= \vec{p}_0 + \tau\vec{\delta}.
\end{aligned}
$$

An important observation is the vector $\vec{r}(t) - \vec{p}(\tau)$ must be perpendicular to both lines at the points where the minimum distance is attained. This means this vector must be perpendicular to both the direction vectors $\vec{d}$ and $\vec{\delta}$

$$
\begin{aligned}
(\vec{r}(t) - \vec{p}(\tau)) \bullet \vec{d} &= 0 \text{ and} \\
(\vec{r}(t) - \vec{p}(\tau)) \bullet \vec{\delta} &= 0.
\end{aligned}
$$

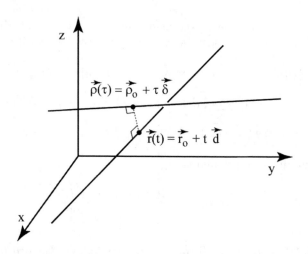

Figure 2.3.2: Distance between Two Lines

Since $\overrightarrow{r}(t) - \overrightarrow{\rho}(\tau) = (\overrightarrow{r}_0 - \overrightarrow{\rho}_0) + t\overrightarrow{d} - \tau\overrightarrow{\delta}$, these two dot product equations give two scalar equations for the two unknown parameters t and τ

$$(\overrightarrow{r}_0 - \overrightarrow{\rho}_0) \bullet \overrightarrow{d} + \overrightarrow{d} \bullet \overrightarrow{d} t - \overrightarrow{\delta} \bullet \overrightarrow{d} \tau = 0 \text{ and}$$
$$(\overrightarrow{r}_0 - \overrightarrow{\rho}_0) \bullet \overrightarrow{\delta} + \overrightarrow{d} \bullet \overrightarrow{\delta} t - \overrightarrow{\delta} \bullet \overrightarrow{\delta} \tau = 0.$$

The matrix form of this is

$$\begin{bmatrix} \overrightarrow{d} \bullet \overrightarrow{d} & -\overrightarrow{\delta} \bullet \overrightarrow{d} \\ \overrightarrow{d} \bullet \overrightarrow{\delta} & -\overrightarrow{\delta} \bullet \overrightarrow{\delta} \end{bmatrix} \begin{bmatrix} t \\ \tau \end{bmatrix} = \begin{bmatrix} -(\overrightarrow{r}_0 - \overrightarrow{\rho}_0) \bullet \overrightarrow{d} \\ -(\overrightarrow{r}_0 - \overrightarrow{\rho}_0) \bullet \overrightarrow{\delta} \end{bmatrix}. \qquad (2.3.2)$$

By Cramer's rule as discussed in Section 2.4 and in Theorem 2.4.1, the solution of this system will exist if the determinant of the 2×2 coefficient matrix is not zero

$$\det\left(\begin{bmatrix} \overrightarrow{d} \bullet \overrightarrow{d} & -\overrightarrow{\delta} \bullet \overrightarrow{d} \\ \overrightarrow{d} \bullet \overrightarrow{\delta} & -\overrightarrow{\delta} \bullet \overrightarrow{\delta} \end{bmatrix} \right) = -\left\| \overrightarrow{d} \right\|^2 \left\| \overrightarrow{\delta} \right\|^2 + (\overrightarrow{d} \bullet \overrightarrow{\delta})^2.$$

If the direction vectors are not parallel, then the cross product will not be zero. But, by the Lagrange identity in equation (2.3.1) with $\overrightarrow{a} = \overrightarrow{d}$ and $\overrightarrow{b} = \overrightarrow{\delta}$ this determinant is $-\left\| \overrightarrow{d} \times \overrightarrow{\delta} \right\|^2 \neq 0$. Once equation (2.3.2) is solved for $t = t_0$ and $\tau = \tau_0$, the points on the two lines given by the position vectors $\overrightarrow{r}(t_0)$ and $\overrightarrow{\rho}(\tau_0)$ can be computed as well as the minimum distance $\left\| \overrightarrow{r}(t_0) - \overrightarrow{\rho}(\tau_0) \right\|$.

The cross product can be used to compute the minimum distance without having to solve the algebraic system in (2.3.2). Observe the cross product $\overrightarrow{d} \times \overrightarrow{\delta}$ must be perpendicular to both $\overrightarrow{d}$ and $\overrightarrow{\delta}$ and, hence, it must also be parallel to $\vec{r}(t_0) - \vec{\rho}(\tau_0)$. This means the angle between $\vec{d} \times \vec{\delta}$ and $\vec{r}(t_0) - \vec{\rho}(\tau_0)$ must be either zero or 180 degrees so that by the cosine identity

$$(\vec{r}(t_0) - \vec{\rho}(\tau_0)) \bullet (\vec{d} \times \vec{\delta}) = \|\vec{r}(t_0) - \vec{\rho}(\tau_0)\| \left\|\vec{d} \times \vec{\delta}\right\| (\pm 1)$$

$$((\vec{r}_0 - \vec{\rho}_0) + \vec{d} t_0 - \vec{\delta} \tau_0) \bullet (\vec{d} \times \vec{\delta}) = \|\vec{r}(t_0) - \vec{\rho}(\tau_0)\| \left\|\vec{d} \times \vec{\delta}\right\| (\pm 1)$$

$$(\vec{r}_0 - \vec{\rho}_0) \bullet (\vec{d} \times \vec{\delta}) + 0 + 0 = \|\vec{r}(t_0) - \vec{\rho}(\tau_0)\| \left\|\vec{d} \times \vec{\delta}\right\| (\pm 1).$$

The minimum distance squared is

$$\|\vec{r}(t_0) - \vec{\rho}(\tau_0)\|^2 = \frac{((\vec{r}_0 - \vec{\rho}_0) \bullet (\vec{d} \times \vec{\delta}))^2}{\left\|\vec{d} \times \vec{\delta}\right\|^2}. \qquad (2.3.3)$$

Theorem 2.3.2 *(Minimum Distance Between Lines) Let two lines be given by the vector equations* $\vec{r}(t) = \vec{r}_0 + t\vec{d}$ *and* $\vec{\rho}(\tau) = \vec{\rho}_0 + \tau\vec{\delta}$. *If the two lines are not parallel and do not intersect, then* $\vec{d} \times \vec{\delta}$ *is not the zero vector and the solution of equation (2.3.2) gives the points on the lines with minimum distance. Moreover, the minimum distance may be computed by (2.3.3).*

Example 2.3.3. Let the two lines be given by $\vec{r}(t) = [1\ 2\ 1] + t[2\ 0\ 1]$ and $\vec{\rho}(\tau) = [0\ 2\ -1] + \tau[1\ -1\ 1]$.

$$\vec{d} \times \vec{\delta} = \det\left(\begin{bmatrix} \vec{i} & \vec{j} & \vec{k} \\ 2 & 0 & 1 \\ 1 & -1 & 1 \end{bmatrix}\right)$$

$$= 1\vec{i} - 1\vec{j} - 2\vec{k}$$

$$= [1\ -1\ -2].$$

Equation (2.3.3) is used to compute the minimum distance squared

$$\frac{((\vec{r}_0 - \vec{\rho}_0) \bullet (\vec{d} \times \vec{\delta}))^2}{\left\|\vec{d} \times \vec{\delta}\right\|^2} = \frac{([1\ 0\ 2] \bullet [1\ -1\ -2])^2}{6} = \frac{3}{2}.$$

Thus, the minimum distance is $\sqrt{3/2}$.

2.3.3   Curves and Projectiles

Lines in space are given by three linear polynomials of the parameter t: $x(t) = x_0 + d_1 t$, $y(t) = y_0 + d_2 t$ and $z(t) = z_0 + d_3 t$. If the linear polynomials are replaced by more complicated functions, then the curve will not be a straight line. An example is the *helix*. Two illustrations are the outer edges of a spiral staircase or a circular car ramp in a parking deck. The x and y components must be projected onto a circle and the z component should be proportional to

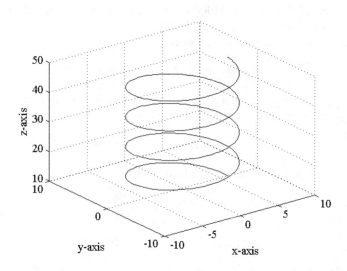

Figure 2.3.3: Helix

the parameter. Consider a circle with radius r, and assume the z component increases a height h for every revolution. One parametric equation for this is

$$
\begin{aligned}
x(t) &= r\cos(2\pi t), \\
y(t) &= r\sin(2\pi t) \text{ and} \\
z(t) &= ht.
\end{aligned}
$$

The MATLAB code helix.m was used to generate Figure 2.3.3.

Another interesting curve is a projectile's path as it moves through the air and is subjected to gravity, air resistance and wind forces. The projectile may be an artillery shell directed at a target or a sport ball moving towards a goal post. In order to model the motion, we apply Newton's law of motion to each of the three directions. Mass times acceleration is equal to the sum of the forces. Suppose the wind velocity is given by speeds in the three directions $\overrightarrow{w} = [w_1\ w_2\ w_3]$. Assume the force acting on the mass from air resistance is proportional to the velocity relative to the air, and let c be the proportionality constant. (A more realistic model for air resistance or drag is a quadratic function of the velocity.) The velocity vector of the mass is $\overrightarrow{v} = [x'\ y'\ z']$ and is derived from the position vector $\overrightarrow{r} = [x\ y\ z]$. The three equations are

$$
\begin{aligned}
mx'' &= cw_1 - cx' \\
my'' &= cw_2 - cy' \text{ and} \\
mz'' &= cw_3 - cz' - mg.
\end{aligned}
$$

Let the initial position be at the origin so that $x(0) = 0$, $y(0) = 0$ and $z(0) =$

0 and for ease of notation consider a unit mass $m = 1$. The initial velocity components are given $x'(0)$, $y'(0)$ and $z'(0)$.

Each of these equations has the form

$$v' = d - cv \text{ and } v(0) = given.$$

If d and c are constants, then this can be solved by making the change of variables $w \equiv d - cv$ and, thus, $w' = 0 - cv'$ and

$$w' = -cw \text{ and } w(0) = d - cv(0).$$

By the basic properties of exponential functions

$$
\begin{aligned}
w(t) &= w(0)e^{-ct} \\
d - cv(t) &= (d - cv(0))e^{-ct} \\
v(t) &= d/c - (d/c - v(0))e^{-ct}. \qquad (2.3.4)
\end{aligned}
$$

The differential equation for $v = x'$ uses $m = 1$, $d = cw_1$, $x'(0)$ given and $x(0) = 0$. Then equation (2.3.4) gives

$$
\begin{aligned}
x'(t) &= w_1 - (w_1 - x'(0))e^{-ct} \text{ and} \\
x(t) &= w_1 t + ((w_1 - x'(0))/c)e^{-ct} - (w_1 - x'(0))/c.
\end{aligned}
$$

The solution for $v = y'$ is similar

$$
\begin{aligned}
y'(t) &= w_2 - (w_2 - y'(0))e^{-ct} \text{ and} \\
y(t) &= w_2 t + ((w_1 - y'(0))/c)e^{-ct} - (w_2 - y'(0))/c.
\end{aligned}
$$

For $v = z'$ use $m = 1$, $d = cw_3 - mg = c0 - 32 = -32$, $z'(0)$ given and $z(0) = 0$

$$
\begin{aligned}
z'(t) &= -32/c - (-32/c - z'(0))e^{-ct} \text{ and} \\
z(t) &= -(32/c)t + ((-(32/c) - z'(0))/c)e^{-ct} + ((32/c) + z'(0))/c.
\end{aligned}
$$

The MATLAB code proj3d.m was used to create the graph for this curve in Figure 2.3.4.

2.3.4   Curves in Space Using MATLAB

The MATLAB command plot3(x,y,z) can be used to graph a curve in space where x, y and z are lists (vectors) of the components on the three axes, respectively. The previous mentioned MATLAB codes helix.m and proj3d.m use this command. The following illustrates how it can be used to create a graph of one of the lines in Example 2.3.3.

```
>> t = -3:.1:3;        % creates row vector with 61 numbers
>> x = 1+2*t;          % computes 1 + 2t for all entries in t
>> y = 2+0*t;          % computes 2 + 0t for all entries in t
```

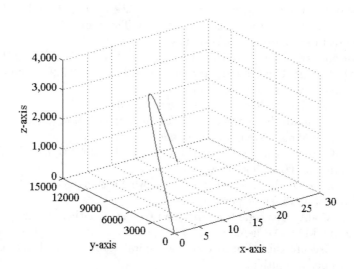

Figure 2.3.4: Projectile in Space

>> z = 1+1*t;   % computes $1 + 1t$ for all entries in t
>> plot3(x,y,z)

The command plot3() can be used to create more than one curve on the graph by simply listing a second triple of coordinate points. A simple illustration is the MATLAB code lines3d.m, which generates the lines in Example 2.3.3. Try the above codes, experiment with the input data and be sure to "rotate" the figures, that is, change perspectives or points of view.

2.3.5 Exercises

1.   Let a line be given by the two points $(1, 0, 3)$ and $(4, 1, 4)$. Find the vector equation for this line.

2.   Let a line be given by the point $(1, 2, -1)$ and the direction vector $\vec{d} = [1\ 1\ 2]$. Find the parametric equations for this line.

3.   Suppose a line has vector equation $\vec{r}(t) = [1\ 1\ 2] + t[-1\ 1\ 1]$.

  (a).   Find the point on the line that is closest to the point $(2, 4, 7)$.

  (b).   Compute this distance.

  (c).   Use the cross product as in Theorem 2.3.1 to compute the distance.

4.   Suppose a line has vector equation $\vec{r}(t) = [-1\ 2\ 2] + t[-1\ 2\ 1]$.

  (a).   Find the point on the line that is closest to the point whose position vector is $\vec{p} = [1\ 5\ 3]$.

  (b).   Compute this distance.

  (c).   Use the cross product as in Theorem 2.3.1 to compute the distance.

5.   Consider two lines given by $\overrightarrow{r}(t) = [1\ \ 1\ \ 2] + t[-1\ \ 1\ \ 1]$ and $\overrightarrow{\rho}(t) = [-1$
$2\ \ 2] + \tau[-1\ \ 2\ \ 1]$. Use the cross product as in Theorem 2.3.2 to compute the
minimum distance between the two lines.

6.   Consider two lines given by $\overrightarrow{r}(t) = [1\ \ 0\ \ -1] + t[1\ \ 1\ \ 3]$ and $\overrightarrow{\rho}(t) = [-1$
$-2\ \ 2] + \tau[-1\ \ 2\ \ 1]$. Use the cross product as in Theorem 2.3.2 to compute the
minimum distance between the two lines.

7.   Consider the two lines given in Example 2.3.3.

   (a).   Compute the components in the matrix equation (2.3.2).

   (b).   Solve this system for $t = t_0$ and $\tau = \tau_0$.

   (c).   Compute the minimum distance directly by $\|\overrightarrow{r}(t_0) - \overrightarrow{\rho}(\tau_0)\|$.

8.   Use the MATLAB lines3d.m.

   (a).   Execute this code to view the two lines in Example 2.3.3.   Use
"rotate" to obtain different perspectives.

   (b).   Modify this code to graph the two lines in either exercise 5 or 6.

9.   Use the MATLAB helix.m.

   (a).   Execute this code to view the "spiral staircase" with radius 6 and
height of 10 per revolution.

   (b).   Modify the code to consider a "spiral ramp" with radius 60 and
height of 30 per revolution.

10.   Use the MATLAB proj3d.m.

   (a).   Execute this code with the wind speeds $\overrightarrow{w} = [5\ \ -10\ \ 0]$ and $\overrightarrow{w} = [5$
$10\ \ 0]$.

   (b).   Execute this code with different air resistance coefficients $c = 0.01$
and $c = 0.005$ (drier air!).

2.4   Planes in $\mathbb{R}^3$

This section will give four descriptions of a plane in space.   The minimum
distance from a point to a plane with an application to curve fitting to data will
be presented. Determinants will be used to solve resulting algebraic systems.

2.4.1   Four Plane Descriptions

The first description uses a *single equation* with three variables x, y and z

$$ax + by + cz = d.$$

For example, if the four constants were positive, then the intercepts with the
coordinate axes will be positive and the graph would look similar to that given
in Figure 2.4.1. A point (x_0, y_0, z_0) is in the plane if and only if it satisfies the
equation

$$ax_0 + by_0 + cz_0 = d.$$

Therefore, any point (x, y, z) in the plane must satisfy

$$ax + by + cz \quad = \quad ax_0 + by_0 + cz_0 \text{ or}$$
$$a(x - x_0) + b(y - y_0) + c(z - z_0) \quad = \quad 0.$$

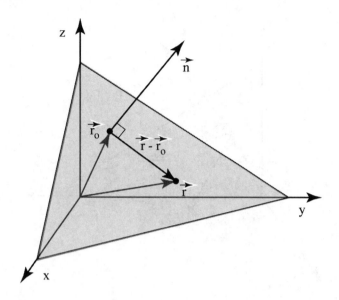

Figure 2.4.1: Normal and Point

The second description of a plane requires a *normal* vector $\overrightarrow{n}$ and a given point (x_0, y_0, z_0). Every vector in the plane must be perpendicular to the normal vector, see Figure 2.4.1. Let any point in the plane be given by the position vector $\overrightarrow{r} = [x \ y \ z]$ and let the given point have position vector $\overrightarrow{r}_0 = [x_0 \ y_0 \ z_0]$. Since $\overrightarrow{r} - \overrightarrow{r}_0$ is a vector in the plane,

$$\overrightarrow{n} \bullet (\overrightarrow{r} - \overrightarrow{r}_0) = 0.$$

The normal vector can be derived from algebraic equation formula from

$$
\begin{aligned}
a(x - x_0) + b(y - y_0) + c(z - z_0) &= 0 \\
[a \ b \ c] \bullet [x - x_0 \ y - y_0 \ z - z_0] &= 0.
\end{aligned}
$$

Thus, one can choose $\overrightarrow{n} = [a \ b \ c]$.

Example 2.4.1. Consider the algebraic equation $6x + 2y + 3z = 6$. The x, y and z intercepts are $x = 1$, $y = 3$ and $z = 2$. If $x_0 = 1/2$ and $y_0 = 1/2$, then z_0 must be $(6 - 6(1/2) - 2(1/2))/3 = 2/3$. The normal vector is $\overrightarrow{n} = [a \ b \ c] = [6 \ 2 \ 3]$ and we can use $\overrightarrow{r}_0 = [1/2 \ 1/2 \ 2/3]$ so that

$$
\begin{aligned}
[a \ b \ c] \bullet [x - x_0 \ y - y_0 \ z - z_0] &= 0 \\
[6 \ 2 \ 3] \bullet [x - 1/2 \ y - 1/2 \ z - 2/3] &= 0.
\end{aligned}
$$

The third description uses *three distinct points* that are not on the same line. Let the position vectors $\overrightarrow{r}_0 = [x_0 \ y_0 \ z_0]$, $\overrightarrow{r}_1 = [x_1 \ y_1 \ z_1]$ and $\overrightarrow{r}_2 = [x_2$

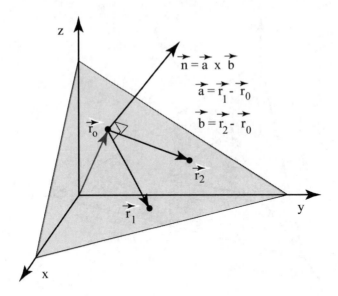

Figure 2.4.2:   Three Points

y_2 z_2] be associated with these points. Then the vectors $\vec{a} = \vec{r}_1 - \vec{r}_0$ and $\vec{b} = \vec{r}_2 - \vec{r}_0$ beginning at the first point and ending at the second and third points are in the plane. Their cross product will be perpendicular to both, as is illustrated in Figure 2.4.2. Since any vector in the plane is a linear combination of $\vec{a}$ and $\vec{b}$, the normal vector can be chosen to be this cross product

$$\vec{n} = \vec{a} \times \vec{b} = (\vec{r}_1 - \vec{r}_0) \times (\vec{r}_2 - \vec{r}_0).$$

The fourth description requires a *point and two non-parallel vectors* in the plane. The two non-parallel vectors could be calculated from three points as above. Any point in the plane can be found by adding the position vector of the given point, a scaled first vector and a scaled second vector

$$\vec{r} = \vec{r}_0 + r\vec{a} + s\vec{b}.$$

The scaling factors are real numbers r and s such that the linear combination $r\vec{a} + s\vec{b}$ is a vector beginning at the point (x_0, y_0, z_0) and ending at a desired point in the plane (x, y, z). See Figure 2.4.3 for an illustration of the linear combination.

Example 2.4.2. Find the plane that contains the points $(1, 1, 3), (2, 2, 1)$ and $(0, 3, 2)$. Choose the position vectors $\vec{r}_0 = [1 \ 1 \ 3], \vec{r}_1 = [2 \ 2 \ 1]$ and $\vec{r}_2 = [0 \ 3 \ 2]$ and compute

$$\vec{a} = [2 \ 2 \ 1] - [1 \ 1 \ 3] = [1 \ 1 \ -2] \text{ and}$$
$$\vec{b} = [0 \ 3 \ 2] - [1 \ 1 \ 3] = [-1 \ 2 \ -1].$$

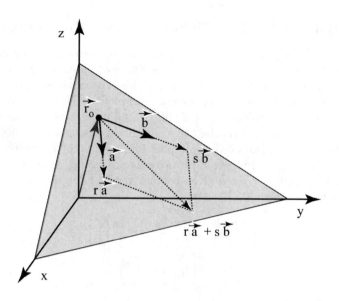

Figure 2.4.3: Linear Combination of Vectors

The normal vector is computed by using the cross product

$$\vec{n} = \vec{a} \times \vec{b} = \det\left(\begin{bmatrix} \vec{i} & \vec{j} & \vec{k} \\ 1 & 1 & -2 \\ -1 & 2 & -1 \end{bmatrix}\right)$$
$$= 3\vec{i} + 3\vec{j} + 3\vec{k} = [3\ 3\ 3].$$

Use this normal vector in the dot product

$$\vec{n} \bullet (\vec{r} - \vec{r}_0) = 0$$
$$[3\ 3\ 3] \bullet [x-1\ \ y-1\ \ z-3] = 0$$
$$3(x-1) + 3(y-1) + 3(z-3) = 0.$$

Thus the algebraic description is $x + y + z = 5$. A linear combination representation of this plane is

$$\vec{r} = \vec{r}_0 + r\vec{a} + s\vec{b}$$
$$[x\ y\ z] = [1\ 1\ 3] + r[1\ 1\ -2] + s[-1\ 2\ -1].$$

The column version of this is

$$\begin{bmatrix} x \\ y \\ z \end{bmatrix} = \begin{bmatrix} 1 \\ 1 \\ 3 \end{bmatrix} + r\begin{bmatrix} 1 \\ 1 \\ -2 \end{bmatrix} + s\begin{bmatrix} -1 \\ 2 \\ -1 \end{bmatrix}. \qquad (2.4.1)$$

The three scalar equations are $x = 1 + 1r - 1s$, $y = 1 + 1r + 2s$ and $z = 3 - 2r - 1s$, which satisfy $x + y + z = 5$.

2.4.2   Solving for Two Unknowns with Determinants

In Example 2.4.2 suppose a particular point on the plane is given and we want
to find the parameters r and s so that the linear combination gives the point.
This will require solving for two unknowns. This type of problem also appeared
in the previous section where we were trying to find the points on two lines
that were closest. A similar problem will be finding a point in a plane that is
closest to a given point not on the plane. For linear systems with small number
of unknowns, determinants give a compact formula for the solutions.

Let a 2×2 matrix A be given by two column vectors $\mathbf{a}$ and $\mathbf{b}$, and let the
unknowns be r and s. With a given column vector $\mathbf{f}$ for the right side of the
vector equation we have

$$\begin{bmatrix} a_1 & b_1 \\ a_2 & b_2 \end{bmatrix} \begin{bmatrix} r \\ s \end{bmatrix} = \begin{bmatrix} f_1 \\ f_2 \end{bmatrix} \text{ or } \qquad (2.4.2)$$

$$r \begin{bmatrix} a_1 \\ a_2 \end{bmatrix} + s \begin{bmatrix} b_1 \\ b_2 \end{bmatrix} = \begin{bmatrix} f_1 \\ f_2 \end{bmatrix}.$$

The two equivalent scalar equations are

$$a_1 r + b_1 s = f_1 \text{ and } a_2 r + b_2 s = f_2.$$

Multiply the second equation by a_1, multiply the first equation by a_2 and subtract the two to get

$$(a_1 b_2 - a_2 b_1)s = a_1 f_2 - a_2 f_1.$$

A similar calculation eliminates r and gives

$$(a_1 b_2 - a_2 b_1)r = f_1 b_2 - f_2 b_1.$$

The coefficient on the left side is the definition of the determinant of the 2×2
matrix A. By moving the column vector $\mathbf{f}$ into the first or second column of A,
one can express the right sides using determinants. This is *Cramer's rule* for
two unknowns.

Theorem 2.4.1 *(Cramer's Rule) Let A be the 2×2 matrix in equation (2.4.2).
If $\det(A) \neq 0$, then the solution is*

$$r = \det\left(\begin{bmatrix} f_1 & b_1 \\ f_2 & b_2 \end{bmatrix}\right) / \det(A) \text{ and } s = \det\left(\begin{bmatrix} a_1 & f_1 \\ a_2 & f_2 \end{bmatrix}\right) / \det(A).$$

Example 2.4.3. Consider Example 2.4.2 where the plane has representation
in equation (2.4.1) or $x + y + z = 5$. The point $(x, y, z) = (4, 3, -2)$ is in this
plane. The problem is to find r and s so that equation (2.4.1) holds. Since
$(4, 3, -2)$ is in this plane, we only need to consider the first two of the scalar
equations $x = 4 = 1 + 1r - 1s$, $y = 3 = 1 + 1r + 2s$

$$r - s = 3$$
$$r + 2s = 2.$$

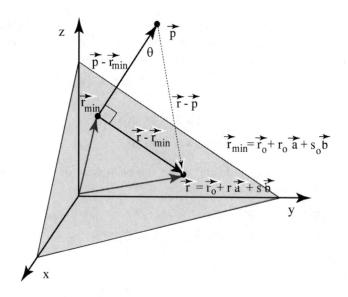

Figure 2.4.4: Minimum Distance to a Plane

The vector form of this is

$$\begin{bmatrix} 1 & -1 \\ 1 & 2 \end{bmatrix} \begin{bmatrix} r \\ s \end{bmatrix} = \begin{bmatrix} 3 \\ 2 \end{bmatrix}.$$

The determinant of the matrix A is $1(2) - (-1)1 = 3$ and the solution is

$$r = \det\left(\begin{bmatrix} 3 & -1 \\ 2 & 2 \end{bmatrix}\right)/3 = 8/3 \text{ and } s = \det\left(\begin{bmatrix} 1 & 3 \\ 1 & 2 \end{bmatrix}\right)/3 = -1/3.$$

It is easy to see that all three scalar equations hold.

2.4.3 Minimum Distance to a Plane

This section extends the study of minimum distance from a given point to a line in a plane to a given point in space to a plane in space. As in Section 1.4, one can approach the problem either using calculus or geometry. The calculus approach requires minimizing a function of two variables, and we will defer this to the multivariable calculus course. The geometric approach gives the same result and relies on the visualization of the problem as depicted in Figure 2.4.4.

Let the plane be given by a point and two non-parallel vectors in the plane

$$\overrightarrow{r} = \overrightarrow{r}_0 + r\overrightarrow{a} + s\overrightarrow{b}.$$

If the given point not in the plane is represented by the position vector $\overrightarrow{p}$, then we must find the real numbers $r = r_0$ and $s = s_0$ so that

$$\overrightarrow{r}_{\min} \equiv \overrightarrow{r}_0 + r_0\overrightarrow{a} + s_0\overrightarrow{b}$$

is closest to the $\vec{p}$. In this case, the vectors $\vec{p} - \vec{r}_{\min}$ and $\vec{r} - \vec{r}_{\min}$ are perpendicular for all points in the plane represented by the position vector. Since $\vec{a}$ and $\vec{b}$ are in the plane, we may choose either $\vec{r} = \vec{r}_{\min} + \vec{a}$ or $\vec{r} = \vec{r}_{\min} + \vec{b}$. In these cases, $\vec{p} - \vec{r}_{\min}$ must be perpendicular to both $\vec{a}$ and $\vec{b}$

$$(\vec{p} - \vec{r}_{\min}) \bullet \vec{a} = 0 \text{ and}$$
$$(\vec{p} - \vec{r}_{\min}) \bullet \vec{b} = 0.$$

Since $\vec{r}_{\min} \equiv \vec{r}_0 + r_0 \vec{a} + s_0 \vec{b}$ and using the properties of the dot product, we obtain the following scalar equations

$$(\vec{p} - \vec{r}_0) \bullet \vec{a} - \vec{a} \bullet \vec{a} r_0 - \vec{b} \bullet \vec{a} s_0 = 0 \text{ and}$$
$$(\vec{p} - \vec{r}_0) \bullet \vec{b} - \vec{a} \bullet \vec{b} r_0 - \vec{b} \bullet \vec{b} s_0 = 0.$$

The matrix form of this is

$$\begin{bmatrix} \vec{a} \bullet \vec{a} & \vec{b} \bullet \vec{a} \\ \vec{a} \bullet \vec{b} & \vec{b} \bullet \vec{b} \end{bmatrix} \begin{bmatrix} r_0 \\ s_0 \end{bmatrix} = \begin{bmatrix} (\vec{p} - \vec{r}_0) \bullet \vec{a} \\ (\vec{p} - \vec{r}_0) \bullet \vec{b} \end{bmatrix}. \tag{2.4.3}$$

The determinant of the 2×2 coefficient matrix is

$$(\vec{a} \bullet \vec{a})(\vec{b} \bullet \vec{b}) - (\vec{a} \bullet \vec{b})^2 = \|\vec{a}\|^2 \|\vec{b}\|^2 - (\vec{b} \bullet \vec{a})^2$$
$$= \|\vec{a} \times \vec{b}\|^2.$$

Since the vectors $\vec{a}$ and $\vec{b}$ are required not to be parallel, this determinant is not zero and equation (2.4.3) must have a solution. Once the solution is found, $\vec{r}_{\min}$ and the desired distance $\|\vec{p} - \vec{r}_{\min}\|$ can be computed.

An alternative way to compute $\|\vec{p} - \vec{r}_{\min}\|$ is to use a normal vector $\vec{n}$ to the plane. Then the cosine of the angle between $\vec{p} - \vec{r}_{\min}$ and $\vec{r}_0 - \vec{p}$ is

$$\cos(\theta) = \frac{\vec{n} \bullet (\vec{r}_0 - \vec{p})}{\|\vec{n}\| \|\vec{r}_0 - \vec{p}\|}.$$

The angle θ is indicated in Figure 2.4.4 with $\vec{r}$ set equal to $\vec{r}_0$ so that

$$\|\vec{p} - \vec{r}_{\min}\| = \|\vec{r}_0 - \vec{p}\| |\cos(\theta)|$$
$$= \left| \frac{\vec{n} \bullet (\vec{r}_0 - \vec{p})}{\|\vec{n}\|} \right|. \tag{2.4.4}$$

Theorem 2.4.2 *(Minimum Distance to a Plane) Let a plane be given by $\vec{r} = \vec{r}_0 + r\vec{a} + s\vec{b}$ where $\vec{r}_0$ is a position vector of a given point in the plane and $\vec{a}$ and $\vec{b}$ are not parallel vectors in the plane. If $\vec{p}$ is a position vector of a point not in the plane, then the solution of equation (2.4.3) gives the point in the plane closest to the point associated with $\vec{p}$. Moreover, the minimum distance may also be calculated by equation (2.4.4).*

Example 2.4.4. Return to the Example 2.4.2 where the plane is given by

$$\overrightarrow{r} = \overrightarrow{r}_0 + r\overrightarrow{a} + s\overrightarrow{b}$$
$$[x\ y\ z] = [1\ 1\ 3] + r[1\ 1\ -2] + s[-1\ 2\ -1].$$

Let the point not on this plane be given by the position vector $\overrightarrow{p} = [10\ 11\ 12]$ and compute $\overrightarrow{p} - \overrightarrow{r}_0 = [10\ 11\ 12] - [1\ 1\ 3] = [9\ 10\ 9]$. Then equation (2.4.3) is

$$\begin{bmatrix} 6 & 3 \\ 3 & 6 \end{bmatrix}\begin{bmatrix} r_0 \\ s_0 \end{bmatrix} = \begin{bmatrix} 1 \\ 2 \end{bmatrix}.$$

Apply Cramer's rule to obtain $r_0 = 0/27$ and $s_0 = 9/27$. Then $\overrightarrow{r}_{min} \equiv \overrightarrow{r}_0 + r_0\overrightarrow{a} + s_0\overrightarrow{b} = [2/3\ 5/3\ 8/3]$ and $\|\overrightarrow{p} - \overrightarrow{r}_{min}\| = 28/\sqrt{3}$. This is consistent with equation (2.4.4)

$$\|\overrightarrow{p} - \overrightarrow{r}_{min}\| = \frac{\overrightarrow{n} \bullet (\overrightarrow{r}_0 - \overrightarrow{p})}{\|\overrightarrow{n}\|}$$
$$= \frac{[1\ 1\ 1] \bullet [9\ 10\ 9]}{\sqrt{3}} = 28/\sqrt{3}.$$

2.4.4 Application to Curve Fitting to Data

A consumer would like to predict the price of a high definition television based upon prices from the previous three months of $2000, \$1950$ and $\$1910$ for the first, second and third months of the year. One way to view this problem is to plot the prices on the vertical axis versus the times $1, 2$ and 3. Then try to draw a line "closest" to the data, estimate the slope m and the y-intercept c, and use these in the formula $y = mx + c$ where x is the time in months and y is the possible price of the television. Here we are trying to "fit" a line to the three points of price data.

A more precise way of doing this is to view m and c as unknowns in the following three scalar equations

$$m1 + c = 2000$$
$$m2 + c = 1950 \text{ and}$$
$$m3 + c = 1910.$$

This is an over determined system with three equations and two unknowns, and in general does not have a solution to all three equations. In terms of matrices this is

$$\begin{bmatrix} 1 & 1 \\ 2 & 1 \\ 3 & 1 \end{bmatrix}\begin{bmatrix} m \\ c \end{bmatrix} = \begin{bmatrix} 2000 \\ 1950 \\ 1910 \end{bmatrix} \qquad (2.4.5)$$

$$m\begin{bmatrix} 1 \\ 2 \\ 3 \end{bmatrix} + c\begin{bmatrix} 1 \\ 1 \\ 1 \end{bmatrix} = \begin{bmatrix} 2000 \\ 1950 \\ 1910 \end{bmatrix}.$$

Consider the plane in space determined by the linear combination of the columns of the 3×2 matrix and find the point in the plane that is closest to the point given by the price column vector on the right side. Apply the results in the previous section with $\vec{r}_0 = [0\ \ 0\ \ 0]$, $\vec{p} = [2000\ \ 1950\ \ 1910]$, $\vec{a} = [1\ \ 2\ \ 3]$ and $\vec{b} = [1\ \ 1\ \ 1]$. Equation (2.4.3) gives the solution of the minimum distance problem for $m = r_0$ and $c = s_0$

$$\begin{bmatrix} 14 & 6 \\ 6 & 3 \end{bmatrix} \begin{bmatrix} m \\ c \end{bmatrix} = \begin{bmatrix} 11630 \\ 5860 \end{bmatrix}.$$

Apply Cramer's rule to solve for $m = -270/6 = -45.00$ and $c = 12260/6 \approx 2043.33$. Use these to predict the price in the fourth month

$$\begin{aligned} y &= mx + c \\ &\approx (-45.00)4 + 2043.33 = 1863.33. \end{aligned}$$

2.4.5   Solving Systems Using MATLAB

The MATLAB command A\d can be used to solve $Ax = d$ where the matrix A has n rows and n columns. In the above curve fitting problem $n = 2$ and the following simple MATLAB computations produce the same results as Cramer's rule.

```
>> A =[14 6;6 3]
        A =
                14  6
                 6  3
>> d =[11630 5860]'
        d =
                11630
                 5860
>> A\d
        ans =
        1.0e+003 *
            -0.0450
             2.0433
```

The MATLAB command A\d can also be used to solve certain over determined systems such as in equation (2.4.5).

```
>> A_big = [1 1;2 1;3 1]
        A_big =
                1  1
                2  1
                3  1
>> d_big = [2000; 1950; 1910]
        d_big =
```

$$2000$$
$$1950$$
$$1910$$
$$>> A\_big\backslash d\_big$$
$$\text{ans} =$$
$$1.0e+003 \ *$$
$$-0.0450$$
$$2.0433$$

A more careful study of what is done to generate possible solutions will be given in the next two chapters.

2.4.6 Exercises

1. Consider the plane given by $x + 2y + 4z = 12$.
   (a). Verify the point $(2, 1, 2)$ is in this plane.
   (b). Find the normal to the plane.
   (c). Find the normal equation representation of this plane.
2. Consider the plane given by $5x + y + 4z = 20$.
   (a). Verify the point $(0, 4, 4)$ is in this plane.
   (b). Find the normal to the plane.
   (c). Find the normal equation representation of this plane.
3. Let a plane be given by the three points $(1, 1, 1)$, $(0, 3, 0)$ and $(3, 0, 1)$.
   (a). Find a normal vector to this plane.
   (b). Compute the algebraic equation that describes this plane.
4. Let a plane be given by the three points $(1, 0, 1)$, $(1, 3, 0)$ and $(0, -1, 1)$.
   (a). Find a normal vector to this plane.
   (b). Compute the algebraic equation that describes this plane.
5. Consider a plane given by

$$[x \ \ y \ \ z] = [1 \ \ 1 \ \ 1] + r[2 \ \ 3 \ \ 0] + s[1 \ \ 0 \ \ 0].$$

   (a). Find the algebraic system in equation (2.4.2) such that $[x \ \ y \ \ z] = [2 \ \ 3 \ \ 1]$.
   (b). Use Cramer's rule to find r and s.
6. Consider a plane given by

$$[x \ \ y \ \ z] = [1 \ \ 0 \ \ 1] + r[4 \ \ 0 \ \ 1] + s[1 \ \ 0 \ \ 1].$$

   (a). Find the algebraic system in equation (2.4.2) such that $[x \ \ y \ \ z] = [1 \ \ 0 \ \ 2]$.
   (b). Use Cramer's rule to find r and s.
7. Consider a plane given by

$$[x \ \ y \ \ z] = [1 \ \ 1 \ \ 1] + r[2 \ \ 3 \ \ 1] + s[1 \ \ 0 \ \ 1].$$

Find the point in the plane that is closest to the point $(5, 4, 6)$.

8.   Consider a plane given by

$$[x \ \ y \ \ z] = [1 \ \ 0 \ \ 1] + r[0 \ \ 3 \ \ 1] + s[1 \ \ 10 \ \ 1].$$

Find the point in the plane that is closest to the point $(5, 4, 6)$.

9.   A consumer would like to predict the price of a laptop computer based upon prices from the previous three months of $1000, $960 and $910 for the first, second and third months of the year. Model this by assuming the price decreases as a linear function of time similar to equation (2.4.5).

(a).   Find the slope and the y-intercept of a straight line "closest" to this data.

(b).   Use these to predict the price after four months.

10.   A consumer would like to predict the price of a laptop computer based upon prices from the previous months one, two and four equal to $1000, $960 and $870, respectively. Model this by assuming the price decreases as a linear function of time similar to equation (2.4.5).

(a).   Find the slope and the y-intercept of a straight line "closest" to this data.

(b).   Use these to predict the price after five months.

2.5   Extensions to $\mathbb{R}^n$

This section is an introduction to vectors with more than three components and to the remaining chapters in this book. There are sections with applications to inventory and price vectors, larger algebraic systems of equations, curve fitting with more than three data points, higher dimensional spaces of vectors, systems of ordinary differential equations and matrices used in the context of visualization. Even though we have mostly considered vectors with two or three components, which may represent points in space or physical quantities such as force and velocity vectors, we have used vectors with a number of components to store information. Examples are the spiral and projectile curves in Subsection 2.3.3 and the lines as given by the MATLAB commands in Subsection 2.3.4. In the line example, four vectors with 61 components were created to store the values for the parameter and the x, y and z values of the points on the line. Many of the operations on vectors have natural generalization to higher dimensions, but some of the geometric aspects are lost as we move beyond dimension two and three.

2.5.1   Inventory and Price Vectors

A store carries six different styles of television sets. There are a number of each style in the inventory, and there is a price per set for each style. So, there are a total of 12 numbers to be recorded. These can be stored in two vectors each with six components. The inventory or stock numbers will be denoted by a column vector **s**, and the price numbers will be stored in a column vector **p**.

Example 2.5.1.

$$\mathbf{s} = \begin{bmatrix} 2 \\ 4 \\ 5 \\ 3 \\ 7 \\ 1 \end{bmatrix} \text{ and } \mathbf{p} = \begin{bmatrix} 400 \\ 250 \\ 280 \\ 900 \\ 700 \\ 100 \end{bmatrix}.$$

If the stock is doubled and the price is decreased by 10%, then the new stock and price vectors are

$$2\mathbf{s} = \begin{bmatrix} 4 \\ 8 \\ 10 \\ 6 \\ 14 \\ 2 \end{bmatrix} \text{ and } .9\mathbf{p} = \begin{bmatrix} 360 \\ 225 \\ 252 \\ 810 \\ 630 \\ 90 \end{bmatrix}.$$

The total value of the original stock and price is given by a "dot product" of the stock and price vectors

$$\begin{aligned} \mathbf{s}^T\mathbf{p} &= 2(400) + 4(250) + 5(280) \\ &\quad + 3(900) + 7(700) + 1(100) \\ &= 10900. \end{aligned}$$

If there is a second store with a known stock vector $\widehat{\mathbf{s}}$, then the total stock can be computed by "adding" the vectors. For example, if the second store stock vector is

$$\widehat{\mathbf{s}} = \begin{bmatrix} 1 \\ 2 \\ 2 \\ 5 \\ 3 \\ 4 \end{bmatrix},$$

then the total stock is

$$\widehat{\mathbf{s}} + \mathbf{s} = \begin{bmatrix} 1+2 \\ 2+4 \\ 2+5 \\ 5+3 \\ 3+7 \\ 4+1 \end{bmatrix} = \begin{bmatrix} 3 \\ 6 \\ 7 \\ 8 \\ 10 \\ 5 \end{bmatrix}.$$

The total value of the two stores can be computed in two ways

$$\begin{aligned} \mathbf{s}^T\mathbf{p} + \widehat{\mathbf{s}}^T\mathbf{p} &= 10900 + 8460 \text{ or} \\ (\widehat{\mathbf{s}} + \mathbf{s})^T\mathbf{p} &= 19260. \end{aligned}$$

Large department stores have thousands of products on their shelves, and their inventory numbers and prices can be stored in vectors with large numbers of components.

2.5.2   Algebraic Systems

In the previous sections algebraic systems with two unknowns and two equations were solved using determinants and Cramer's rule. In Subsection 2.2.3 the non-zero volume of a parallelepiped is generated by three column vectors $\mathbf{a}, \mathbf{b}$ and $\mathbf{c}$. Any point in space can be represented by a linear combination of these vectors (see Figure 2.2.4)

$$r\mathbf{a} + s\mathbf{b} + t\mathbf{c} = \mathbf{r}_0$$

$$\begin{bmatrix} a_1 & b_1 & c_1 \\ a_2 & b_2 & c_2 \\ a_3 & b_3 & c_3 \end{bmatrix} \begin{bmatrix} r \\ s \\ t \end{bmatrix} = \begin{bmatrix} x_0 \\ y_0 \\ z_0 \end{bmatrix}.$$

Given the three column vectors and the point, we want to find r, s and t. Determinants and a generalization of Cramer's rule can be used to compute these coefficients. For three unknowns the rule is (see Section 3.7)

$$\begin{aligned} r &= \det([\; \mathbf{r}_0 \quad \mathbf{b} \quad \mathbf{c}\;])/\det([\; \mathbf{a} \quad \mathbf{b} \quad \mathbf{c}\;]), \\ s &= \det([\; \mathbf{a} \quad \mathbf{r}_0 \quad \mathbf{c}\;])/\det([\; \mathbf{a} \quad \mathbf{b} \quad \mathbf{c}\;]) \text{ and} \\ t &= \det([\; \mathbf{a} \quad \mathbf{b} \quad \mathbf{r}_0\;])/\det([\; \mathbf{a} \quad \mathbf{b} \quad \mathbf{c}\;]). \end{aligned}$$

Example 2.5.2. Consider the Example 2.2.5 where $\mathbf{a}^T = [1 \; 2 \; 1]$, $\mathbf{b}^T = [2 \; 0 \; 1]$, $\mathbf{c}^T = [0 \; 1 \; 4]$ and $\det([\; \mathbf{a} \quad \mathbf{b} \quad \mathbf{c}\;]) = -15$. If $\mathbf{r}_0^T = [5 \; 6 \; 4]$, then solve

$$\begin{bmatrix} 1 & 2 & 0 \\ 2 & 0 & 1 \\ 1 & 1 & 4 \end{bmatrix} \begin{bmatrix} r \\ s \\ t \end{bmatrix} = \begin{bmatrix} 5 \\ 6 \\ 4 \end{bmatrix}.$$

The three additional determinants are easy but messy to compute

$$\begin{aligned} r &= \det([\; \mathbf{r}_0 \quad \mathbf{b} \quad \mathbf{c}\;])/(-15) = -45/(-15) = 3, \\ s &= \det([\; \mathbf{a} \quad \mathbf{r}_0 \quad \mathbf{c}\;])/(-15) = -15/(-15) = 1 \text{ and} \\ t &= \det([\; \mathbf{a} \quad \mathbf{b} \quad \mathbf{r}_0\;])/(-15) = 0/(-15) = 0. \end{aligned}$$

For larger algebraic problems determinants are not used and usually row operations are better suited for the solution. An example of a large system would be from models of building whose support is from beams connected at a number of nodes or joints. In order for the building to be in a static state, the sum of the forces at each node must be zero. Each node has a small number of equations, but the building usually will have a large number of nodes. Thus, there are a large number of equations. This will be discussed in chapter three.

2.5.3 Curve Fitting

In Subsection 2.4.4 the minimum distance from a point in space to a plane was used to fit a line to three data points. If there are more than three data points, then the minimization problem will be in a higher dimensional space. For example, consider an additional data point to the data in Subsection 2.4.4

$$
\begin{array}{rcl}
m1 + c &=& 2000 \\
m2 + c &=& 1950 \\
m3 + c &=& 1910 \text{ and} \\
m4 + c &=& 1890.
\end{array}
$$

In terms of matrices this is

$$
\begin{bmatrix} 1 & 1 \\ 2 & 1 \\ 3 & 1 \\ 4 & 1 \end{bmatrix} \begin{bmatrix} m \\ c \end{bmatrix} = \begin{bmatrix} 2000 \\ 1950 \\ 1910 \\ 1890 \end{bmatrix}
$$

$$
m \begin{bmatrix} 1 \\ 2 \\ 3 \\ 4 \end{bmatrix} + c \begin{bmatrix} 1 \\ 1 \\ 1 \\ 1 \end{bmatrix} = \begin{bmatrix} 2000 \\ 1950 \\ 1910 \\ 1890 \end{bmatrix}.
$$

The solution of this requires a notion of perpendicular vectors in four dimensional space! The solution has the same form and is called the *normal equations*. Let $\mathbf{a}^T = [1 \ 2 \ 3 \ 4]$, $\mathbf{b}^T = [1 \ 1 \ 1 \ 1]$ and $\mathbf{p}^T = [2000 \ 1950 \ 1910 \ 1890]$

$$
\begin{bmatrix} \mathbf{a}^T\mathbf{a} & \mathbf{a}^T\mathbf{b} \\ \mathbf{b}^T\mathbf{a} & \mathbf{b}^T\mathbf{b} \end{bmatrix} \begin{bmatrix} m \\ c \end{bmatrix} = \begin{bmatrix} \mathbf{a}^T\mathbf{p} \\ \mathbf{b}^T\mathbf{p} \end{bmatrix}.
$$

These concepts will be developed in chapter four.

2.5.4 Multiple Solution of Algebraic Systems

Consider a single equation with three variables $ax + by + cz = d$. The points in space that satisfy this equation form a plane, which may be represented by a linear combination of two vectors plus a given vector. The plane is a two dimension subset in a three dimensional space. If there are two equations and the corresponding planes are not parallel, then they intersect and the set of points in this intersection is a line, which is a one dimensional subset in a three dimensional space.

Example 2.5.3. Find the line given by the intersection of the planes $x+y+z = 1$ and $6x + 2y + 3z = 6$. The direction vector of the line must be perpendicular to both normal vectors, which are $\vec{n}_1 = [1 \ 1 \ 1]$ and $\vec{n}_2 = [6 \ 2 \ 3]$. Since the cross product of two vectors is perpendicular to both vectors, we can choose the direction vector $\vec{d}$ to be

$$\vec{n}_1 \times \vec{n}_2 = \det\left(\begin{bmatrix} \vec{i} & \vec{j} & \vec{k} \\ 1 & 1 & 1 \\ 6 & 2 & 3 \end{bmatrix}\right)$$
$$= \vec{i}\,(1) - \vec{j}\,(-3) + \vec{k}\,(-4)$$
$$= [1 \ \ 3 \ \ -4].$$

Find a point on both planes by setting $z = 0$ and solve the equations $x+y+0 = 1$ and $6x + 2y + 0 = 6$ to obtain $x = 1$ and $y = 0$. Then the vector equation for the line is $\vec{r} = [1 \ \ 0 \ \ 0] + t[1 \ \ 3 \ \ -4]$.

If there are m equations and n variables with $m < n$, then one would like to be able to describe the set of points satisfying the m equations. Problems of this nature arise as parts of models for structures, circuits and fluids. These problems will be discussed in chapter five.

2.5.5   Systems of Differential Equations

In Subsection 2.3.3 the movement of a projectile was given by three differential equations, which described Newton's law of motion in the three directions. Let the velocity vector be given by three components v_1, v_2 and v_3 and assuming a unit mass so that the three equations are

$$\frac{d}{dt}v_1 = av_1 + d_1,$$
$$\frac{d}{dt}v_2 = bv_2 + d_2 \text{ and}$$
$$\frac{d}{dt}v_3 = cv_3 + d_3.$$

The matrix form is

$$\frac{d}{dt}\begin{bmatrix} v_1 \\ v_2 \\ v_3 \end{bmatrix} = \begin{bmatrix} a & 0 & 0 \\ 0 & b & 0 \\ 0 & 0 & c \end{bmatrix}\begin{bmatrix} v_1 \\ v_2 \\ v_3 \end{bmatrix} + \begin{bmatrix} d_1 \\ d_2 \\ d_3 \end{bmatrix}.$$

A generalization of this is to replace the 3×3 diagonal matrix by any 3×3 matrix. Moreover, applications often have a number of unknowns and differential equations so that the matrix would be replaced by an $n \times n$ matrix where n is the number of unknown functions. Chapters six and seven will start the study of these important time dependent models.

2.5.6   Images and Matrices

Grayscale images are associated with $m \times n$ matrices whose components are integers. For 8-bit images the integers range from 0 to $255 = 2^8 - 1$. For

16-bit images they range from 0 to $65535 = 2^{16} - 1$. The black image pixel is associated with 0, and the white image pixel is associated with 255 (8-bit) or 65535 (16-bit). If these matrices are converted to floating point numbers, then one can use matrix operations to change the image. For example, if the image is too bright, then multiply the corresponding matrix by a constant less than one so that the resulting pixels have smaller values.

In MATLAB one can "view" the image in several ways. First, just look at the matrix components. Second, use the MATLAB command mesh() to generate a surface of the image where the indices of the matrix are on the xy-plane and the intensity of the image is on the z-axis. Third, one can map the matrix into a standard image file such as a *.jpg file. This can be done by the MATLAB command imwrite(). The inverse of the imwrite() is imread(), which generates a matrix from a *.jpg file.

The following MATLAB function file creates a 50×40 matrix that represents the letter N. The input parameter g in the function lettern(g) should be between 0 and 255 and will indicate how bright the letter will be. Other letter generating functions are similar.

MATLAB function lettern()
```
function letn = lettern(g)
    letn = zeros(50,40);
    letn(10:40,4:8) = g;
    letn(10:40,32:36) = g;
    for j = 8:32
        letn(48-j,j) = g;
        letn(47-j,j) = g;
        letn(46-j,j) = g;
        letn(45-j,j) = g;
        letn(44-j,j) = g;
        letn(43-j,j) = g;
        letn(42-j,j) = g;
    end
```

The MATLAB code file imagncsu.m creates the letters NCSU and displays them via mesh(), imwrite() as a *.jpg file and imwrite() for the negative image, see Figures 2.5.1, 2.5.2 and 2.5.3.

MATLAB code file imagncsu.m
```
ncsu = [lettern(5) letterc(7) letters(9) letteru(11)];
mesh(ncsu);
newncsu = 20*ncsu;
newncsu = newncsu(50:-1:1,:);
negncsu = 255*ones(size(newncsu))-newncsu;
newncsu1 = uint8(newncsu);
imwrite(newncsu1, 'ncsu.jpg');
negncsu1 = uint8(negncsu);
imwrite(negncsu1, 'negncsu.jpg');
```

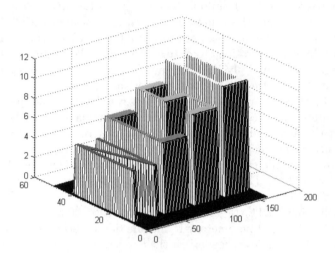

Figure 2.5.1: Mesh of Image Matrix

Figure 2.5.2: Imwrite of Image Matrix

Images and signals often have unwanted noise associated with the transmission and reception. Matrix manipulations can "filter" some of the unwanted attributes. Chapters eight and nine will investigate space and frequency filters. The frequency filters use the "fast Fourier transform." This is associated with matrix products where the Fourier matrix has complex numbers as components.

Figure 2.5.3: Negative Image Matrix

2.5.7 Exercises

1.   A computer store stocks five different laptops and has $5, 6, 2, 1$ and 5 units of the respective models. Suppose the prices per unit are $1000, 850, 1700, 2220$ and 1150, respectively.

(a).   Find the stock and price vectors.

(b).   Compute this store's total value of the laptop stock by using the dot product of these vectors.

2.   Consider the computer store in exercise one. The same owner wishes to open a second store with the same five laptop models and $4, 5, 3, 3$ and 2 units of the respective models.

(a).   Find the stock and price vectors for the second store.

(b).   Compute the total value of the stock in both stores by using the dot product.

3.   Let $r\mathbf{a} + s\mathbf{b} + t\mathbf{c} = \mathbf{r}_0$ where $\mathbf{a}^T = [1\ 0\ 1]$, $\mathbf{b}^T = [2\ 1\ 1]$, $\mathbf{c}^T = [4\ 1\ 4]$ and $\mathbf{r}_0^T = [10\ 1\ 4]$.

(a).   Write this as a matrix equation.

(b).   Use determinants and Cramer's rule to solve the matrix equation.

4.   Let $r\mathbf{a} + s\mathbf{b} + t\mathbf{c} = \mathbf{r}_0$ where $\mathbf{a}^T = [-1\ 0\ 1]$, $\mathbf{b}^T = [3\ 1\ 2]$, $\mathbf{c}^T = [2\ 1\ 4]$ and $\mathbf{r}_0^T = [10\ 1\ 4]$.

(a).   Write this as a matrix equation.

(b).   Use determinants and Cramer's rule to solve the matrix equation.

5.   Two planes are given by $2x + y + 3z = 12$ and $-x + 3y + z = 6$.

(a).   Find the normal vectors to each plane.

(b).   Find the vector equation for the line given by the intersection of the two planes.

6.   Two planes are given by $x + y + z = 3$ and $-x + 3y + z = 12$.

(a).   Find the normal vectors to each plane.

(b).   Find the vector equation for the line given by the intersection of the two planes.

7.   Use the MATLAB command A\d to solve the matrix equation in exercise three.

8.   Use the MATLAB command A\d to solve the matrix equation in exercise four.

9.   Use the MATLAB function file lettern(g) to experiment with different brightness levels. Use mesh(letn) to view the letter n, for example,

>> lettern(50)
>> mesh(letn)

10.   Use the MATLAB function file lettern(g) as a template to create letter matrices for your first name.

11.   Use the MATLAB code file imagncsu.m as a template to create a *.jpg file for your first name.

Chapter 3

Ax = d: Unique Solution

This and the next two chapters contain the basic methods for solving matrix equations of the form $A\mathbf{x} = \mathbf{d}$ where A is a matrix and $\mathbf{x}$ and $\mathbf{d}$ are column vectors. A much more complete description of solution methods and linear algebra can be found in [5]. The possibilities of a unique solution, no solution, multiple solutions and least square solutions are discussed. In this chapter most of the algebraic systems are assumed to be square so that the number rows (equations) and columns (variables) are equal. Row operations and elementary matrices are used to do by-hand computations of the Gauss elimination, LU factorization and inverse matrix methods. These methods are also implemented in MATLAB. Applications to steady state circuits, mixing tanks, heat conduction and support trusses are given.

3.1 Matrix Models

The extension from vectors with two or three components and from 2×2 or 3×3 matrices to higher dimensions will be motivated by graphics and a variety of applications. In this section the simplest model of cost, circuits and trusses will be presented. We will return to these models in the remainder of the text so as to increase their complexity, time dependence and accuracy.

3.1.1 Column Vectors and $\mathbb{R}^n$

In Section 2.5 price, inventory and graphing vectors were introduced. Vectors may either be row or column vectors and usually are an ordered list of real or complex numbers. However, they could have more general components such as a phone book where the components are triples having a name, address and phone number. In the remainder of the text we will assume the vector is a column and will use a bold lower case font to denote vectors.

Definition 3.1.1. An $n \times 1$ *column vector* is an ordered list of n real or complex numbers. This will be described in two ways

$$\mathbf{a} = \begin{bmatrix} a_1 \\ \vdots \\ a_n \end{bmatrix} \text{ or } \mathbf{a} = [a_i] \text{ where } i = 1, \cdots, n.$$

The set of all $n \times 1$ column vectors with real components is denoted by $\mathbb{R}^n$, and $\mathbb{C}^n$ is the set of all $n \times 1$ column vectors with complex components.

Four basic operations with vectors are scalar product, addition, augmentation and transpose.

Definitions 3.1.2. Let $\mathbf{a}$ and $\mathbf{b}$ be $n \times 1$ column vectors and let s be a real or complex number.

Scalar product $s\mathbf{a}$ is another $n \times 1$ column vector whose components are sa_i

$$s\mathbf{a} \equiv \begin{bmatrix} sa_1 \\ \vdots \\ sa_n \end{bmatrix} = [sa_i] \text{ where } i = 1, \cdots, n.$$

Notation: $s\mathbf{a} = [sa_i] = [a_i s] = \mathbf{a}s$.

Vector addition is $\mathbf{a} + \mathbf{b}$ is another $n \times 1$ column vector whose components are $a_i + b_i$

$$\mathbf{a} + \mathbf{b} \equiv \begin{bmatrix} a_1 + b_1 \\ \vdots \\ a_n + b_n \end{bmatrix} = [a_i + b_i] \text{ where } i = 1, \cdots, n.$$

Vector augmentation of $\mathbf{a}$ and $\mathbf{b}$ denoted by $[\mathbf{a} \ \mathbf{b}]$ is an $n \times 2$ matrix with two column vectors where $\mathbf{a}$ is the first column of the matrix

$$[\mathbf{a} \ \mathbf{b}] = \begin{bmatrix} a_1 & b_1 \\ \vdots & \vdots \\ a_n & b_n \end{bmatrix} = [a_i \ b_i] \text{ where } i = 1, \cdots, n.$$

One can augment m $n \times 1$ column vectors to form an $n \times m$ matrix, which will be denoted by upper case fonts.

Transpose of $\mathbf{a}$ is a $1 \times n$ *row vector* $\mathbf{a}^T$

$$\mathbf{a}^T = [a_1 \cdots a_n].$$

Example 3.1.1. Consider the three 4×1 column vectors

$$\mathbf{a} = \begin{bmatrix} 1 \\ 6 \\ 3 \\ 5 \end{bmatrix}, \ \mathbf{b} = \begin{bmatrix} 3 \\ 2 \\ -1 \\ 3 \end{bmatrix} \text{ and } \mathbf{c} = \begin{bmatrix} 0 \\ 1 \\ 4 \\ -2 \end{bmatrix}.$$

One can combine scalar multiplication and vector addition to form a linear combination of these vectors

$$
2\mathbf{a}+3\mathbf{b}-\mathbf{c} = 2\begin{bmatrix} 1 \\ 6 \\ 3 \\ 5 \end{bmatrix} + 3\begin{bmatrix} 3 \\ 2 \\ -1 \\ 3 \end{bmatrix} - \begin{bmatrix} 0 \\ 1 \\ 4 \\ -2 \end{bmatrix}
$$

$$
= \begin{bmatrix} 2(1) + 3(3) - 0 \\ 2(6) + 3(2) - 1 \\ 2(3) + 3(-1) - 4 \\ 2(5) + 3(3) - (-2) \end{bmatrix} = \begin{bmatrix} 11 \\ 17 \\ -1 \\ 21 \end{bmatrix}.
$$

This is in contrast to the augmentation of the three column vectors, which is a 4×3 matrix or array

$$
A = \begin{bmatrix} 1 & 3 & 0 \\ 6 & 2 & 1 \\ 3 & -1 & 4 \\ 5 & 3 & -2 \end{bmatrix}.
$$

The set of $n \times 1$ column vectors $\mathbb{R}^n$ is called a vector space because it has the properties listed in the next theorem.

Theorem 3.1.1 *(Vector Space Properties of $\mathbb{R}^n$) Let $\mathbf{a}$, $\mathbf{b}$ and $\mathbf{c}$ be in $\mathbb{R}^n$ and let s and t be real numbers. Let $\mathbf{a}+\mathbf{b}$ and $s\mathbf{a}$ denote vector addition and scalar product. Then the following hold:*

$\mathbf{a}+\mathbf{b} \in \mathbb{R}^n$, $\mathbf{a}+\mathbf{0} = \mathbf{a}$ *where* $\mathbf{0} \in \mathbb{R}^n$ *has zeros as components*, $\mathbf{a}+(-\mathbf{a}) = \mathbf{0}$,
$\mathbf{a}+\mathbf{b} = \mathbf{b}+\mathbf{a}$, $\mathbf{a}+(\mathbf{b}+\mathbf{c}) = (\mathbf{a}+\mathbf{b})+\mathbf{c}$;
$s\mathbf{a} \in \mathbb{R}^n$, $1\mathbf{a} = \mathbf{a}$, $s(\mathbf{a}+\mathbf{b}) = s\mathbf{a}+s\mathbf{b}$, $(s+t)\mathbf{a} = s\mathbf{a}+t\mathbf{a}$ *and* $s(t\mathbf{a}) = (st)\mathbf{a}$.

The definitions of dot product and norms for vectors in $\mathbb{R}^2$ and $\mathbb{R}^3$, see Sections 1.4 and 2.1, have the following generalizations and analogous properties.

Definitions 3.1.3. Let $\mathbf{a}$ and $\mathbf{b}$ be given vectors in $\mathbb{R}^n$. The *dot product* of $\mathbf{a}$ and $\mathbf{b}$ is a real number defined by

$$
\mathbf{a} \bullet \mathbf{b} \equiv \mathbf{a}^T\mathbf{b} = a_1b_1 + a_2b_2 + \cdots + a_nb_n.
$$

The vectors $\mathbf{a}$ and $\mathbf{b}$ are called *orthogonal* if and only if their dot product in zero

$$
\mathbf{a} \bullet \mathbf{b} = \mathbf{a}^T\mathbf{b} = 0.
$$

The *Euclidean norm* of $\mathbf{a}$ is a real number defined by

$$
\|\mathbf{a}\| \equiv (\mathbf{a} \bullet \mathbf{a})^{1/2} = (\mathbf{a}^T\mathbf{a})^{1/2} = (a_1^2 + a_2^2 + \cdots + a_n^2)^{1/2}.
$$

In order to illustrate the dot product and norm, consider Example 3.1.1 where $n = 4$

$$
\begin{aligned}
\mathbf{a} \bullet \mathbf{b} &= 1(3) + 6(2) + 3(-1) + 5(3) = 27 \text{ and} \\
\|\mathbf{a}\| &= (1^2 + 6^2 + 3^2 + 5^2)^{1/2} = \sqrt{71}.
\end{aligned}
$$

Many properties of the dot product and norm, that hold for vectors in the plane and space, also hold for vectors in $\mathbb{R}^n$. The proofs of these are similar except for the Cauchy-Schwarz inequality.

Theorem 3.1.2 *(Properties of Dot Product)   Let* **a**, **b** *and* **c** *be given vectors in* $\mathbb{R}^n$ *and let* s *be a real number. Then the following hold*

$$\mathbf{a} \bullet \mathbf{b} = \mathbf{b} \bullet \mathbf{a}, \ s(\mathbf{a} \bullet \mathbf{b}) = (s\mathbf{a}) \bullet \mathbf{b},$$
$$\mathbf{a} \bullet (\mathbf{b} + \mathbf{c}) = \mathbf{a} \bullet \mathbf{b} + \mathbf{a} \bullet \mathbf{c} \ and$$
$$|\mathbf{a} \bullet \mathbf{b}| \leq \|\mathbf{a}\| \, \|\mathbf{b}\| \ (Cauchy\text{-}Schwarz \ inequality); \qquad (3.1.1)$$

Theorem 3.1.3 *(Properties of Norm)   Let* **a** *and* **b** *be given vectors in* $\mathbb{R}^n$ *and let* s *be a real number. Then the following hold*

$$\|\mathbf{a}\| \geq 0; \ \mathbf{a} = \mathbf{0} \ if \ and \ only \ if \ \|\mathbf{a}\| = 0,$$
$$\|s\mathbf{a}\| = |s| \, \|\mathbf{a}\| \ and$$
$$\|\mathbf{a} + \mathbf{b}\| \leq \|\mathbf{a}\| + \|\mathbf{b}\| \ (triangle \ inequality).$$

The *Cauchy-Schwarz inequality* in (3.1.1) for vectors in $\mathbb{R}^2$ and $\mathbb{R}^3$ follows from the cosine identity in (2.1.2). The proof for vectors in $\mathbb{R}^n$ makes use of the dot product properties and a clever observation. Let t be a real number and define the following function with $\mathbf{b} \neq \mathbf{0}$

$$f(t) \equiv (\mathbf{a} + t\mathbf{b}) \bullet (\mathbf{a} + t\mathbf{b}) \geq 0.$$

Note, it can be written as a quadratic function of t

$$f(t) \equiv \mathbf{a} \bullet \mathbf{a} + 2t\mathbf{a} \bullet \mathbf{b} + t^2 \mathbf{b} \bullet \mathbf{b}.$$

The first and second derivatives are

$$\frac{df}{dt} = 0 + 2\mathbf{a} \bullet \mathbf{b} + 2t\mathbf{b} \bullet \mathbf{b} \ and$$
$$\frac{d^2 f}{dt^2} = 0 + 2\mathbf{b} \bullet \mathbf{b}.$$

Choose t_0 so that $f(t_0)$ is a minimum, that is,

$$\frac{df}{dt} = 0 + 2\mathbf{a} \bullet \mathbf{b} + 2t_0 \mathbf{b} \bullet \mathbf{b} = 0.$$

Thus,

$$t_0 = \frac{-\mathbf{a} \bullet \mathbf{b}}{\mathbf{b} \bullet \mathbf{b}} \ and$$
$$f(t_0) = \mathbf{a} \bullet \mathbf{a} + 2t_0 \mathbf{a} \bullet \mathbf{b} + t_0^2 \mathbf{b} \bullet \mathbf{b}$$
$$= \mathbf{a} \bullet \mathbf{a} - \frac{(\mathbf{a} \bullet \mathbf{b})^2}{\mathbf{b} \bullet \mathbf{b}} \geq 0.$$

The inequality is equivalent to the *Cauchy-Schwarz* inequality.

3.1.2 Matrices

In Section 2.5 matrices were used to store the intensity at each point or pixel in an image. Matrices can be used to store a variety of information, which often requires visualizing or modifications. There are four element-wise or array operations that will be very useful.

Definitions 3.1.4. An *augmentation* of n $m \times 1$ column vectors with $1 \le j \le n$ of the form

$$\begin{bmatrix} a_{1j} \\ \vdots \\ a_{mj} \end{bmatrix}$$

is an $m \times n$ matrix A where the first index $1 \le i \le m$ is the row number, the second index $1 \le j \le n$ is the column number and

$$A = \begin{bmatrix} a_{11} & a_{12} & \cdots & a_{1n} \\ \vdots & \vdots & \ddots & \vdots \\ a_{m1} & a_{n2} & \cdots & a_{mn} \end{bmatrix} = [a_{ij}].$$

Let A and B be $m \times n$ matrices. Define the following four *array operations*, which generate additional $m \times n$ matrices:

$$\begin{aligned} f(A) &\equiv [f(a_{ij})], \\ A.*B &\equiv [a_{ij}b_{ij}], \\ A./B &\equiv [a_{ij}/b_{ij}] \text{ and} \\ A.\hat{}k &\equiv [a_{ij}\hat{}k]. \end{aligned}$$

Example 3.1.2. Consider the 4×3 matrix in Example 3.1.1 formed by augmenting the three column vectors. The component in the third row and first column is $a_{31} = 3$, and $a_{23} = 1$ is the component in the second row and third column. Array operations with exponents are not the same as matrix products

$$A.\hat{}2 = \begin{bmatrix} 1^2 & 3^2 & 0^2 \\ 6^2 & 2^2 & 1^2 \\ 3^2 & (-1)^2 & 4^2 \\ 5^2 & 3^2 & (-2)^2 \end{bmatrix} = \begin{bmatrix} 1 & 9 & 0 \\ 36 & 4 & 1 \\ 9 & 1 & 16 \\ 25 & 9 & 4 \end{bmatrix}.$$

Example 3.1.3. This example illustrates how array operations and matrices can be used to create a graph in MATLAB of a surface given by $z = f(x, y) = 100 - x^2 - 4y^2$. In order to graph the surface, a collection of points in the xy-plane must be selected and then the values of $f(x, y)$ at these points must be computed and stored. Suppose three points in the x-direction are $x = 0, 4$ and 8, and four points in the y-direction are $y = 0, 2, 4$ and 6. There are a total of 12 points in the xy-plane; the 12 values of x can be stored in a 4×3 matrix X, and the 12 values of y can be stored in a 4×3 matrix Y. The corresponding 12 values of $f(x, y)$ can be computed and stored in a 4×3 matrix

F. The following MATLAB commands do this. Note, the row 4 and column 3 component of F is $100 - 8^2 - 4(6^2) = -108$. By using smaller increments in the MATLAB command meshgrid(), one can generate much larger matrices and a more accurate depiction of the surface, which is graphed by the MATLAB command mesh(X,Y,F).

```
>> [X Y] = meshgrid(0:4:8,0:2:6)
      X =
            0 4 8
            0 4 8
            0 4 8
            0 4 8
      Y =
            0 0 0
            2 2 2
            4 4 4
            6 6 6
>> F = 100 - X.^2 - 4*Y.^2
      F =
            100   84   36
            084   68   20
            036   20  -28
           -44  -60  -108
>> mesh(X,Y,F)
```

3.1.3   Application to Visualization of Minimum Cost

The objective is to find the dimensions of a 3D box with fixed volume and with minimum cost. In order to approximate the solution, the cost function will be graphed, and we will inspect the graph for a minimum cost. Suppose the volume must be 1000 ft^3, the bottom cost 3 \$/ft^2, the four sides cost 1 \$/ft^2 and there is no top. Let the bottom have edges equal to x and y, and let the height of the box be z. This is illustrated by Figure 3.1.1. The cost of the bottom is $3xy$, and the cost of the sides is $1(yz + yz + xz + xz)$. Since the volume must be 1000, $1000 = xyz$ or $z = 1000/(xy)$. Because the total cost equals the sum of the cost of the bottom and the four sides,

$$
\begin{aligned}
C(x,y) &= 3xy + 2(yz + xz) \\
&= 3xy + 2(y1000/(xy) + x1000/(xy)) \\
&= 3xy + 2(1000/x + 1000/y).
\end{aligned}
$$

The following MATLAB commands create the surface and contour plots of this cost function as given in Figure 3.1.2. The three matrices X, Y and C are 20×20 and note how array operations are used to compute the possible costs in the matrix C. An estimate of the minimum cost is given by $x = 9$ and $y = 9$

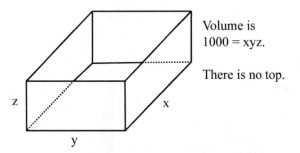

Volume is
1000 = xyz.

There is no top.

Figure 3.1.1: Box with Fixed Volume

so that $C(9,9) = 243 + 4000/9 \approx 687.4444$. A better estimate is given by using multivariable calculus where one sets the partial derivatives equal to zero and solve for x and y to attain $x = y = 8.7358$ and minimum cost equal to 686.8285.

```
>> [X Y] = meshgrid(1:1:20,1:1:20);
>> C = 3*X.*Y + 2000./Y + 2000./X;
>> subplot(1,2,1)
>> mesh(X,Y,C)
>> subplot(1,2,2)
>> contour(X,Y,C, 650:25:900)
```

3.1.4 Application to Two-bar Truss

Consider the two-bar truss in Figure 3.1.3. This truss is attached to a wall on the left and can be used to support a mass on the node joining the two bars. In the truss the bar ends are always pin connected. Consequently, the force in the bars are always along the axis. Assume there is no deformation and the bars are in a static state. The horizontal bar will have compression force $f_2 < 0$, and the diagonal bar will have a tension force $f_1 > 0$. At the joining node there are three force vectors whose sum must be the zero vector to satisfy the equilibrium equations

$$[-f_1 \cos(\theta) \quad f_1 \sin(\theta)] + [-f_2 \quad 0] + [0 \quad -w] = [0 \quad 0] \text{ or}$$
$$[-f_1 \cos(\theta) - f_2 \quad f_1 \sin(\theta) - w] = [0 \quad 0].$$

By equating the first and second components of the vector equation, we obtain two scalar equations

$$-f_1 \cos(\theta) - f_2 = 0 \text{ and}$$
$$f_1 \sin(\theta) = w.$$

The matrix version of this is

$$\begin{bmatrix} -\cos(\theta) & -1 \\ \sin(\theta) & 0 \end{bmatrix} \begin{bmatrix} f_1 \\ f_2 \end{bmatrix} = \begin{bmatrix} 0 \\ w \end{bmatrix}.$$

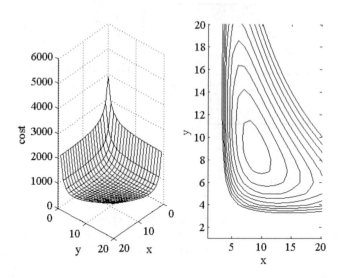

Figure 3.1.2:   Cost of a Box

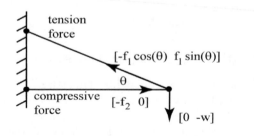

Figure 3.1.3:   Two-bar Truss

The solution is easy to compute. For example, if $w = 100$ and $\theta = \pi/6$, then

$$\left[\begin{array}{c} f_1 \\ f_2 \end{array} \right] = \left[\begin{array}{c} 200 \\ -100\sqrt{3} \end{array} \right].$$

If θ decreases, then the tension and compressive force magnitudes will increase. If there are more than two bars, then the number of joints will increase as well as the number of force vectors and force vector equations. This eventually leads to much larger algebraic systems as demonstrated in Subsections 3.4.4.

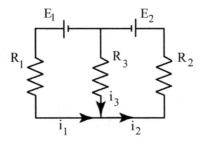

Figure 3.1.4: Two-loop Circuit

3.1.5 Application to Two-loop Circuit

Consider the two-loop circuit in Figure 3.1.4 with three resistors and two batteries. The current going through resistors R_1, R_2 and R_3 will be denoted by i_1, i_2 and i_3 and will have directions indicated by the arrows in the figure. Kirchhoff's current law requires the sum of the currents at any node be zero

$$i_1 - i_2 + i_3 = 0.$$

Ohm's law states that the voltage drop across each resistor is the resistance times the current. Kirchhoff's voltage law requires the sum of the voltage drops in each loop be equal to zero

$$E_1 - R_1 i_1 + R_3 i_3 = 0$$
$$-E_2 - R_2 i_2 - R_3 i_3 = 0.$$

The matrix version for the three unknown currents and the above three scalar equations is

$$\begin{bmatrix} 1 & -1 & 1 \\ R_1 & 0 & -R_3 \\ 0 & -R_2 & -R_3 \end{bmatrix} \begin{bmatrix} i_1 \\ i_2 \\ i_3 \end{bmatrix} = \begin{bmatrix} 0 \\ E_1 \\ E_2 \end{bmatrix}.$$

The solution of this 3×3 algebraic system can be found in a number of ways. For example, if $R_1 = 1$, $R_2 = 2$, $R_3 = 3$, $E_1 = 10$ and $E_2 = 20$, then one could use Cramer's rule, see Sections 2.5 and 3.7, to find the currents. Let

$$A = \begin{bmatrix} 1 & -1 & 1 \\ 1 & 0 & -3 \\ 0 & -2 & -3 \end{bmatrix} \text{ and } \mathbf{d} = \begin{bmatrix} 0 \\ 10 \\ 20 \end{bmatrix} \text{ and note}$$

$$\det(A) = 1(0 - (-3)(-2)) - (-1)(1(-3) - 0) + 1(1(-2) - 0) = -11.$$

The solutions are

$$
i_1 = \det\left(\begin{bmatrix} 0 & -1 & 1 \\ 10 & 0 & -3 \\ 20 & -2 & -3 \end{bmatrix} \right) / \det(A) = 10/(-11),
$$

$$
i_2 = \det\left(\begin{bmatrix} 1 & 0 & 1 \\ 1 & 10 & -3 \\ 0 & 20 & -3 \end{bmatrix} \right) / \det(A) = 50/(-11) \text{ and }
$$

$$
i_3 = \det\left(\begin{bmatrix} 1 & -1 & 0 \\ 1 & 0 & 10 \\ 0 & -2 & 20 \end{bmatrix} \right) / \det(A) = 40/(-11).
$$

Most circuits have many more loops so that the resulting algebraic systems are very large. In these cases it is not practical to solve them using Cramer's rule. In the following sections viable alternative methods will be developed.

3.1.6   Exercises

1.   Let $a^T = [2\ 5\ 7\ 1\ -1]$, $b^T = [0\ 1\ 2\ 7\ -4]$ and $c^T = [1\ 4\ 2\ 1\ -3]$.
   (a).   Find $3a - 4b + 2c$.
   (b).   Find the augmentations of a, b and c.
   (c).   Find $a \bullet b$.
   (d).   Find $\|a\|$.
2.   Let $a^T = [0\ 8\ -3\ 1\ -2]$, $b^T = [1\ 1\ 2\ -4\ -4]$ and $c^T = [2\ 1\ -5\ 1\ -3]$.
   (a).   Find $2a + 4b - 3c$.
   (b).   Find the augmentations of a, b and c.
   (c).   Find $b \bullet c$.
   (d).   Find $\|c\|$.
3.   Consider Example 3.1.2.
   (a).   Find $A.\hat{\ }3$.
   (b).   Find $a./b$ where a and b are the first and second columns of A.
   (c).   Find $sin(A)$.
4.   Consider Example 3.1.2.
   (a).   Find $A.\hat{\ }(1/2)$.
   (b).   Find $b. * c$ where b and c are the second and third columns of A.
   (c).   Find $b \bullet c$.
5.   Consider the surface $z = f(x, y) = 400 - 2x^2 - y^2$.
   (a).   By hand compute the 2×3 matrix

$$
F = \begin{bmatrix} f(1,1) & f(1,2) & f(1,3) \\ f(2,1) & f(2,2) & f(2,3) \end{bmatrix}.
$$

   (b).   Use MATLAB and mesh(X,Y,F) to compute F and graph.
6.   Consider the surface $z = f(x, y) = 200 - x^2 - 3y^2$.

(a).   By hand compute the 2×3 matrix

$$F = \left[\begin{array}{ccc} f(1,1) & f(1,1.5) & f(1,2) \\ f(2,1) & f(2,1.5) & f(2,2) \end{array} \right].$$

(b).   Use MATLAB and mesh(X,Y,F) to compute F and graph.

7.   Consider the minimum cost of the box with no top, volume equals 200, bottom costs 4 $/ft^2 and sides cost 1 $/ft^2.

(a).   Find the cost function.

(b).   Use MATLAB and mesh(X,Y,C) to approximate the minimum cost.

8.   Consider the minimum cost of the box with a top costing 2 $/ft^2, volume equals 200, bottom costs 4 $/ft^2 and sides cost 1 $/ft^2.

(a).   Find the cost function.

(b).   Use MATLAB and mesh(X,Y,C) to approximate the minimum cost.

9.   Consider the two-bar truss with $w = 100$ and variable $\theta = \pi/12, 2\pi/12, 3\pi/12$ and $4\pi/12$. Find the forces on the two bars.

10.   Consider the two-bar truss with $\theta = \pi/6$ and variable $w = 100, 200, 300$ and 400. Find the forces on the two bars.

11.   Consider the two-loop circuit with $E_1 = 10$, $E_2 = 20$, $R_1 = 1$, $R_2 = 2$ and $R_3 = 2$. Find the three currents.

12.   Consider the two-loop circuit with arbitrary batteries and resistors. Find the three currents and show the center current must be negative.

3.2 Matrix Products

In the previous sections matrices were used to describe systems of equations with two or three unknowns. Most realistic models have many more unknowns, and the resulting matrices are much larger. The general matrix-vector and matrix-matrix products will be defined and their properties will described. An application to heat conduction in a thin wire will be introduced, and this model will be enhanced in subsequent sections.

An algebraic system for three unknowns x_1, x_2 and x_3 can be listed either as three scalar equations or as a vector equation

$$\begin{aligned} a_{11}x_1 + a_{12}x_2 + a_{13}x_3 &= d_1, \\ a_{21}x_1 + a_{22}x_2 + a_{23}x_3 &= d_2 \text{ and} \\ a_{31}x_1 + a_{32}x_2 + a_{33}x_3 &= d_3. \end{aligned}$$

The vector equation has the form

$$\left[\begin{array}{ccc} a_{11} & a_{12} & a_{13} \\ a_{21} & a_{22} & a_{23} \\ a_{31} & a_{32} & a_{33} \end{array} \right] \left[\begin{array}{c} x_1 \\ x_2 \\ x_3 \end{array} \right] = \left[\begin{array}{c} d_1 \\ d_2 \\ d_3 \end{array} \right] \text{ or}$$

$$A\mathbf{x} = \mathbf{d} \text{ where } A \text{ is } 3 \times 3.$$

If there were two equations with four unknowns, then

$$a_{11}x_1 + a_{12}x_2 + a_{13}x_3 + a_{14}x_4 = d_1 \text{ and}$$
$$a_{21}x_1 + a_{22}x_2 + a_{23}x_3 + a_{24}x_4 = d_2.$$

The vector form is

$$\begin{bmatrix} a_{11} & a_{12} & a_{13} & a_{14} \\ a_{21} & a_{22} & a_{23} & a_{24} \end{bmatrix} \begin{bmatrix} x_1 \\ x_2 \\ x_3 \\ x_4 \end{bmatrix} = \begin{bmatrix} d_1 \\ d_2 \\ d_3 \\ d_4 \end{bmatrix} \text{ or}$$

$$A\mathbf{x} = \mathbf{d} \text{ where } A \text{ is } 2 \times 4.$$

3.2.1   Matrix-vector Products

In the product of a 3×3 matrix times a 3×1 column vector we previously noted that this can be done by either products of rows in the matrix and the column vector, or by linear combinations of the columns of the matrix. The row version is

$$\begin{bmatrix} a_{11} & a_{12} & a_{13} \\ a_{21} & a_{22} & a_{23} \\ a_{31} & a_{32} & a_{33} \end{bmatrix} \begin{bmatrix} x_1 \\ x_2 \\ x_3 \end{bmatrix} = \begin{bmatrix} a_{11}x_1 + a_{12}x_2 + a_{13}x_3 \\ a_{21}x_1 + a_{22}x_2 + a_{23}x_3 \\ a_{31}x_1 + a_{32}x_2 + a_{33}x_3 \end{bmatrix}.$$

The column version is

$$\begin{bmatrix} a_{11} & a_{12} & a_{13} \\ a_{21} & a_{22} & a_{23} \\ a_{31} & a_{32} & a_{33} \end{bmatrix} \begin{bmatrix} x_1 \\ x_2 \\ x_3 \end{bmatrix} = x_1 \begin{bmatrix} a_{11} \\ a_{21} \\ a_{31} \end{bmatrix} + x_2 \begin{bmatrix} a_{12} \\ a_{22} \\ a_{32} \end{bmatrix} + x_3 \begin{bmatrix} a_{13} \\ a_{23} \\ a_{33} \end{bmatrix}.$$

This important observation also holds for general matrix-vector products.

Definitions 3.2.1. Let A and B be $m \times n$ matrices, $\mathbf{a}$ and $\mathbf{x}$ be $n \times 1$ column vectors and let s be a real number. The ij-component of A is a_{ij} where i is the row number and j is the column number.

   Row vector $\mathbf{a}^T$ *times a column vector* $\mathbf{x}$ is a real number equal to the sum of the products of the components (also called the dot product of $\mathbf{a}$ and $\mathbf{x}$)

$$\mathbf{a}^T\mathbf{x} \equiv a_1x_1 + \cdots + a_nx_n.$$

Matrix A times a column vector $\mathbf{x}$ is a $n \times 1$ column vector whose i^{th} component is row i of A times the column vector $\mathbf{x}$

$$A\mathbf{x} \equiv [a_{i1}x_1 + \cdots + a_{in}x_n] \text{ where } i = 1, \cdots, m.$$

   The *scalar s times a matrix A* is another $m \times n$ matrix whose ij-component is s times the ij-component of A

$$sA \equiv [sa_{ij}].$$

Notation: $sA = [sa_{ij}] = [a_{ij}s] = As$.

The *addition of two matrices A and B* is another $m \times n$ matrix whose ij-component is the sum of the ij-components of A and B

$$A + B \equiv [a_{ij} + b_{ij}].$$

Example 3.2.1. Let A and B be 4×2 and $\mathbf{x}$ be a 2×1 column vector

$$A = \begin{bmatrix} 1 & 2 \\ -1 & 0 \\ 7 & 4 \\ 2 & 3 \end{bmatrix}, \quad B = \begin{bmatrix} -1 & 3 \\ 2 & 1 \\ 0 & 5 \\ -2 & 4 \end{bmatrix} \text{ and } \mathbf{x} = \begin{bmatrix} 3 \\ 4 \end{bmatrix}.$$

Then $A\mathbf{x}$ is defined to be a 4×1 column vector

$$A\mathbf{x} = \begin{bmatrix} 1(3) + 2(4) \\ -1(3) + 0(4) \\ 7(3) + 4(4) \\ 2(3) + 3(4) \end{bmatrix} = \begin{bmatrix} 11 \\ -3 \\ 37 \\ 18 \end{bmatrix}.$$

The matrix vector product could also have been computed by a linear combination of the two column vectors of A

$$A\mathbf{x} = 3 \begin{bmatrix} 1 \\ -1 \\ 7 \\ 2 \end{bmatrix} + 4 \begin{bmatrix} 2 \\ 0 \\ 4 \\ 3 \end{bmatrix} = \begin{bmatrix} 3 \\ -3 \\ 21 \\ 6 \end{bmatrix} + \begin{bmatrix} 8 \\ 0 \\ 16 \\ 12 \end{bmatrix} = \begin{bmatrix} 11 \\ -3 \\ 37 \\ 18 \end{bmatrix}.$$

If

$$C = \begin{bmatrix} 2 & 4 \\ -1 & 2 \end{bmatrix},$$

then $A+C$ is not defined because their row and column numbers are not equal. If $s = 7$, then sA is defined to be the 4×2 matrix

$$sA = \begin{bmatrix} 7(1) & 7(2) \\ 7(-1) & 7(0) \\ 7(7) & 7(4) \\ 7(2) & 7(3) \end{bmatrix} = \begin{bmatrix} 7 & 14 \\ -7 & 0 \\ 49 & 28 \\ 14 & 21 \end{bmatrix} \text{ and }$$

$$A + B = \begin{bmatrix} 1-1 & 2+3 \\ -1+2 & 0+1 \\ 7+0 & 4+5 \\ 2-2 & 3+4 \end{bmatrix} = \begin{bmatrix} 0 & 5 \\ 1 & 1 \\ 7 & 9 \\ 0 & 7 \end{bmatrix}.$$

Theorem 3.2.1 *(Matrix-vector Products Properties) Let A, B and C be $m \times n$ matrices. Let Z be a $m \times n$ matrix with all components equal to zero, called the zero matrix. Let $\mathbf{x}$ and $\mathbf{y}$ be $n \times 1$ column vectors. Then the following are true:*

1. $A + (B + C) = (A + B) + C,\ A + B = B + C\ \text{and}\ A + Z = A.$

2. $A(s\mathbf{x}) = (sA)\mathbf{x}$ where s is a real number, $A(\mathbf{x} + \mathbf{y}) = A\mathbf{x} + A\mathbf{y}.$

3.   The matrix-vector product can be computed as a linear combination of the columns. Let column j of A be denoted by $\mathbf{a}_{:j}$

$$Ax = x_1\mathbf{a}_{:1} + \cdots + x_n\mathbf{a}_{:n}.$$

The proofs of the above six properties are routine. The last property is the column version of a matrix-vector product whose proof is

$$
\begin{aligned}
A\mathbf{x} &= \begin{bmatrix} a_{11}x_1 + \cdots + a_{1m}x_m \\ \vdots \\ a_{n1}x_1 + \cdots + a_{nm}x_m \end{bmatrix} \\
&= \begin{bmatrix} a_{11}x_1 \\ \vdots \\ a_{n1}x_1 \end{bmatrix} + \cdots + \begin{bmatrix} a_{1m}x_n \\ \vdots \\ a_{nm}x_m \end{bmatrix} \\
&= x_1 \begin{bmatrix} a_{11} \\ \vdots \\ a_{m1} \end{bmatrix} + \cdots + x_n \begin{bmatrix} a_{1n} \\ \vdots \\ a_{mn} \end{bmatrix}.
\end{aligned}
$$

Example 3.2.2. Verify the matrix addition properties in the above theorem for

$$A = \begin{bmatrix} 1 & 2 \\ 2 & 0 \\ 4 & 6 \end{bmatrix},\ B = \begin{bmatrix} -1 & 3 \\ 2 & 1 \\ 2 & 5 \end{bmatrix}\ \text{and}\ C = \begin{bmatrix} 1 & -5 \\ 2 & 3 \\ 4 & 3 \end{bmatrix}.$$

$$
\begin{aligned}
A + (B + C) &= \begin{bmatrix} 1 & 2 \\ 2 & 0 \\ 4 & 6 \end{bmatrix} + \left(\begin{bmatrix} -1 & 3 \\ 2 & 1 \\ 2 & 5 \end{bmatrix} + \begin{bmatrix} 1 & -5 \\ 2 & 3 \\ 4 & 3 \end{bmatrix} \right) \\
&= \begin{bmatrix} 1 + (-1 + 1) & 2 + (3 - 5) \\ 2 + (2 + 2) & 0 + (1 + 3) \\ 4 + (2 + 4) & 6 + (5 + 3) \end{bmatrix} \\
&= \begin{bmatrix} (1 - 1) + 1 & (2 + 3) - 5 \\ (2 + 2) + 2 & (0 + 1) + 3 \\ (4 + 2) + 4 & (6 + 5) + 3 \end{bmatrix} = (A + B) + C.
\end{aligned}
$$

$$A + B = \begin{bmatrix} 1 & 2 \\ 2 & 0 \\ 4 & 6 \end{bmatrix} + \begin{bmatrix} -1 & 3 \\ 2 & 1 \\ 2 & 5 \end{bmatrix}$$

$$= \begin{bmatrix} 1 - 1 & 2 + 3 \\ 2 + 2 & 0 + 1 \\ 4 + 2 & 6 + 5 \end{bmatrix}$$

$$= \begin{bmatrix} 1 - 1 & 3 + 2 \\ 2 + 2 & 1 + 0 \\ 2 + 4 & 5 + 6 \end{bmatrix} = B + A.$$

$$A + Z = \begin{bmatrix} 1 & 2 \\ 2 & 0 \\ 4 & 6 \end{bmatrix} + \begin{bmatrix} 0 & 0 \\ 0 & 0 \\ 0 & 0 \end{bmatrix}$$

$$= \begin{bmatrix} 1 + 0 & 2 + 0 \\ 2 + 0 & 0 + 0 \\ 4 + 0 & 6 + 0 \end{bmatrix} = A.$$

3.2.2 Matrix-matrix Products

Matrix-vector products may be extended to matrix-matrix products as long as the number of columns in the left matrix is the same as the number of rows in the right matrix. Another important fact is that the order of matrix-matrix product is important, that is, AB may not be the same as BA! Like the matrix-vector product there is a row version and a column version. For example, consider the product of two 2×2 matrices

$$\begin{bmatrix} 2 & 3 \\ 4 & 5 \end{bmatrix} \begin{bmatrix} 6 & 7 \\ 8 & 9 \end{bmatrix} = \begin{bmatrix} 2(6) + 3(8) & 2(7) + 3(9) \\ 4(6) + 5(8) & 4(7) + 5(9) \end{bmatrix}$$

$$= \begin{bmatrix} 2 & 3 \\ 4 & 5 \end{bmatrix} \begin{bmatrix} 6 \\ 8 \end{bmatrix} + \begin{bmatrix} 2 & 3 \\ 4 & 5 \end{bmatrix} \begin{bmatrix} 7 \\ 9 \end{bmatrix}.$$

This product is either computed by rows times columns or by matrix times columns. Initially, we use the rows times columns approach for the general definition of a matrix-matrix product.

Definition 3.2.2. Let A be an $m \times n$ and let B be an $n \times p$ matrices. The *product* AB is an $m \times p$ matrix whose ij-component is the product of row i of A times column j of B

$$AB \equiv [a_{i1}b_{1j} + \cdots + a_{in}b_{nj}].$$

Example 3.2.3. First, let

$$A = \begin{bmatrix} 1 & 2 \\ 3 & 4 \end{bmatrix} \text{ and } B = \begin{bmatrix} 0 & 1 \\ 1 & 0 \end{bmatrix} \text{ and note}$$

$$AB = \begin{bmatrix} 2 & 1 \\ 4 & 3 \end{bmatrix} \neq \begin{bmatrix} 3 & 4 \\ 1 & 2 \end{bmatrix} = BA.$$

Second, note AB may be defined while BA is not defined. Let A be 2×3 and B be 3×3

$$A = \begin{bmatrix} 2 & 4 & 7 \\ 8 & 9 & 10 \end{bmatrix} \text{ and } B = \begin{bmatrix} 2 & 1 & 0 \\ 1 & 2 & 3 \\ 0 & 1 & 1 \end{bmatrix}.$$

$$AB = \begin{bmatrix} 2(2)+4(1)+7(0) & 2(1)+4(2)+7(1) & 2(0)+4(3)+7(1) \\ 8(2)+9(1)+10(0) & 8(1)+9(2)+10(1) & 8(0)+9(3)+10(1) \end{bmatrix}$$

$$= \begin{bmatrix} 8 & 17 & 19 \\ 25 & 36 & 37 \end{bmatrix}.$$

This product also can be written as columns of the product given by A times the column of B

$$AB = \begin{bmatrix} A\mathbf{b}_1 & A\mathbf{b}_2 & A\mathbf{b}_3 \end{bmatrix} \text{ where } B = \begin{bmatrix} \mathbf{b}_1 & \mathbf{b}_2 & \mathbf{b}_3 \end{bmatrix}$$

$$A\mathbf{b}_1 = \begin{bmatrix} 2 & 4 & 7 \\ 8 & 9 & 10 \end{bmatrix} \begin{bmatrix} 2 \\ 1 \\ 0 \end{bmatrix} = \begin{bmatrix} 8 \\ 25 \end{bmatrix}$$

$$A\mathbf{b}_2 = \begin{bmatrix} 2 & 4 & 7 \\ 8 & 9 & 10 \end{bmatrix} \begin{bmatrix} 1 \\ 2 \\ 1 \end{bmatrix} = \begin{bmatrix} 17 \\ 36 \end{bmatrix}$$

$$A\mathbf{b}_3 = \begin{bmatrix} 2 & 4 & 7 \\ 8 & 9 & 10 \end{bmatrix} \begin{bmatrix} 0 \\ 3 \\ 1 \end{bmatrix} = \begin{bmatrix} 19 \\ 37 \end{bmatrix}.$$

Theorem 3.2.2 *(Matrix-matrix Products Properties) Let* $A, \widehat{A}$ *be* $m \times n$, B *be* $n \times p$, C *be* $p \times q$ *and* s *be a real number. Then the following are true:*

1. $A(BC) = (AB)C, s(AB) = (sA)B = A(sB)$ *and*

$$AB = A\begin{bmatrix} \mathbf{b}_1 & \cdots & \mathbf{b}_p \end{bmatrix} = \begin{bmatrix} A\mathbf{b}_1 & \cdots & A\mathbf{b}_p \end{bmatrix}.$$

2. $(A + \widehat{A})B = AB + \widehat{A}B.$

The proof of $(A + \widehat{A})B = AB + \widehat{A}B$ follows from the distributive property of real numbers. Let $1 \leq i \leq m$ and $1 \leq j \leq p$

$$\begin{aligned}
(A + \widehat{A})B &= [(a_{i1} + \widehat{a}_{i1})b_{1j} + \cdots + (a_{in} + \widehat{a}_{in})b_{nj}] \\
&= [(a_{i1}b_{1j} + \widehat{a}_{i1}b_{1j}) + \cdots + (a_{in}b_{nj} + \widehat{a}_{in}b_{nj})] \\
&= [a_{i1}b_{1j} + \cdots + a_{in}b_{nj}] + [\widehat{a}_{i1}b_{1j} + \cdots + \widehat{a}_{in}b_{nj}] \\
&= AB + \widehat{A}B.
\end{aligned}$$

Example 3.2.4. Verify the associative rule $A(BC) = (AB)C$ for the following 3×3 matrices

$$
A = \begin{bmatrix} 1 & 0 & 0 \\ 0 & 1 & 0 \\ 0 & 2/3 & 1 \end{bmatrix}, \quad B = \begin{bmatrix} 1 & 0 & 0 \\ 1/2 & 1 & 0 \\ 0 & 0 & 1 \end{bmatrix} \quad \text{and } C = \begin{bmatrix} 2 & -1 & 0 \\ -1 & 2 & -1 \\ 0 & -1 & 2 \end{bmatrix}.
$$

$$
\begin{aligned}
A(BC) &= \begin{bmatrix} 1 & 0 & 0 \\ 0 & 1 & 0 \\ 0 & 2/3 & 1 \end{bmatrix} \left(\begin{bmatrix} 1 & 0 & 0 \\ 1/2 & 1 & 0 \\ 0 & 0 & 1 \end{bmatrix} \begin{bmatrix} 2 & -1 & 0 \\ -1 & 2 & -1 \\ 0 & -1 & 2 \end{bmatrix} \right) \\
&= \begin{bmatrix} 1 & 0 & 0 \\ 0 & 1 & 0 \\ 0 & 2/3 & 1 \end{bmatrix} \left(\begin{bmatrix} 2 & -1 & 0 \\ 0 & 3/2 & -1 \\ 0 & -1 & 2 \end{bmatrix} \right) = \begin{bmatrix} 2 & -1 & 0 \\ 0 & 3/2 & -1 \\ 0 & 0 & 4/3 \end{bmatrix}.
\end{aligned}
$$

$$
\begin{aligned}
(AB)C &= \left(\begin{bmatrix} 1 & 0 & 0 \\ 0 & 1 & 0 \\ 0 & 2/3 & 1 \end{bmatrix} \begin{bmatrix} 1 & 0 & 0 \\ 1/2 & 1 & 0 \\ 0 & 0 & 1 \end{bmatrix} \right) \begin{bmatrix} 2 & -1 & 0 \\ -1 & 2 & -1 \\ 0 & -1 & 2 \end{bmatrix} \\
&= \left(\begin{bmatrix} 1 & 0 & 0 \\ 1/2 & 1 & 0 \\ 1/3 & 2/3 & 1 \end{bmatrix} \right) \begin{bmatrix} 2 & -1 & 0 \\ -1 & 2 & -1 \\ 0 & -1 & 2 \end{bmatrix} = \begin{bmatrix} 2 & -1 & 0 \\ 0 & 3/2 & -1 \\ 0 & 0 & 4/3 \end{bmatrix}.
\end{aligned}
$$

3.2.3 Application to Heat Conduction

The objective is to predict the temperature of a mass given an initial temperature and the temperature on the boundary. A very complicated example is the temperature inside a spacecraft as it enters the earth's atmosphere. Such models have a long history of previous less sophisticated models. In this section we restrict the mass to a long thin wire so that one may assume the diffusion of thermal energy is only along the length of the wire. For the present model, we assume no heat is lost through the lateral surface area. Heat may enter the wire through the left or right ends of the wire. Thus, the temperature will only vary continuously with time and position along the wire. More complicated heat models can be found in later sections of this text and in [12].

The model will be a discrete approximation of both time and space, that is, the time and space variables will be broken into a finite number of segments and the temperature will be assumed to be constant within each segment. This is depicted in Figure 3.2.1 where there are four space segments with length $\Delta x = L/4$ and cross sectional area A. The temperatures in the three interior segments are u_1, u_2 and u_3; the temperatures at the left u_0 and right u_4 are given. The heat energy in the ith segment is

$$
\rho c u_i (A\Delta x)
$$

where ρ is the density, $A\Delta x$ is the volume and c is the specific heat of the wire.

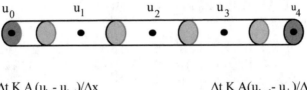

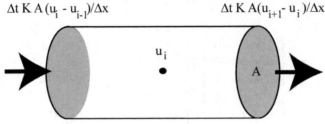

Figure 3.2.1: Heat Conduction in a Wire

The *Fourier heat law* states that the heat flows from hot to cold, and the amount of heat moving normal to a cross section is directly proportional to the product of the time interval, cross section area and the derivative of the temperature normal to the cross section area. Any change in the heat energy of the segment is from diffusion from left or right sides. The proportionality constant is called the *thermal conductivity* K so that the heat entering from the right is

$$\Delta t \, A \, K \frac{u_{i+1} - u_i}{\Delta x}$$

and the heat entering from the left is

$$-\Delta t \, A \, K \frac{u_i - u_{i-1}}{\Delta x}.$$

The new heat of the ith segment is modeled or approximated by

$$\rho c(u_i^{new})(A\Delta x) = \rho c u_i(A\Delta x) - \Delta t \, A \, K \frac{u_i - u_{i-1}}{\Delta x} + \Delta t \, A \, K \frac{u_{i+1} - u_i}{\Delta x}.$$

Divide by $\rho c(A\Delta x)$ and let $\alpha \equiv (\Delta t/\Delta x^2)(K/\rho c)$ to get for $i = 1, 2$ and 3

$$u_i^{new} = u_i - \alpha(-u_{i-1} + 2u_i - u_{i+1}).$$

One may write this as three scalar equations

$$
\begin{aligned}
u_1^{new} &= u_1 - \alpha(-u_0 + 2u_1 - u_2), \\
u_2^{new} &= u_2 - \alpha(-u_1 + 2u_2 - u_3) \text{ and} \\
u_3^{new} &= u_3 - \alpha(-u_2 + 2u_2 - u_4).
\end{aligned}
$$

Or, as one vector equation

$$
\begin{bmatrix} u_1^{new} \\ u_2^{new} \\ u_3^{new} \end{bmatrix} = \begin{bmatrix} u_1 \\ u_2 \\ u_3 \end{bmatrix} - \begin{bmatrix} 2\alpha & -\alpha & 0 \\ -\alpha & 2\alpha & -\alpha \\ 0 & -\alpha & 2\alpha \end{bmatrix} \begin{bmatrix} u_1 \\ u_2 \\ u_3 \end{bmatrix} + \begin{bmatrix} \alpha u_0 \\ 0 \\ \alpha u_4 \end{bmatrix}
$$

$$
\mathbf{u}^{new} = \mathbf{u} - A\mathbf{u} + \mathbf{d}.
$$

The initial temperature is given and one can compute the temperature at the next time step by using the above.

In order for the discrete model to give accurate approximations of the continuous heat conduction process, the step sizes in both time and space must be suitably small, and for the above model they must satisfy the stability constraint $1 - 2\alpha > 0$. Since $\alpha \equiv (\Delta t/\Delta x^2)(K/\rho c)$, changing either the step sizes or the physical properties will alter $1 - 2\alpha$. If diffusion of heat occurs in all three directions, then the discrete model can have a large number of unknown temperatures. For example, if each direction has 100 unknown temperatures, then there will be 100^3 unknown temperatures in a three dimensional box!

Example 3.2.5. Consider the above with $L = 10$, $\Delta x = 10/4$, $\rho = c = 1$, $K = 1/1000$, $\Delta t = 1000$ so that $\alpha \equiv (\Delta t/\Delta x^2)(K/\rho c) = 16/100 = 0.16$. If the initial temperature is zero in the interior, the left side is 200 and the right side is 70, then

$$u_0 = 200, \ u_1 = u_2 = u_3 = 0 \text{ and } u_4 = 70.$$

The approximation for the temperature at time equal to 1000 is $\mathbf{u}^1$ where the superscript denotes the first time step

$$
\begin{aligned}
\mathbf{u}^1 &= \mathbf{u} - A\mathbf{u} + \mathbf{d} \\
&= \begin{bmatrix} 0 \\ 0 \\ 0 \end{bmatrix} - (0.16) \begin{bmatrix} 2 & -1 & 0 \\ -1 & 2 & -1 \\ 0 & -1 & 2 \end{bmatrix} \begin{bmatrix} 0 \\ 0 \\ 0 \end{bmatrix} + \begin{bmatrix} (0.16)200 \\ 0 \\ (0.16)70 \end{bmatrix} \\
&= \begin{bmatrix} 32.0 \\ 0 \\ 11.2 \end{bmatrix}.
\end{aligned}
$$

The approximate temperature at the second time step equal to 2000 is $\mathbf{u}^2$ where the superscript denotes the second time step

$$
\begin{aligned}
\mathbf{u}^2 &= \mathbf{u}^1 - A\mathbf{u}^1 + \mathbf{d} \\
&= \begin{bmatrix} 32.0 \\ 0 \\ 11.2 \end{bmatrix} - (0.16) \begin{bmatrix} 2 & -1 & 0 \\ -1 & 2 & -1 \\ 0 & -1 & 2 \end{bmatrix} \begin{bmatrix} 32.0 \\ 0 \\ 11.2 \end{bmatrix} + \begin{bmatrix} (0.16)200 \\ 0 \\ (0.16)70 \end{bmatrix} \\
&= \begin{bmatrix} 53.7600 \\ 6.9120 \\ 18.8160 \end{bmatrix}.
\end{aligned}
$$

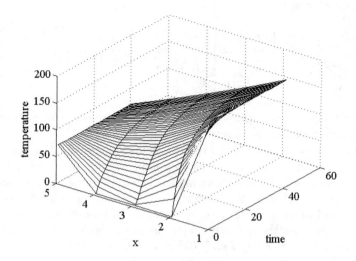

Figure 3.2.2:   Steady State Heat Diffusion

Note $\mathbf{u}^2$ can be represented by matrix-matrix products

$$
\begin{aligned}
\mathbf{u}^2 &= (\mathbf{u} - A\mathbf{u} + \mathbf{d}) - A(\mathbf{u} - A\mathbf{u} + \mathbf{d}) + \mathbf{d} \\
&= \mathbf{u} - 2A\mathbf{u} + A^2\mathbf{u} + 2\mathbf{d} - A\mathbf{d}.
\end{aligned}
$$

One can continue solving for the temperature at the next time step. This is illustrated in Figure 3.2.2, which was generated by the MATLAB code wire_time.m. As the time steps increase, the temperature vector reaches a steady state, which is a straight line. The means for larger time steps

$$\mathbf{u} \approx \mathbf{u} - A\mathbf{u} + \mathbf{d}.$$

Hence, the *steady state solution* is defined to be the solution of $A\mathbf{u} = \mathbf{d}$

$$
(0.16)
\begin{bmatrix}
2 & -1 & 0 \\
-1 & 2 & -1 \\
0 & -1 & 2
\end{bmatrix}
\begin{bmatrix}
u_1 \\
u_2 \\
u_3
\end{bmatrix}
=
\begin{bmatrix}
(0.16)200 \\
0 \\
(0.16)70
\end{bmatrix}.
$$

The solution is $u_1 = 167.5$, $u_2 = 135.0$ and $u_3 = 102.5$.

3.2.4   Matrix Computations Using MATLAB

The calculations for the heat conduction application can easily be done by MATLAB. Here we have computed the first three time steps and the steady state solution. The interested reader should examine the MATLAB code wire_time.m to see how this can be done using for-loops.

```
>> A = .16*[2 -1 0;-1 2 -1;0 -1 2]        % creates a 3x3 matrix
   A =
           0.3200 -0.1600  0.0000
          -0.1600  0.3200 -0.1600
           0.0000 -0.1600  0.3200
>> B = eye(3) - A                 % eye(3) is 3x3 identity matrix
   B =
           0.6800  0.1600  0.0000
           0.1600  0.6800  0.1600
           0.0000  0.1600  0.6800
>> d = [.16*200 0 .16*70]'
   d =
          32.0000
           0
          11.2000
>> u = zeros(3,1)
   u =
           0
           0
           0
>> newu1 = B*u+d                          % first time step
   newu1 =
          32.0000
           0
          11.2000
>> newu2 = B*newu1+d                       % second time step
   newu2 =
          53.7600
           6.9120
          18.8160
>> newu3 = B*newu2+d                       % third time step
   newu3 =
          69.6627
          16.3123
          25.1008
>> SSu = A\d                               % solves Au = d
   SSu =
         167.5000
         135.0000
         102.5000
```

3.2.5 Exercises

1.   Compute $A\mathbf{x}$ (both row and column versions), $A + B$, $A + C$ and sA where

$$A = \begin{bmatrix} 1 & 2 & 3 & 4 \\ 4 & -1 & 2 & 5 \end{bmatrix}, \; C = \begin{bmatrix} 1 & -3 & 3 & 4 \\ 2 & -1 & 12 & 5 \end{bmatrix},$$

$$B = \begin{bmatrix} 0 & 22 & 2 & 4 \\ 1 & -1 & 3 & 7 \end{bmatrix}, \; \mathbf{x} = \begin{bmatrix} 2 \\ 1 \\ 5 \\ 4 \end{bmatrix} \text{ and } s = 9.$$

2.   Compute $A\mathbf{x}$ (both row and column versions), $A + B$, $A + C$ and sA where

$$A = \begin{bmatrix} 1 & 5 & 3 \\ 8 & -1 & 2 \end{bmatrix}, \; C = \begin{bmatrix} 3 & 7 \\ 0 & 2 \end{bmatrix}, \; s = -7,$$

$$B = \begin{bmatrix} 1 & 0 & 10 \\ 2 & -3 & 1 \end{bmatrix} \text{ and } \mathbf{x} = \begin{bmatrix} 2 \\ 11 \\ 5 \end{bmatrix}.$$

3.   Verify $A + (B + C) = (A + B) + C$ where A, B and C are the same as in exercise one.

4.   Verify $A(2\mathbf{x} + 3\mathbf{y}) = 2A\mathbf{x} + 3A\mathbf{y}$ where A and $\mathbf{x}$ are the same as in exercise two and $\mathbf{y}^T = [1 \; 3 \; 2 \; -2]$.

5.   Compute AB (use both row and column versions) where

$$A = \begin{bmatrix} 1 & 5 & 3 \\ 8 & -1 & 2 \end{bmatrix} \text{ and } B = \begin{bmatrix} 1 & 0 & 2 & -1 \\ 5 & 4 & 7 & 1 \\ -1 & 2 & 0 & 1 \end{bmatrix}.$$

6.   Compute BA (use both row and column versions) where

$$A = \begin{bmatrix} 1 & 2 \\ 5 & -2 \\ -1 & 0 \\ 7 & 8 \end{bmatrix} \text{ and } B = \begin{bmatrix} 1 & 0 & 2 & -1 \\ 5 & 4 & 7 & 1 \\ -1 & 2 & 0 & 1 \end{bmatrix}.$$

7.   Verify the $A(BC) = (AB)C$ where

$$A = \begin{bmatrix} 1 & 5 & 3 \\ 8 & -1 & 2 \end{bmatrix}, \; B = \begin{bmatrix} 1 & 2 \\ -3 & 0 \\ 1 & 3 \end{bmatrix} \text{ and } C = \begin{bmatrix} 3 & 7 \\ 0 & 2 \end{bmatrix}.$$

8.   Verify the $(A + \widehat{A})B = AB + \widehat{A}B$ where

$$A = \begin{bmatrix} 1 & 5 & 3 \\ 8 & -1 & 2 \end{bmatrix}, \; \widehat{A} = \begin{bmatrix} -1 & 0 & 1 \\ 3 & -11 & 0 \end{bmatrix} \text{ and } B = \begin{bmatrix} 1 & 2 \\ -3 & 0 \\ 1 & 3 \end{bmatrix}.$$

9.   Consider Example 3.2.5 of heat conduction. Change the thermal conductivity from $K = 1/1000$ to $K = 2/1000$ and leave the other parameters the same. Compute α, $\mathbf{u}^1$, $\mathbf{u}^2$ and the steady state solution.

10.   Consider Example 3.2.5 of heat conduction. Change the number of segments from $n = 4$ to $n = 5$ and leave the other parameters the same. Compute α, $\mathbf{u}^1$, $\mathbf{u}^2$ and the steady state solution. Note there are now four unknowns, the vectors have four components, the matrix is 4×4, the left temperature is $u_0 = 200$ and the right temperature is $u_5 = 70$.

11.   Use MATLAB to do the computations in exercise 9.

12.   Use MATLAB to do the computations in exercise 10.

13.   Modify the MATLAB code wire_time.m to do exercise 9 and observe convergence to the steady state solution.

14.   Modify the MATLAB code wire_time.m to do exercise 10 and observe convergence to the steady state solution.

3.3   Special Cases of Ax = d

In the previous sections we have solved algebraic systems with two and three unknowns. These could be written as a vector equation $A\mathbf{x} = \mathbf{d}$ where A is an $n \times n$ matrix with $n = 2$ or 3, $\mathbf{d}$ is a given $n \times 1$ column vector and $\mathbf{x}$ is an unknown $n \times 1$ column vector. The overall objective is be able to systematically solve problems with much larger n. In this section some important special matrices will be considered and an application to heat conduction in a wire will continue to be studied.

3.3.1   Four Possible Classes of "Solutions"

Algebraic systems may have a solution, no solution or multiple solutions. Moreover, any solution may be very sensitive to the data in A or $\mathbf{d}$. These will be illustrated by four simple 2×2 systems. One can gain insight to these examples by viewing them from both an algebraic and geometric perspective. Consider two equations

$$
\begin{aligned}
a_{11}x_1 + a_{12}x_2 &= d_1 \text{ and} \\
a_{21}x_1 + a_{22}x_2 &= d_2.
\end{aligned}
$$

Or, view this as a single vector equation

$$
\begin{bmatrix} a_{11} & a_{12} \\ a_{21} & a_{22} \end{bmatrix} \begin{bmatrix} x_1 \\ x_2 \end{bmatrix} = \begin{bmatrix} d_1 \\ d_2 \end{bmatrix}.
$$

Plot x_1 on the horizontal axis and x_2 on the vertical axis. The first scalar equation may be written in slope-intercept form of a line

$$
x_2 = (-a_{11}/a_{12})x_1 + d_1/a_{12}, \text{ if } a_{12} \neq 0.
$$

Do the same for the second scalar equation. The two lines are either not parallel, parallel and do not overlap, parallel and overlap, or they are "nearly" parallel. If they are not parallel, then there is exactly one solution. If they are parallel and do not overlap, then there is no solution. If they are parallel and overlap, then there are multiple solutions given by the points on the single line. If the two lines are "nearly" parallel, then small changes in the data can cause large changes in the intersection, that is, the solution. In the last case the small changes in the data can be caused by measurement, roundoff or human errors.

Case One. If the lines are not parallel, then

$$\det(A) = \det\left(\begin{bmatrix} a_{11} & a_{12} \\ a_{21} & a_{22} \end{bmatrix} \right) \neq 0.$$

Cramer's rule can be used to find the unique solution

$$x_1 = \det\left(\begin{bmatrix} d_1 & a_{12} \\ d_2 & a_{22} \end{bmatrix} \right)/\det(A) \text{ and } x_2 = \det\left(\begin{bmatrix} a_{11} & d_1 \\ a_{21} & d_2 \end{bmatrix} \right)/\det(A).$$

Case Two. The lines are parallel and do not overlap. For example,

$$\begin{aligned} x_1 + x_2 &= 1 \text{ and} \\ 2x_1 + 2x_2 &= 4. \end{aligned}$$

The second equation can be divided by 2 to get $x_1 + x_2 = 2$, which contradicts the first equation! So, there is no solution.

Case Three. The lines are parallel and do overlap. For example,

$$\begin{aligned} x_1 + x_2 &= 1 \text{ and} \\ 2x_1 + 2x_2 &= 2. \end{aligned}$$

The second equation can be divided by 2 so that $x_1 + x_2 = 1$, which is the same as the first equation. Any point on this line will be a solution of algebraic system.

Case Four. This example should be of great concern because it raises the issue of sensitivity of the solution to the data. Let ϵ represent a small variation in the data for $d_1 = 1$ and consider the system

$$\begin{aligned} x_1 + x_2 &= 1 + \epsilon \text{ and} \\ (1 - 10^{-6})x_1 + x_2 &= 1. \end{aligned}$$

If $\epsilon = 0$, then the unique solution is $x_1 = 0$ and $x_2 = 1$. If $\epsilon = 10^{-3}$, then the unique solution is extremely different and is $x_1 = 1000$ and $x_2 = -998.999$. Moreover, for small variations in the right side the residuals, $\mathbf{r} = \mathbf{d} - A\mathbf{x}$, appear to be small

$$\begin{aligned} 1 - (x_1 + x_2) &= 1 - (1000 - 998.999) = -.001 \text{ and} \\ 1 - ((1 - 10^{-6})x_1 + x_2) &= 1 - ((1 - 10^{-6})1000 - 998.999) = 0! \end{aligned}$$

Problems whose solutions are very sensitive to small changes in the data are called *ill-conditioned.*

The challenge is to detect these possibilities for large systems. This chapter is restricted to the unique solution case, and chapter five studies the multiple solution case. Another possibility is having more equations than unknowns, which will be investigated in the fourth chapter.

3.3.2 Triangular Matrices

The simplest system to solve is the diagonal system where the equations are for $i = 1, \cdots, n$

$$a_{ii}x_i = d_i.$$

The matrix version of this uses the *diagonal* matrix $A = D$

$$\begin{bmatrix} a_{11} & 0 & \cdots & 0 \\ 0 & a_{22} & \ddots & \vdots \\ \vdots & \ddots & \ddots & 0 \\ 0 & \cdots & 0 & a_{nn} \end{bmatrix} \begin{bmatrix} x_1 \\ x_2 \\ \vdots \\ x_n \end{bmatrix} = \begin{bmatrix} d_1 \\ d_2 \\ \vdots \\ d_n \end{bmatrix}$$

$$D\mathbf{x} = \mathbf{d}.$$

If each component on the diagonal is not zero ($a_{ii} \neq 0$ for all i), then the solution is $x_i = d_i/a_{ii}$.

More general $n \times n$ matrices are *lower triangular* $A = L$ and *upper triangular* $A = U$

$$\begin{bmatrix} a_{11} & 0 & \cdots & 0 \\ a_{21} & a_{22} & \ddots & \vdots \\ \vdots & \ddots & \ddots & 0 \\ a_{n1} & \cdots & a_{n,n-1} & a_{nn} \end{bmatrix} \begin{bmatrix} x_1 \\ x_2 \\ \vdots \\ x_n \end{bmatrix} = \begin{bmatrix} d_1 \\ d_2 \\ \vdots \\ d_n \end{bmatrix}$$

$$L\mathbf{x} = \mathbf{d} \text{ and}$$

$$\begin{bmatrix} a_{11} & a_{12} & \cdots & a_{1n} \\ 0 & a_{22} & \ddots & \vdots \\ \vdots & \ddots & \ddots & a_{n-1,n} \\ 0 & \cdots & 0 & a_{nn} \end{bmatrix} \begin{bmatrix} x_1 \\ x_2 \\ \vdots \\ x_n \end{bmatrix} = \begin{bmatrix} d_1 \\ d_2 \\ \vdots \\ d_n \end{bmatrix}$$

$$U\mathbf{x} = \mathbf{d}.$$

Both of these are relatively easy to solve provided all the diagonal components a_{ii} are not zero. The lower triangular system is solved by first finding x_1 and then x_2 and so on until x_n is found; this is called *forward substitution.* The upper triangular system is solved in reverse order by first finding x_n and then x_{n-1} and so on until x_1 is determined; this is called *backward substitution.*

Example 3.3.1. Let $n = 3$ and solve the following lower triangular system

$$\begin{bmatrix} 1 & 0 & 0 \\ 2 & 3 & 0 \\ 4 & 5 & 6 \end{bmatrix} \begin{bmatrix} x_1 \\ x_2 \\ x_3 \end{bmatrix} = \begin{bmatrix} 11 \\ 28 \\ 60 \end{bmatrix}.$$

The system is equivalent to the following three scalar equations

$$\begin{aligned} 1x_1 &= 11, \\ 2x_1 + 3x_2 &= 28 \text{ and} \\ 4x_1 + 5x_2 + 6x_3 &= 60. \end{aligned}$$

The first equation gives $x_1 = 11$. Put this into the second equation $2(11) + 3x_2 = 28$ to compute $x_2 = (28 - 2(11))/3 = 2$. Finally, put $x_1 = 11$ and $x_2 = 2$ into the third equation and find $x_3 = (60 - 4(11) - 5(2))/6 = 1$.

Another way to solve this is by using the column version of the matrix-vector product

$$x_1 \begin{bmatrix} 1 \\ 2 \\ 4 \end{bmatrix} + x_2 \begin{bmatrix} 0 \\ 3 \\ 5 \end{bmatrix} + x_3 \begin{bmatrix} 0 \\ 0 \\ 6 \end{bmatrix} = \begin{bmatrix} 11 \\ 28 \\ 60 \end{bmatrix}.$$

The first equation gives $x_1 = 11$. Now, put this into the left column on the left side and move it to the right side

$$x_2 \begin{bmatrix} 0 \\ 3 \\ 5 \end{bmatrix} + x_3 \begin{bmatrix} 0 \\ 0 \\ 6 \end{bmatrix} = \begin{bmatrix} 11 \\ 28 \\ 60 \end{bmatrix} - 11 \begin{bmatrix} 1 \\ 2 \\ 4 \end{bmatrix} = \begin{bmatrix} 0 \\ 6 \\ 16 \end{bmatrix}.$$

The second equation gives $x_2 = 2$. The last step is to put $x_2 = 2$ into the left column and move it to the right side

$$x_3 \begin{bmatrix} 0 \\ 0 \\ 6 \end{bmatrix} = \begin{bmatrix} 0 \\ 6 \\ 16 \end{bmatrix} - 2 \begin{bmatrix} 0 \\ 3 \\ 5 \end{bmatrix} = \begin{bmatrix} 0 \\ 0 \\ 6 \end{bmatrix}.$$

The third equation yields $x_3 = 1$.

Example 3.3.2. Let $n = 4$ and solve the following upper triangular system

$$\begin{bmatrix} 2 & 3 & 0 & 1 \\ 0 & 1 & -2 & 3 \\ 0 & 0 & 3 & -1 \\ 0 & 0 & 0 & 2 \end{bmatrix} \begin{bmatrix} x_1 \\ x_2 \\ x_3 \\ x_4 \end{bmatrix} = \begin{bmatrix} 25 \\ 1 \\ 9 \\ 6 \end{bmatrix}.$$

The four equivalent scalar equations are

$$\begin{aligned} 2x_1 + 3x_2 + 0x_3 + 1x_4 &= 25, \\ 1x_2 - 2x_3 + 3x_4 &= 1, \\ 3x_3 - 1x_4 &= 9 \text{ and} \\ 2x_4 &= 6. \end{aligned}$$

The solution of the last equation is $x_4 = 6/2 = 3$. Put this into the third equation and solve for $x_3 = (9 + 1(3))/3 = 4$. The second equation becomes $1x_2 - 2(4) + 3(3) = 1$ so that $x_2 = 0$. The first equation gives $2x_1 + 3(0) + 0(4) + 1(3) = 25$ so that $x_1 = 11$.

This can also be solved by using the column version of the matrix-vector product

$$x_1 \begin{bmatrix} 2 \\ 0 \\ 0 \\ 0 \end{bmatrix} + x_2 \begin{bmatrix} 3 \\ 1 \\ 0 \\ 0 \end{bmatrix} + x_3 \begin{bmatrix} 0 \\ -2 \\ 3 \\ 0 \end{bmatrix} + x_4 \begin{bmatrix} 1 \\ 3 \\ -1 \\ 2 \end{bmatrix} = \begin{bmatrix} 25 \\ 1 \\ 9 \\ 6 \end{bmatrix}.$$

The last equation gives $x_4 = 3$. Put this into the right column on the left side and move it to the right side

$$x_1 \begin{bmatrix} 2 \\ 0 \\ 0 \\ 0 \end{bmatrix} + x_2 \begin{bmatrix} 3 \\ 1 \\ 0 \\ 0 \end{bmatrix} + x_3 \begin{bmatrix} 0 \\ -2 \\ 3 \\ 0 \end{bmatrix} = \begin{bmatrix} 25 \\ 1 \\ 9 \\ 6 \end{bmatrix} - 3 \begin{bmatrix} 1 \\ 3 \\ -1 \\ 2 \end{bmatrix} = \begin{bmatrix} 22 \\ -8 \\ 12 \\ 0 \end{bmatrix}.$$

The third equation gives $x_3 = 4$. Put $x_3 = 4$ into the right column on the left side and move it to the right side

$$x_1 \begin{bmatrix} 2 \\ 0 \\ 0 \\ 0 \end{bmatrix} + x_2 \begin{bmatrix} 3 \\ 1 \\ 0 \\ 0 \end{bmatrix} = \begin{bmatrix} 22 \\ -8 \\ 12 \\ 0 \end{bmatrix} - 4 \begin{bmatrix} 0 \\ -2 \\ 3 \\ 0 \end{bmatrix} = \begin{bmatrix} 22 \\ 0 \\ 0 \\ 0 \end{bmatrix}.$$

The second and first equations give $x_2 = 0$ and then $x_1 = 11$.

The solution process in all of the above cases requires that the diagonal components not be zero. In this case, there is one and only one solution, which can be found by using either the row or column versions of the matrix-vector product. For large systems there may be some advantages to using one of the two matrix-vector products. This is often dependent on the computing tools used to do the solves.

Theorem 3.3.1 *(Triangular Solves) Consider a lower or an upper triangular matrix $A = L$ or U. If each diagonal component of the matrix is not zero, then the triangular systems $Lx = d$ or $Ux = d$ have one and only one solution.*

3.3.3   A Has LU Factorization

If the given matrix can be written as a product of a lower and upper triangular matrices, then one may be able to solve $Ax = d$. Suppose $A = LU$ and both L and U have non-zero diagonal components. By the associative property

$$\begin{aligned} Ax &= d \\ (LU)x &= d \\ L(Ux) &= d. \end{aligned}$$

So, solve $L\mathbf{y} = \mathbf{d}$ by forward substitution, and then solve $U\mathbf{x} = \mathbf{y}$ by backward substitution.

Example 3.3.3. Consider the steady state heat problem in Example 3.2.5 where

$$\begin{bmatrix} 2 & -1 & 0 \\ -1 & 2 & -1 \\ 0 & -1 & 2 \end{bmatrix} \begin{bmatrix} x_1 \\ x_2 \\ x_3 \end{bmatrix} = \begin{bmatrix} 200 \\ 0 \\ 70 \end{bmatrix}.$$

Observe the matrix can be factored as follows

$$\begin{bmatrix} 2 & -1 & 0 \\ -1 & 2 & -1 \\ 0 & -1 & 2 \end{bmatrix} = \begin{bmatrix} 1 & 0 & 0 \\ -1/2 & 1 & 0 \\ 0 & -2/3 & 1 \end{bmatrix} \begin{bmatrix} 2 & -1 & 0 \\ 0 & 3/2 & -1 \\ 0 & 0 & 4/3 \end{bmatrix}$$

$$A \quad = \quad LU.$$

Later we will learn two methods for finding these factors. First, solve $L\mathbf{y} = \mathbf{d}$

$$\begin{bmatrix} 1 & 0 & 0 \\ -1/2 & 1 & 0 \\ 0 & -2/3 & 1 \end{bmatrix} \begin{bmatrix} y_1 \\ y_2 \\ y_3 \end{bmatrix} = \begin{bmatrix} 200 \\ 0 \\ 70 \end{bmatrix}.$$

The equation associated with the first row gives $y_1 = 200$. The equation associated with the second row is $(-1/2)200 + 1y_2 = 0$ implies $y_2 = 100$. The equation associated with the last row is $(-2/3)100 + 1y_3 = 70$ implies $y_3 = 410/3$. Second, solve $U\mathbf{x} = \mathbf{y}$

$$\begin{bmatrix} 2 & -1 & 0 \\ 0 & 3/2 & -1 \\ 0 & 0 & 4/3 \end{bmatrix} \begin{bmatrix} x_1 \\ x_2 \\ x_3 \end{bmatrix} = \begin{bmatrix} 200 \\ 100 \\ 410/3 \end{bmatrix}.$$

The last row implies $x_3 = (410/3)/(4/3) = 102.5$. The second row gives $(3/2)x_2 - 102.5 = 100$ or $x_2 = 202.5/(3/2) = 135.0$. The top row gives $2x_1 - 135 = 200$ or $x_1 = 335/2 = 167.5$.

If the matrix is *tridiagonal*, a matrix whose only non-zero components are in the diagonal, sub-diagonal and super-diagonal, then one may be able to find the LU factors by a process similar to the following. Let A be 3×3 tridiagonal matrix and assume the L and U factors have the special forms

$$\begin{bmatrix} a_{11} & a_{12} & 0 \\ a_{21} & a_{22} & a_{23} \\ 0 & a_{32} & a_{33} \end{bmatrix} = \begin{bmatrix} 1 & 0 & 0 \\ \alpha_1 & 1 & 0 \\ 0 & \alpha_2 & 1 \end{bmatrix} \begin{bmatrix} \beta_1 & a_{12} & 0 \\ 0 & \beta_2 & a_{23} \\ 0 & 0 & \beta_3 \end{bmatrix}$$

$$= \begin{bmatrix} \beta_1 & a_{12} & 0 \\ \alpha_1\beta_1 & \alpha_1 a_{12} + \beta_2 & a_{23} \\ 0 & \alpha_2\beta_2 & \alpha_2 a_{23} + \beta_3 \end{bmatrix}.$$

The matrices are equal if and only if they have the same number of rows and columns and their ij-components are all equal. This gives us five non-trivial scalar equations in the above 3×3 matrix equation: $a_{11} = \beta_1$, $a_{21} = \alpha_1\beta_1$, $a_{22} = \alpha_1 a_{12} + \beta_2$, $a_{32} = \alpha_2\beta_2$ and $a_{33} = \alpha_2 a_{23} + \beta_3$. If each of the β_1, β_2 and β_3 are non-zero, then we can solve for α_1 and α_2 as well as do the upper triangular solve. As an illustration consider Example 3.3.3 where the diagonal components are all equal to 2 and the other non-zero components are equal to -1 :

$$
\begin{aligned}
2 &= \beta_1, -1 = \alpha_1\beta_1 = \alpha_1 2 \text{ and so } \alpha_1 = -1/2, \\
2 &= -\alpha_1 + \beta_2 = -(-1/2) + \beta_2 \text{ and then } \beta_2 = 3/2, \\
-1 &= \alpha_2\beta_2 \text{ so that } \alpha_2 = -2/3 \text{ and} \\
2 &= -\alpha_2 + \beta_3 \text{ gives } \beta_3 = 4/3.
\end{aligned}
$$

3.3.4 Application to Heat in Wire with Current

Consider heat conduction in a long thin wire as in the previous section. Assume there are five segments, not four, so that the unknown temperatures are u_1, u_2, u_3 and u_4 with $u_0 = 70$ at the left end and $u_5 = 70$ at the right end. Also, assume there is an electrical current, which generates heat at a rate of f per unit volume, per unit time. The heat in a small i^{th} segment is now modeled or approximated by

$$
\rho c(u_i^{new})(A\Delta x) = \rho c u_i(A\Delta x) - \Delta t\, A\, K\frac{u_i - u_{i-1}}{\Delta x} +
$$
$$
\Delta t\, A\, K\frac{u_{i+1} - u_i}{\Delta x} + f\,\Delta t\,(A\Delta x).
$$

Divide by $\rho c(A\Delta x)$ and let $\alpha \equiv (\Delta t/\Delta x^2)(K/\rho c)$ to get for $i = 1, 2, 3$ and 4

$$
u_i^{new} = u_i - \alpha(-u_{i-1} + 2u_i - u_{i+1}) + f\,\Delta t/(\rho c).
$$

One may write this as four scalar equations

$$
\begin{aligned}
u_1^{new} &= u_1 - \alpha(-u_0 + 2u_1 - u_2) + f\,\Delta t/(\rho c), \\
u_2^{new} &= u_2 - \alpha(-u_1 + 2u_2 - u_3) + f\,\Delta t/(\rho c), \\
u_3^{new} &= u_3 - \alpha(-u_2 + 2u_2 - u_4) + f\,\Delta t/(\rho c) \text{ and} \\
u_4^{new} &= u_4 - \alpha(-u_3 + 2u_4 - u_5) + f\,\Delta t/(\rho c).
\end{aligned}
$$

Or, view it as one vector equation

$$
\begin{bmatrix} u_1^{new} \\ u_2^{new} \\ u_3^{new} \\ u_4^{new} \end{bmatrix}
=
\begin{bmatrix} u_1 \\ u_2 \\ u_3 \\ u_4 \end{bmatrix}
-
\begin{bmatrix} 2\alpha & -\alpha & 0 & 0 \\ -\alpha & 2\alpha & -\alpha & 0 \\ 0 & -\alpha & 2\alpha & -\alpha \\ 0 & 0 & -\alpha & 2\alpha \end{bmatrix}
\begin{bmatrix} u_1 \\ u_2 \\ u_3 \\ u_4 \end{bmatrix}
+
$$
$$
\begin{bmatrix} \alpha u_0 \\ 0 \\ 0 \\ \alpha u_5 \end{bmatrix}
+
\begin{bmatrix} f\,\Delta t/(\rho c) \\ f\,\Delta t/(\rho c) \\ f\,\Delta t/(\rho c) \\ f\,\Delta t/(\rho c) \end{bmatrix}
$$
$$
\mathbf{u}^{new} = \mathbf{u} - A\mathbf{u} + \mathbf{d}.
$$

The time dependent solution can be generated by the MATLAB code wire _time _cur.m, and the graphical solution is illustrated in Figure 3.3.1. The steady state solution is given by $\mathbf{u} = \mathbf{u} - A\mathbf{u} + \mathbf{d}$ so that one must solve $A\mathbf{u} = \mathbf{d}$

$$\alpha \begin{bmatrix} 2 & -1 & 0 & 0 \\ -1 & 2 & -1 & 0 \\ 0 & -1 & 2 & -1 \\ 0 & 0 & -1 & 2 \end{bmatrix} \begin{bmatrix} u_1 \\ u_2 \\ u_3 \\ u_4 \end{bmatrix} = \begin{bmatrix} \alpha u_0 \\ 0 \\ 0 \\ \alpha u_5 \end{bmatrix} + \begin{bmatrix} f \Delta t/(\rho c) \\ f \Delta t/(\rho c) \\ f \Delta t/(\rho c) \\ f \Delta t/(\rho c) \end{bmatrix}.$$

Example 3.3.4. Consider the above with $L = 10$, $n = 5$, $\Delta x = 10/5$, $\rho = c = 1$, $K = 1/1000$, $\Delta t = 1000$ so that $\alpha \equiv (\Delta t/\Delta x^2)(K/\rho c) = 25/100 = 0.25$. If the electrical current in the wire generates heat at a rate equal to $f = 0.1$, then $f \Delta t/(\rho c) = 100$ and the above steady state model becomes

$$\begin{bmatrix} 2 & -1 & 0 & 0 \\ -1 & 2 & -1 & 0 \\ 0 & -1 & 2 & -1 \\ 0 & 0 & -1 & 2 \end{bmatrix} \begin{bmatrix} u_1 \\ u_2 \\ u_3 \\ u_3 \end{bmatrix} = \begin{bmatrix} 70 \\ 0 \\ 0 \\ 70 \end{bmatrix} + \begin{bmatrix} 400 \\ 400 \\ 400 \\ 400 \end{bmatrix}.$$

The reader should verify $A = LU$ where L and U are given below

$$\begin{bmatrix} 1 & 0 & 0 & 0 \\ -1/2 & 1 & 0 & 0 \\ 0 & -2/3 & 1 & 0 \\ 0 & 0 & -3/4 & 1 \end{bmatrix} \begin{bmatrix} 2 & -1 & 0 & 0 \\ 0 & 3/2 & -1 & 0 \\ 0 & 0 & 4/3 & -1 \\ 0 & 0 & 0 & 5/4 \end{bmatrix} \begin{bmatrix} u_1 \\ u_2 \\ u_3 \\ u_3 \end{bmatrix} = \begin{bmatrix} 470 \\ 400 \\ 400 \\ 470 \end{bmatrix}.$$

One must solve $A\mathbf{u} = L(U\mathbf{u}) = \mathbf{d}$. The solution of $L\mathbf{y} = \mathbf{d}$ is $y_1 = 470.0$, $y_2 = 635.0$, $y_3 = 823.3$ and $y_4 = 1087.5$. The solution of $U\mathbf{u} = \mathbf{y}$ is $u_1 = 870.0$, $u_2 = 1270.0$, $u_3 = 1270.0$ and $u_4 = 870.0$. This is consistent with the graphical solution for larger times steps in Figure 3.3.1.

3.3.5 Matrix Computations Using MATLAB

The calculations for the heat conduction with electrical current application can easily be done by MATLAB. The steady state solution has been computed several ways. The interested reader should examine the MATLAB code wire _time _cur.m to see how the time dependent solution can be computed using for-loops.

```
>> A = [ 2 -1  0  0;
        -1  2 -1  0;
         0 -1  2 -1;
         0  0 -1  2];
>> d = [470 400 400 470]';
>> A\d                          % solves Au = d
        ans =
```

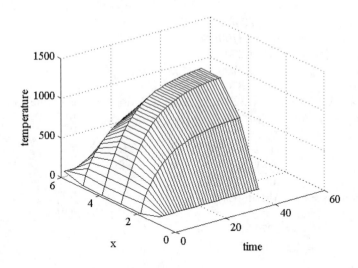

Figure 3.3.1:   Temperature in Wire with Current

```
                    1.0e+003 *
                       0.8700
                       1.2700
                       1.2700
                       0.8700
>> [L U] = lu(A)                        % finds the LU factorization of A
         L =
                 1.0000  0.0000  0.0000  0.0000
                -0.5000  1.0000  0.0000  0.0000
                 0.0000 -0.6667  1.0000  0.0000
                 0.0000  0.0000 -0.7500  1.0000
         U =
                 2.0000 -1.0000  0.0000  0.0000
                 0.0000  1.5000 -1.0000  0.0000
                 0.0000  0.0000  1.3333 -1.0000
                 0.0000  0.0000  0.0000  1.2500
>> y = L\d                              % first step using LU
         y =
                    1.0e+003 *
                       0.4700
                       0.6350
                       0.8233
                       1.0875
>> u = U\y                              % second step using LU
```

u =

   1.0e+003 *

    0.8700

    1.2700

    1.2700

    0.8700

 >> A3 = [A(:,1) A(:,2) d A(:,4)] % use Cramer's Rule to find u3

  A3 =

    2 -1   470   0

    -1   2   400   0

    0 -1   400 -1

    0   0   470   2

 >> u3 = det(A3)/det(A)

  u3 =

    1270

3.3.6   Exercises

1. Find the solution of

$$\begin{bmatrix} 2 & 0 & 0 \\ 1 & 3 & 0 \\ 3 & 4 & -1 \end{bmatrix} \begin{bmatrix} x_1 \\ x_2 \\ x_3 \end{bmatrix} = \begin{bmatrix} 4 \\ 14 \\ 19 \end{bmatrix}.$$

2. Find the solution of

$$\begin{bmatrix} 2 & 0 & 0 \\ -1 & \pi & 0 \\ 2 & 4 & 3 \end{bmatrix} \begin{bmatrix} x_1 \\ x_2 \\ x_3 \end{bmatrix} = \begin{bmatrix} e \\ 2 \\ 10 \end{bmatrix}.$$

3. Find the solution of

$$\begin{bmatrix} 1 & 7 & 8 & 6 \\ 0 & 2 & 3 & 4 \\ 0 & 0 & 1 & 2 \\ 0 & 0 & 0 & 3 \end{bmatrix} \begin{bmatrix} x_1 \\ x_2 \\ x_3 \\ x_4 \end{bmatrix} = \begin{bmatrix} 10 \\ 23 \\ 7 \\ 3 \end{bmatrix}.$$

4. Find the solution of

$$\begin{bmatrix} 2 & -3 & 7 & 6 \\ 0 & 5 & 0 & 1 \\ 0 & 0 & 7 & 2 \\ 0 & 0 & 0 & 4 \end{bmatrix} \begin{bmatrix} x_1 \\ x_2 \\ x_3 \\ x_4 \end{bmatrix} = \begin{bmatrix} \pi \\ \sqrt{2} \\ 15 \\ 16 \end{bmatrix}.$$

5. Use the LU factors in Example 3.3.3 to solve

$$\begin{bmatrix} 2 & -1 & 0 \\ -1 & 2 & -1 \\ 0 & -1 & 2 \end{bmatrix} \begin{bmatrix} x_1 \\ x_2 \\ x_3 \end{bmatrix} = \begin{bmatrix} 210 \\ 10 \\ 80 \end{bmatrix}.$$

6.   Solve

$$
\begin{bmatrix} 1 & 0 & 0 \\ -1 & 2 & 0 \\ 2 & 0 & 3 \end{bmatrix} \begin{bmatrix} 2 & 7 & 8 \\ 0 & 9 & 10 \\ 0 & 0 & 1 \end{bmatrix} \begin{bmatrix} x_1 \\ x_2 \\ x_3 \end{bmatrix} = \begin{bmatrix} 1 \\ 2 \\ 3 \end{bmatrix}.
$$

7.   Consider the heat conduction with current in Example 3.3 4. Verify the solutions of $L\mathbf{y} = \mathbf{d}$ and $U\mathbf{u} = \mathbf{y}$.

8.   Consider the heat conduction with current in Example 3.3 4. Replace $f = 0.1$ by $f = 0.05$, the left end temperature from 70 to 80 and the right end temperature from 70 to 75. Find the steady state solution.

9.   Use MATLAB to do the computations in exercise 7.

10.   Use MATLAB to do the computations in exercise 8.

11.   Use the MATLAB code wire_time_cur.m to do the time dependent computations in exercise 8.

3.4 Row Operations and Gauss Elimination

The objective of this section is to formulate a systematic way of transforming $A\mathbf{x} = \mathbf{d}$ to an equivalent upper triangular system $U\mathbf{x} = \widehat{\mathbf{d}}$. The solution of the upper triangular system can be found provided the diagonal components of the upper triangular matrix are all not zero. This method is called *Gauss elimination* and requires about n^2 storage and $n^3/3$ operations for an $n \times n$ matrix A. This systematic approach can be coded so that a computer can do these operations. An application to a six-bar truss will be given.

3.4.1 Introductory Illustration

Consider the following algebraic system for three unknowns

$$
\begin{cases} x_1 - x_2 + x_3 = 5 \\ -x_1 + x_2 + x_3 = 1 \\ 4x_2 + x_3 = -1. \end{cases}
$$

The matrix version $A\mathbf{x} = \mathbf{d}$ is not in upper triangular form

$$
\begin{bmatrix} 1 & -1 & 1 \\ -1 & 1 & 1 \\ 0 & 4 & 1 \end{bmatrix} \begin{bmatrix} x_1 \\ x_2 \\ x_3 \end{bmatrix} = \begin{bmatrix} 5 \\ 1 \\ -1 \end{bmatrix}.
$$

If one adds equation one (row one) to equation two (row two), then the new algebraic system is

$$
\begin{cases} x_1 - x_2 + x_3 = 5 \\ 2x_3 = 6 \\ 4x_2 + x_3 = -1. \end{cases}
$$

This is equivalent to multiplying the matrix equation by an elementary matrix $E_{21}(1)$

$$E_{21}(1)\,A\mathbf{x} \;=\; E_{21}(1)\,\mathbf{d}$$

$$\begin{bmatrix} 1 & 0 & 0 \\ 1 & 1 & 0 \\ 0 & 0 & 1 \end{bmatrix} \begin{bmatrix} 1 & -1 & 1 \\ -1 & 1 & 1 \\ 0 & 4 & 1 \end{bmatrix} \begin{bmatrix} x_1 \\ x_2 \\ x_3 \end{bmatrix} = \begin{bmatrix} 1 & 0 & 0 \\ 1 & 1 & 0 \\ 0 & 0 & 1 \end{bmatrix} \begin{bmatrix} 5 \\ 1 \\ -1 \end{bmatrix}$$

$$\begin{bmatrix} 1 & -1 & 1 \\ 0 & 0 & 2 \\ 0 & 4 & 1 \end{bmatrix} \begin{bmatrix} x_1 \\ x_2 \\ x_3 \end{bmatrix} = \begin{bmatrix} 5 \\ 6 \\ -1 \end{bmatrix}.$$

Next interchange equation two (row two) and equation three (row three)

$$\begin{cases} x_1 - x_2 + x_3 = 5 \\ \quad\;\; 4x_2 + x_3 = -1 \\ \quad\qquad\;\; 2x_3 = 6. \end{cases}$$

The interchange is equivalent to multiplying the matrix equation by an elementary permutation matrix P_{23}

$$P_{23}\,E_{21}(1)\,A\mathbf{x} \;=\; P_{23}\,E_{21}(1)\,\mathbf{d}$$

$$\begin{bmatrix} 1 & 0 & 0 \\ 0 & 0 & 1 \\ 0 & 1 & 0 \end{bmatrix} \begin{bmatrix} 1 & -1 & 1 \\ 0 & 0 & 2 \\ 0 & 4 & 1 \end{bmatrix} \begin{bmatrix} x_1 \\ x_2 \\ x_3 \end{bmatrix} = \begin{bmatrix} 1 & 0 & 0 \\ 0 & 0 & 1 \\ 0 & 1 & 0 \end{bmatrix} \begin{bmatrix} 5 \\ 6 \\ -1 \end{bmatrix}$$

$$\begin{bmatrix} 1 & -1 & 1 \\ 0 & 4 & 1 \\ 0 & 0 & 2 \end{bmatrix} \begin{bmatrix} x_1 \\ x_2 \\ x_3 \end{bmatrix} = \begin{bmatrix} 5 \\ -1 \\ 6 \end{bmatrix}.$$

Note this is in upper triangular form $U\mathbf{x} = \widehat{\mathbf{d}}$ where $U = P_{23}\,E_{21}(1)\,A$ and $\widehat{\mathbf{d}} = P_{23}\,E_{21}(1)\,\mathbf{d}$. Since adding and interchanging equations can be reversed, the solution of the upper triangular system must also be a solution of the original system. The solution is $x_3 = 6/2 = 3$, $x_2 = (-1 - 3)/4 = -1$ and $x_1 = (5 - 3 + (-1)) = 1$.

3.4.2   Three Types of Row Operations

In order to minimize the repeated writing of the equations, the augmented matrix notation will be used. Let A be an $n \times n$ matrix and let $\mathbf{d}$ be an $n \times 1$ column vector. The *augmented matrix* $[A\ \mathbf{d}]$ is an $n \times (n+1)$ matrix with the column vector augmented to the matrix. In the above example

$$[A\ \mathbf{d}] = \begin{bmatrix} 1 & -1 & 1 & 5 \\ -1 & 1 & 1 & 1 \\ 0 & 4 & 1 & -1 \end{bmatrix}.$$

The two row operations can be written as

$$P_{23}\,E_{21}(1)\,[A\ \mathbf{d}] = [U\ \widehat{\mathbf{d}}].$$

There are three *row operations*, which are used to transform matrices to upper triangular matrices. They can be used to solve algebraic systems, find the *LU* factors, evaluate determinants, find multiple solutions and eigenvectors. The row operations can be represented by *elementary matrices* $E_{ij}(a)$, P_{ij} and $E_i(c)$.

Add a **times** (*row_j*) **to** *row_i*. This can be represented by an $n \times n$ elementary matrix $E_{ij}(a)$ with the only non-zero components being ones on the diagonal and a equal to the ij-component. For example, for $n = 3$, $a = -2$, $i = 3$ and $j = 2$

$$E_{32}(-2) = \begin{bmatrix} 1 & 0 & 0 \\ 0 & 1 & 0 \\ 0 & -2 & 1 \end{bmatrix}.$$

Note, $E_{ij}(-a)$ is the inverse operation to $E_{ij}(a)$, that is, $E_{ij}(-a)\, E_{ij}(a) = I$. For example,

$$E_{32}(2)\, E_{32}(-2) = \begin{bmatrix} 1 & 0 & 0 \\ 0 & 1 & 0 \\ 0 & 2 & 1 \end{bmatrix} \begin{bmatrix} 1 & 0 & 0 \\ 0 & 1 & 0 \\ 0 & -2 & 1 \end{bmatrix} = \begin{bmatrix} 1 & 0 & 0 \\ 0 & 1 & 0 \\ 0 & 0 & 1 \end{bmatrix}.$$

Interchange *row_i* **and** *row_j*. This can be represented by an $n \times n$ matrix P_{ij} with the only non-zero components being ones on the diagonal except for *row_i* and *row_j* where the ij-component and ji-component equal one, respectively. For example, for $n = 4$, $i = 2$ and $j = 3$

$$P_{23} = \begin{bmatrix} 1 & 0 & 0 & 0 \\ 0 & 0 & 1 & 0 \\ 0 & 1 & 0 & 0 \\ 0 & 0 & 0 & 1 \end{bmatrix}.$$

Note that $P_{ij}\, P_{ij} = I$.

Multiply *row_i* **by** c. The is represented by an $n \times n$ matrix $E_i(c)$ with the only non-zero components being ones on the diagonal except for *row_i* where the diagonal component is c. For example, for $n = 4$, $c = 9$ and $i = 3$

$$E_3(9) = \begin{bmatrix} 1 & 0 & 0 & 0 \\ 0 & 1 & 0 & 0 \\ 0 & 0 & 9 & 0 \\ 0 & 0 & 0 & 1 \end{bmatrix}.$$

It is easy to verify that $E_i(1/c)\, E_i(c) = I$.

Row operations are used, starting with the left columns in the augmented matrix, to transform it to an upper triangular matrix. Once the upper triangular matrix is found, the solution may be computed by backward substitution as long as the diagonal components of the upper triangular matrix are not zero. The following 2×2 example illustrates this may not always be possible

$$
\begin{aligned}
x_1 + x_2 &= 1 \text{ and} \\
2x_1 + 2x_2 &= 3.
\end{aligned}
$$

The augmented matrix is

$$
[A \ \mathbf{d}] = \begin{bmatrix} 1 & 1 & 1 \\ 2 & 2 & 3 \end{bmatrix}.
$$

In order to obtain a zero in the 21-component add (-2) times *row*_1 to *row*_2

$$
E_{21}(-2)[A \ \mathbf{d}] = \begin{bmatrix} 1 & 0 \\ -2 & 1 \end{bmatrix} \begin{bmatrix} 1 & 1 & 1 \\ 2 & 2 & 3 \end{bmatrix} = \begin{bmatrix} 1 & 1 & 1 \\ 0 & 0 & 1 \end{bmatrix} = [U \ \widehat{\mathbf{d}}].
$$

The 22-component of U is zero! Moreover, the second row is a shorthand way of writing $0x_1 + 0x_2 = 1$, which is impossible. Such algebraic systems are called *inconsistent* and have no solution.

3.4.3   Gauss Elimination for Solving $Ax = d$

Assume $Ax = d$ has a solution and it is unique. The Gauss elimination method is defined so that it can be implemented for larger systems and using computers. The reader will observe the by-hand calculations can become a little more than tiresome, but the objective here is to illustrate how the method can be used to solve larger systems.

Definition 3.4.1. Consider the $n \times n$ algebraic system $Ax = d$. The *Gauss elimination method* for solving this system has two stages:
Stage 1.   Transform $Ax = d$ to upper triangular form $Ux = \widehat{d}$, that is, transform the augmented matrix $[A \ d]$ to an upper triangular matrix $[U \ \widehat{d}]$.
    (a). start with the left column, column $j = 1$, and use row operations to transform column $j = 1$ to zeros below row $i = 1$,
    (b). move to the next column, column $j = 2$, and use row operations to transform column $j = 2$ to zeros below row $i = 2$ and
    (c). repeat this until column $j = n - 1$ has been done.
Stage 2.   Solve the upper triangular system $Ux = \widehat{d}$ by backward substitution.
    (a). start with the bottom row, row $i = n$, and solve the corresponding equation for x_n,
    (b). put x_n into the equations corresponding to rows $i = n - 1, \cdots, 1$,
    (c). solve equation $i = n - 1$ for x_{n-1} and
    (d). repeat this until x_1 has been computed.

Example 3.4.1. Consider the 3×3 algebraic system

$$
\begin{cases}
5x_2 + 6x_3 = 6 \\
x_1 + 3x_2 + x_3 = 2 \\
2x_1 + x_2 + x_3 = 3.
\end{cases}
$$

The augmented matrix is

$$[A \ \mathbf{d}] = \begin{bmatrix} 0 & 5 & 6 & 6 \\ 1 & 3 & 1 & 2 \\ 2 & 1 & 1 & 3 \end{bmatrix}.$$

Stage 1. In order to get zeros in column $j = 1$, the 11-component must not be zero. So, first interchange row_1 and row_2

$$P_{12} \ [A \ \mathbf{d}] = \begin{bmatrix} 0 & 1 & 0 \\ 1 & 0 & 0 \\ 0 & 0 & 1 \end{bmatrix} \begin{bmatrix} 0 & 5 & 6 & 6 \\ 1 & 3 & 1 & 2 \\ 2 & 1 & 1 & 3 \end{bmatrix} = \begin{bmatrix} 1 & 3 & 1 & 2 \\ 0 & 5 & 6 & 6 \\ 2 & 1 & 1 & 3 \end{bmatrix}.$$

Add (-2) times row_1 to row_3

$$E_{31}(-2) \ P_{12} \ [A \ \mathbf{d}] = \begin{bmatrix} 1 & 0 & 0 \\ 0 & 1 & 0 \\ -2 & 0 & 1 \end{bmatrix} \begin{bmatrix} 1 & 3 & 1 & 2 \\ 0 & 5 & 6 & 6 \\ 2 & 1 & 1 & 3 \end{bmatrix} = \begin{bmatrix} 1 & 3 & 1 & 2 \\ 0 & 5 & 6 & 6 \\ 0 & -5 & -1 & -1 \end{bmatrix}.$$

Move to the second column and add row_2 to row_3

$$E_{32}(1) \ E_{31}(-2) \ P_{12} \ [A \ \mathbf{d}] = \begin{bmatrix} 1 & 0 & 0 \\ 0 & 1 & 0 \\ 0 & 1 & 1 \end{bmatrix} \begin{bmatrix} 1 & 3 & 1 & 2 \\ 0 & 5 & 6 & 6 \\ 0 & -5 & -1 & -1 \end{bmatrix}$$

$$= \begin{bmatrix} 1 & 3 & 1 & 2 \\ 0 & 5 & 6 & 6 \\ 0 & 0 & 5 & 5 \end{bmatrix}.$$

Stage 2. The last row corresponds to the equation $5x_3 = 5$ and so $x_3 = 1$. Put this into the equations corresponding to the rows 1 and 2 to get

$$\begin{aligned} x_1 + 3x_2 + 1(1) &= 2 \text{ and} \\ 5x_2 + 6(1) &= 6. \end{aligned}$$

Then $x_2 = 0$ and $x_1 = 1$.

Example 3.4.2. Consider another 3×3 algebraic system

$$\begin{cases} 2x_1 + 6x_2 + 8x_3 = 16 \\ 4x_1 + 15x_2 + 19x_3 = 38 \\ 2x_1 + 3x_3 = 8. \end{cases}$$

The augmented matrix is

$$[A \ \mathbf{d}] = \begin{bmatrix} 2 & 6 & 8 & 16 \\ 4 & 15 & 19 & 38 \\ 2 & 0 & 3 & 8 \end{bmatrix}.$$

Stage 1. Get zeros in the first column by adding (-2) times row_1 to row_2, and then adding (-1) times row_1 to row_3

$$E_{31}(-1)\, E_{21}(-2)\, [A\ \mathbf{d}] = \begin{bmatrix} 2 & 6 & 8 & 16 \\ 0 & 3 & 3 & 6 \\ 0 & -6 & -5 & -8 \end{bmatrix}.$$

By adding (2) times row_2 to row_3 we get a zero in the 32-component

$$E_{32}(2)\, E_{31}(-1)\, E_{21}(-2)\, [A\ \mathbf{d}] = \begin{bmatrix} 2 & 6 & 8 & 16 \\ 0 & 3 & 3 & 6 \\ 0 & 0 & 1 & 4 \end{bmatrix}.$$

Stage 2. The last row corresponds to the equation $1x_3 = 4$ and so $x_3 = 4$. Put this into the equations corresponding to the rows 1 and 2 to get

$$\begin{aligned} 2x_1 + 6x_2 + 8(4) &= 16 \text{ and} \\ 3x_2 + 3(4) &= 6. \end{aligned}$$

Then $x_2 = -2$ and $x_1 = -2$.

Example 3.4.3. Consider the two-loop circuit problem that was discussed in Subsection 3.1.4

$$\begin{cases} i_1 - i_2 + i_3 = 0 \\ R_1 i_1 - R_3 i_3 = E_1 \\ -R_2 i_2 - R_3 i_3 = E_2. \end{cases}$$

Let $R_1 = 1$, $R_2 = 2$, $R_3 = 3$, $E_1 = 10$ and $E_2 = 20$ so that the augmented matrix is

$$[A\ \mathbf{d}] = \begin{bmatrix} 1 & -1 & 1 & 0 \\ 1 & 0 & -3 & 10 \\ 0 & -2 & -3 & 20 \end{bmatrix}.$$

Stage 1. Obtain zeros in the 21-component and 32-component by using

$$E_{32}(2)\, E_{21}(-1)\, [A\ \mathbf{d}] = \begin{bmatrix} 1 & -1 & 1 & 0 \\ 0 & 1 & -4 & 10 \\ 0 & 0 & -11 & 40 \end{bmatrix}.$$

Stage 2. The last row corresponds to the equation $-11i_3 = 40$ and so $i_3 = 40/(-11)$. Put this into the equations corresponding to the rows 1 and 2 to get

$$\begin{aligned} i_1 - i_2 + 40/(-11) &= 0 \text{ and} \\ i_2 - 4(40/(-11)) &= 10. \end{aligned}$$

This gives $i_2 = 50/(-11)$ and $i_1 = 10/(-11)$.

Example 3.4.4. Consider the steady state heat wire in Example 3.3.4 where there were five segments and electrical current generating heat within the wire. The algebraic problem for the temperature in each interior segment is

$$
\begin{bmatrix}
2 & -1 & 0 & 0 \\
-1 & 2 & -1 & 0 \\
0 & -1 & 2 & -1 \\
0 & 0 & -1 & 2
\end{bmatrix}
\begin{bmatrix}
u_1 \\
u_2 \\
u_3 \\
u_3
\end{bmatrix}
=
\begin{bmatrix}
470 \\
400 \\
400 \\
470
\end{bmatrix}
$$

and the augmented matrix is

$$
[A \; \mathbf{d}] =
\begin{bmatrix}
2 & -1 & 0 & 0 & 470 \\
-1 & 2 & -1 & 0 & 400 \\
0 & -1 & 2 & -1 & 400 \\
0 & 0 & -1 & 2 & 470
\end{bmatrix}.
$$

Stage 1. Obtain zeros in the subdiagonal by using

$$
E_{43}(3/4) \, E_{32}(2/3) \, E_{21}(1/2) \, [A \; \mathbf{d}] =
\begin{bmatrix}
2 & -1 & 0 & 0 & 470.00 \\
0 & 3/2 & -1 & 0 & 635.00 \\
0 & 0 & 4/3 & -1 & 823.33 \\
0 & 0 & 0 & 5/4 & 1087.50
\end{bmatrix}.
$$

Stage 2. The last row corresponds to the equation $(5/4)u_4 = 1087.5$ and so $u_4 = 1087.50/1.25 = 870$. Put this into the equations corresponding the rows 1, 2 and 3 to get $u_3 = u_2 = 1270$ and $u_1 = 870$. The reader should note the similarities of these calculations with the LU factorization method as given in Subsection 3.3.4.

3.4.4   Application to Six-bar Truss

In Subsection 3.1.4 an application to a two-bar truss was presented. There was one stationary node where a balance of 2D forces gave two scalar equations for the forces within the two bars. If a bar is under tension, then the force within the bar will be positive. If the bar is under compression, then the force will be negative. The two-bar truss will be enhanced to a six-bar truss with three stationary nodes, see Figure 3.4.1. Unlike the two-bar truss, it is not possible to assume the bars are under compression or tension forces. Hence, all the unknown bar forces in Figure 3.4.1 are initially displayed as under tension.

In order to derive an algebraic system for the six forces, use at each node the equilibrium condition that the sum of the force vectors must be the zero force vector. Let the angle between bars one and two be equal to θ and use the short notation for sine and cosine $s = \sin(\theta)$ and $c = \cos(\theta)$. Assume the mass at the right end of the support structure has a weight equal to 10.

Node 1: $[-f_1 \; 0] + [-f_2 c \; f_2 s] + [0 \; -10] = [0 \; 0],$

Node 2: $[f_2 c \; -f_2 s] + [0 \; -f_3] + [-f_4 \; 0] = [0 \; 0]$ and

Node 3: $[f_1 \; 0] + [0 \; f_3] + [-f_5 c \; f_5 s] + [-f_6 \; 0] = [0 \; 0].$

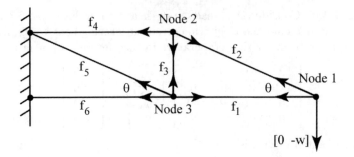

Figure 3.4.1: Six-bar Truss

Now equate the first and second components in each vector equation to obtain a system with six scalar equations

$$\begin{cases} -f_1 - f_2 c = 0 \\ f_2 s = 10 \\ f_2 c - f_4 = 0 \\ -f_2 s - f_3 = 0 \\ f_1 - f_5 c - f_6 = 0 \\ f_3 + f_5 s = 0. \end{cases}$$

The vector version of these scalar equations is

$$\begin{bmatrix} -1 & -c & 0 & 0 & 0 & 0 \\ 0 & s & 0 & 0 & 0 & 0 \\ 0 & c & 0 & -1 & 0 & 0 \\ 0 & -s & -1 & 0 & 0 & 0 \\ 1 & 0 & 0 & 0 & -c & -1 \\ 0 & 0 & 1 & 0 & s & 0 \end{bmatrix} \begin{bmatrix} f_1 \\ f_2 \\ f_3 \\ f_4 \\ f_5 \\ f_6 \end{bmatrix} = \begin{bmatrix} 0 \\ 10 \\ 0 \\ 0 \\ 0 \\ 0 \end{bmatrix}$$

and the augmented matrix is

$$[A \ \mathbf{d}] = \begin{bmatrix} -1 & -c & 0 & 0 & 0 & 0 & 0 \\ 0 & s & 0 & 0 & 0 & 0 & 10 \\ 0 & c & 0 & -1 & 0 & 0 & 0 \\ 0 & -s & -1 & 0 & 0 & 0 & 0 \\ 1 & 0 & 0 & 0 & -c & -1 & 0 \\ 0 & 0 & 1 & 0 & s & 0 & 0 \end{bmatrix}.$$

In order to do the first stage of Gauss elimination, one must use row operation to transform the lower portion of the augmented matrix to zero components. Fortunately, this matrix has many zero components. The required row operations are indicated by groups of operations on columns:

$$E_{51}(1) [A \ \mathbf{d}] = \begin{bmatrix} -1 & -c & 0 & 0 & 0 & 0 & 0 \\ 0 & s & 0 & 0 & 0 & 0 & 10 \\ 0 & c & 0 & -1 & 0 & 0 & 0 \\ 0 & -s & -1 & 0 & 0 & 0 & 0 \\ 0 & -c & 0 & 0 & -c & -1 & 0 \\ 0 & 0 & 1 & 0 & s & 0 & 0 \end{bmatrix},$$

$$E_{52}(c/s) \ E_{42}(1) \ E_{32}(-c/s) \ E_{51}(1) [A \ \mathbf{d}] =$$

$$\begin{bmatrix} -1 & -c & 0 & 0 & 0 & 0 & 0 \\ 0 & s & 0 & 0 & 0 & 0 & 10 \\ 0 & 0 & 0 & -1 & 0 & 0 & -10c/s \\ 0 & 0 & -1 & 0 & 0 & 0 & 10 \\ 0 & 0 & 0 & 0 & -c & -1 & 10c/s \\ 0 & 0 & 1 & 0 & s & 0 & 0 \end{bmatrix},$$

$$E_{63}(1) \ P_{43} \ E_{52}(c/s) \ E_{42}(1) \ E_{32}(-c/s) \ E_{51}(1) [A \ \mathbf{d}] =$$

$$\begin{bmatrix} -1 & -c & 0 & 0 & 0 & 0 & 0 \\ 0 & s & 0 & 0 & 0 & 0 & 10 \\ 0 & 0 & -1 & 0 & 0 & 0 & 10 \\ 0 & 0 & 0 & -1 & 0 & 0 & -10c/s \\ 0 & 0 & 0 & 0 & -c & -1 & 10c/s \\ 0 & 0 & 0 & 0 & s & 0 & 10 \end{bmatrix} \quad \text{and}$$

$$E_{65}(s/c) \ E_{63}(1) \ P_{43} \ E_{52}(c/s) \ E_{42}(1) \ E_{32}(-c/s) \ E_{51}(1) [A \ \mathbf{d}] =$$

$$\begin{bmatrix} -1 & -c & 0 & 0 & 0 & 0 & 0 \\ 0 & s & 0 & 0 & 0 & 0 & 10 \\ 0 & 0 & -1 & 0 & 0 & 0 & 10 \\ 0 & 0 & 0 & -1 & 0 & 0 & -10c/s \\ 0 & 0 & 0 & 0 & -c & -1 & 10c/s \\ 0 & 0 & 0 & 0 & 0 & -s/c & 20 \end{bmatrix}.$$

The second stage is a backwards substitution:

$$\begin{aligned} f_6 &= -20c/s = -20\cos(\theta)/\sin(\theta), \\ f_5 &= (10c/s - 20c/s)/(-c) = 10/s = 10/\sin(\theta), \\ f_4 &= 10c/s = 10\cos(\theta)/\sin(\theta), \\ f_3 &= -10, \\ f_2 &= 10/s = 10/\sin(\theta) \text{ and} \\ f_1 &= -c(10/s) = -10c/s = -10\cos(\theta)/\sin(\theta). \end{aligned}$$

The negative signs indicate bars 1, 3 and 6 are under compression, and the other bars 2, 4 and 5 are under tension. As the angle θ decreases, the magnitudes of the forces increase, except for bar 3.

3.4.5 Gauss Elimination Using MATLAB

Row operations using MATLAB are illustrated in the MATLAB code gauss_el.m. Both stages of Gauss elimination are combined in the single command MATLAB command A\d. Consider the six-beam support structure with $\theta = \pi/6$.

```
>>c = cos(pi/6)
>> s = sin(pi/6);
>> A = [-1 -c  0 0 0 0 ;
         0 s  0  0 0  0 ;
         0 c  0 -1  0 0 ;
         0 -s -1  0  0  0 ;
         1  0  0  0 -c -1 ;
         0  0  1  0  s 0];
>> d = [0 10 0 0 0 0]';
>> A\d
      ans =
         -17.3205          % = f1
          20.0000          % = f2
         -10.0000          % = f3
          17.3205          % = f4
          20.0000          % = f5
         -34.6410          % = f6
```

The reader will find it interesting to experiment with different angles and weights, which can easily be done using the MATLAB code support.m.

3.4.6 Exercises

1. Find the augmented matrix for

$$\begin{cases} 2x_1 - 2x_2 + x_3 = 7 \\ -4x_1 + x_2 + x_3 = 0 \\ 3x_2 + x_3 = -2. \end{cases}$$

2. Find the augmented matrix for

$$\begin{cases} x_1 - 2x_2 + x_3 - 5x_4 = 1 \\ x_1 - 2x_2 + 5x_4 = 1 \\ -3x_1 - 2x_2 + x_3 + 2x_4 = 0 \\ 4x_1 + x_3 - x_4 = 5. \end{cases}$$

3. Let $n = 4$ and find the elementary matrices $E_{43}(-7)$, P_{23} and $E_4(10)$.
4. Let $n = 3$ and find the elementary matrices $E_{21}(2)$, P_{13} and $E_3(\pi)$.
5. Consider the algebraic system

$$\begin{cases} x_1 + 7x_2 + x_3 = 1 \\ 2x_1 + 0x_2 + 7x_3 = 2 \\ 14x_2 + x_3 = 6. \end{cases}$$

(a).   Find the augmented matrix.

(b).   Use row operations to transform the augmented matrix to an upper triangular matrix.

(c).   Find the solution by using the upper triangular matrix.

6.   As in exercise 5 use Gauss elimination to solve the system in exercise 1.

7.   Consider the algebraic system

$$\begin{cases} x_1 - x_2 + x_4 = 6 \\ x_1 - x_3 - x_4 = 7 \\ 2x_2 + 7x_3 = 4 \\ 2x_1 - x_4 = 8. \end{cases}$$

(a).   Find the augmented matrix.

(b).   Use row operations to transform the augmented matrix to an upper triangular matrix.

(c).   Find the solution by using the upper triangular matrix.

8.   As in exercise 7 use Gauss elimination to solve the system in exercise 2.

9.   Consider the two-loop circuit in Example 3.4.3 with slightly different $R_1 = 1$, $R_2 = 2$, $R_3 = 4$, $E_1 = 20$ and $E_2 = 10$. Use Gauss elimination to find the solution.

10.   Consider the two-loop circuit in Example 3.4.3 with arbitrary symbols for R_1, R_2, R_3, E_1 and E_2. Use Gauss elimination to find the solution.

11.   Consider the steady state heat conduction within a wire in Example 3.4.4. Verify the row operations to find the indicated upper triangular matrix.

12.   Consider the steady state heat conduction within a wire in Example 3.4.4. Let the left end temperature be $u_0 = 80$, the right end temperature be $u_5 = 75$ and the heat generated by the current be $f = 0.05$ per unit time per unit volume. Use Gauss elimination to solve the steady state problem

$$\begin{bmatrix} 2 & -1 & 0 & 0 \\ -1 & 2 & -1 & 0 \\ 0 & -1 & 2 & -1 \\ 0 & 0 & -1 & 2 \end{bmatrix} \begin{bmatrix} u_1 \\ u_2 \\ u_3 \\ u_3 \end{bmatrix} = \begin{bmatrix} 280 \\ 200 \\ 200 \\ 275 \end{bmatrix}.$$

13.   Consider the six-bar truss.

(a).   Verify the operations and the solution.

(b).   Change the weight from $w = 10$ to $w = 20$ and compute the new forces.

(c).   Show the solution for $w = c10$ is always $c\mathbf{f}$ where $\mathbf{f}$ is the solution in the text.

14.   Consider the six-bar truss. Suppose the support structure has weights at both nodes one $[0 \ -w_1]$ and three $[0 \ -w_2]$.

(a).   Find the model for this new problem.

(b).   Find the solution when $w_1 = 10$ and $w_2 = 20$.

15.   Use MATLAB to confirm any of your calculations in exercises 5-14.

16.   Use MATLAB to find the solutions in the six-bar truss problem for $\theta = \pi/6$, $\pi/4$ and $2\pi/6$. You may use the MATLAB code support.m.

17.   Use MATLAB to find the solutions in the six-bar truss with two weights as specified in exercise 14. You may modify the MATLAB code support.m, and experiment with different weights and angles.

3.5 Inverse Matrices

The three types of elementary row operations can be reversed and be represented by elementary matrices. In terms of matrix operations this means for each elementary matrix there is another elementary matrix such that the product is the identity matrix:

$$
\begin{aligned}
E_{ij}(-a)\, E_{ij}(a) &= I \\
P_{ij}\, P_{ij} &= I \text{ and} \\
E_i(1/c)\, E_i(c) &= I.
\end{aligned}
$$

The objective of this section is to be able to find an $n \times n$ matrix B such that for a given $n \times n$ matrix A we have $AB = I$. B is called an *inverse matrix of A*, is often written as $A^{-1} = B$. An important use is to find the solution $Ax = d$ given by $x = A^{-1}d$ because

$$
Ax = A(A^{-1}d) = (AA^{-1})d = Id = d.
$$

Unfortunately, one can not always find such matrices! We shall see in Section 3.7 A with a non-zero determinant implies that an inverse matrix exists.

3.5.1 Examples of Inverse Matrices

The identity $n \times n$ matrix I has all zero components except on the diagonal where the components are equal to one. The *identity matrix* can be written as a matrix whose j^{th} column vector is the jth unit column vector e_j

$$
\begin{aligned}
I &= \begin{bmatrix} e_1 & e_2 & \cdots & e_n \end{bmatrix} \\
&= \begin{bmatrix} 1 & 0 & \cdots & 0 \\ 0 & 1 & \cdots & 0 \\ \vdots & \vdots & \ddots & \vdots \\ 0 & 0 & \cdots & 1 \end{bmatrix}.
\end{aligned}
$$

Definition 3.5.1. Let A be an $n \times n$ matrix. An $n \times n$ matrix B is called an *inverse matrix of A* if and only if $AB = BA = I$.

Notation. The inverse matrix is usually denoted by A^{-1}. Here one should be careful not to write this as $1/A$ or to have components equal to $1/a_{ij}$!

Basic Examples.

1.   Elementary Matrices.

If $A = E_{ij}(a)$, then $A^{-1} = E_{ij}(-a)$. If $A = P_{ij}$, then $A^{-1} = P_{ij}$. If $A = E_i(c)$, then $A^{-1} = E_i(1/c)$.

2.   Diagonal Matrices.

If A is a diagonal matrix with non-zero diagonal components $a_{ii} \neq 0$, then A^{-1} is a diagonal matrix with diagonal components equal to $1/a_{ii}$. For example, if $n = 3$ and

$$A = \begin{bmatrix} 4 & 0 & 0 \\ 0 & 2 & 0 \\ 0 & 0 & 3 \end{bmatrix},$$

then because $AA^{-1} = I$

$$A^{-1} = \begin{bmatrix} 1/4 & 0 & 0 \\ 0 & 1/2 & 0 \\ 0 & 0 & 1/3 \end{bmatrix}.$$

3.   Block Diagonal Matrices.

If A is a block diagonal matrix with diagonal blocks that have inverses, then A has an inverse matrix which is also a block diagonal matrix. For example, consider a 5×5 matrix with two diagonal blocks

$$A = \begin{bmatrix} A_{11} & \\ & A_{22} \end{bmatrix} = \begin{bmatrix} 1 & 0 & 0 & 0 & 0 \\ 0 & 1 & 0 & 0 & 0 \\ 0 & 4 & 1 & 0 & 0 \\ 0 & 0 & 0 & 2 & 0 \\ 0 & 0 & 0 & 0 & 3 \end{bmatrix} \quad \text{where}$$

$$A_{11} = E_{32}(4) = \begin{bmatrix} 1 & 0 & 0 \\ 0 & 1 & 0 \\ 0 & 4 & 1 \end{bmatrix} \quad \text{and} \quad A_{22} = \begin{bmatrix} 2 & 0 \\ 0 & 3 \end{bmatrix}.$$

Both A_{11} and A_{22} have inverses and, hence, the inverse of A is

$$A^{-1} = \begin{bmatrix} A_{11}^{-1} & \\ & A_{22}^{-1} \end{bmatrix} = \begin{bmatrix} 1 & 0 & 0 & 0 & 0 \\ 0 & 1 & 0 & 0 & 0 \\ 0 & -4 & 1 & 0 & 0 \\ 0 & 0 & 0 & 1/2 & 0 \\ 0 & 0 & 0 & 0 & 1/3 \end{bmatrix}.$$

4. A is a 2×2 Matrix with $det(A) \neq 0$.

The inverse matrix B must satisfy the matrix equation

$$AB = I$$
$$A \begin{bmatrix} \mathbf{b}_1 & \mathbf{b}_2 \end{bmatrix} = \begin{bmatrix} \mathbf{e}_1 & \mathbf{e}_2 \end{bmatrix}.$$

This is equivalent to two vector equations

$$A\mathbf{b}_1 = \mathbf{e}_1 \quad \text{and} \quad A\mathbf{b}_2 = \mathbf{e}_2.$$

One can solve these by either Cramer's rule or Gauss elimination. For example, use Cramer's rule to find the inverse of

$$A = \begin{bmatrix} a_{11} & a_{12} \\ a_{21} & a_{22} \end{bmatrix}.$$

The first column $\mathbf{b}_1$ of the inverse matrix must satisfy

$$A\mathbf{b}_1 \;=\; \mathbf{e}_1$$
$$\begin{bmatrix} a_{11} & a_{12} \\ a_{21} & a_{22} \end{bmatrix} \begin{bmatrix} b_{11} \\ b_{21} \end{bmatrix} = \begin{bmatrix} 1 \\ 0 \end{bmatrix}.$$

Then

$$b_{11} \;=\; \det(\begin{bmatrix} 1 & a_{12} \\ 0 & a_{22} \end{bmatrix}) / \det(A) = a_{22}/\det(A) \text{ and}$$

$$b_{21} \;=\; \det(\begin{bmatrix} a_{11} & 1 \\ a_{21} & 0 \end{bmatrix}) / \det(A) = -a_{21}/\det(A).$$

The second column $\mathbf{b}_2$ of the inverse matrix must satisfy

$$A\mathbf{b}_2 \;=\; \mathbf{e}_2$$
$$\begin{bmatrix} a_{11} & a_{12} \\ a_{21} & a_{22} \end{bmatrix} \begin{bmatrix} b_{12} \\ b_{22} \end{bmatrix} = \begin{bmatrix} 0 \\ 1 \end{bmatrix}.$$

Then

$$b_{12} \;=\; \det(\begin{bmatrix} 0 & a_{12} \\ 1 & a_{22} \end{bmatrix}) / \det(A) = -a_{12}/\det(A) \text{ and}$$

$$b_{22} \;=\; \det(\begin{bmatrix} a_{11} & 0 \\ a_{21} & 1 \end{bmatrix}) / \det(A) = a_{11}/\det(A).$$

Therefore, the inverse of A is

$$A^{-1} = B = \begin{bmatrix} \mathbf{b}_1 & \mathbf{b}_2 \end{bmatrix} = \begin{bmatrix} a_{22} & -a_{12} \\ -a_{21} & a_{11} \end{bmatrix} / \det(A).$$

The alternative is to use row operations and Gauss elimination. The Gauss-Jordan method is a variation of Gauss elimination, which can be adapted to larger matrices and to computers.

3.5.2   Gauss-Jordan Method to Find Inverse Matrices

As motivation consider a special case of the above 2×2 example where

$$A = \begin{bmatrix} 2 & -1 \\ -1 & 2 \end{bmatrix} \text{ and } A^{-1} = \begin{bmatrix} 2/3 & 1/3 \\ 1/3 & 2/3 \end{bmatrix}.$$

First, use row operations and the augmented matrix to solve $A\mathbf{b}_1 = \mathbf{e}_1$

$$[A \ \ \mathbf{e}_1] = \begin{bmatrix} 2 & -1 & 1 \\ -1 & 2 & 0 \end{bmatrix}$$

$$E_{21}(1/2) \ [A \ \ \mathbf{e}_1] = \begin{bmatrix} 1 & 0 \\ 1/2 & 1 \end{bmatrix} \begin{bmatrix} 2 & -1 & 1 \\ -1 & 2 & 0 \end{bmatrix} = \begin{bmatrix} 2 & -1 & 1 \\ 0 & 3/2 & 1/2 \end{bmatrix}.$$

At this point either use backward substitution or more row operation to obtain an identity matrix in the left side of the augmented matrix. The row operations include two row multiples and then adding $(1/2)$ times *row_2* to *row_1* :

$$E_2(2/3) \ E_1(1/2) \ E_{21}(1/2) \ [A \ \ \mathbf{e}_1] = \begin{bmatrix} 1 & -1/2 & 1/2 \\ 0 & 1 & 1/3 \end{bmatrix}$$

$$E_{12}(1/2) \ E_2(2/3) \ E_1(1/2) \ E_{21}(1/2) \ [A \ \ \mathbf{e}_1] = \begin{bmatrix} 1 & 0 & 2/3 \\ 0 & 1 & 1/3 \end{bmatrix}.$$

The right column in the transform augmented matrix is the first column of the inverse matrix. Let E be the product of the four elementary matrices and write the above as

$$E \ [A \ \ \mathbf{e}_1] = [I \ \ \mathbf{b}_1].$$

In order to find the second column of the inverse matrix, the *same* row operations are needed to transform the augmented matrix from $[A \ \ \mathbf{e}_2]$ to $[I \ \ \mathbf{b}_2]$

$$E \ [A \ \ \mathbf{e}_2] = [I \ \ \mathbf{b}_2].$$

An efficient way to do both these solves is to combine the two columns of the inverse matrix into the right side of a larger augmented matrix. That is, use the row operations to transform $[A \ \ \mathbf{e}_1 \ \ \mathbf{e}_2] = [A \ \ I]$ to $[I \ \ \mathbf{b}_1 \ \ \mathbf{b}_2] = [I \ \ B]$

$$E \ [A \ \ \mathbf{e}_1 \ \ \mathbf{e}_2] = [I \ \ \mathbf{b}_2 \ \ \mathbf{b}_2].$$

In order to find the inverse of an $n \times n$ matrix A, one must solve n vector equations

$$AB = I$$
$$A \begin{bmatrix} \mathbf{b}_1 & \mathbf{b}_2 & \cdots & \mathbf{b}_n \end{bmatrix} = \begin{bmatrix} \mathbf{e}_1 & \mathbf{e}_2 & \cdots & \mathbf{e}_n \end{bmatrix}$$
$$\begin{bmatrix} A\mathbf{b}_1 & A\mathbf{b}_2 & \cdots & A\mathbf{b}_n \end{bmatrix} = \begin{bmatrix} \mathbf{e}_1 & \mathbf{e}_2 & \cdots & \mathbf{e}_n \end{bmatrix} \text{ or}$$
$$A\mathbf{b}_j = \mathbf{e}_j \text{ for } j = 1, \ldots, n.$$

Use row operations and elementary matrices to transform the augmented matrix $[A \ \ I]$ to $[I \ \ A^{-1}]$.

Definition 3.5.2. Consider the $n \times n$ matrix A and find the inverse matrix, if it exists. The *Gauss-Jordan method* for solving this system has two stages:
Stage 1. Transform the augmented matrix $[A \ \ I]$ to upper triangular form $[U \ \ L]$.

(a). start with the left column, column $j = 1$, and use row operations
to transform column $j = 1$ to zeros below row $i = 1$,

(b). move to the next column, column $j = 2$, and use row operations
to transform column $j = 2$ to zeros below row $i = 2$ and

(c). repeat this until column $j = n - 1$ has been done.

Stage 2. Transform the upper triangular matrix $[U \ L]$ to $[I \ A^{-1}]$

(a). use elementary matrices so that the diagonal of U is transformed
to have ones on all of the diagonal components,

(b). start with the right column of U, column $j = n$, and use row operations
to transform column $j = n$ to zeros above row $i = n$,

(c). move to the next left column, column $j = n-1$, and use row operations
to transform column $j = n - 1$ to zeros above row $i = n - 1$ and

(d). repeat this until column $j = 2$ has been done.

Example 3.5.1. Use the Gauss-Jordan method to find the inverse of

$$A = \begin{bmatrix} 1 & 2 & 1 \\ 1 & 3 & 2 \\ 1 & 0 & 1 \end{bmatrix}.$$

The augmented matrix is

$$[A \ I] = \begin{bmatrix} 1 & 2 & 1 & 1 & 0 & 0 \\ 1 & 3 & 2 & 0 & 1 & 0 \\ 1 & 0 & 1 & 0 & 0 & 1 \end{bmatrix}.$$

Stage 1. Subtract row_1 from row_2 and row_3

$$E_{31}(-1) \, E_{21}(-1) \, [A \ I] = \begin{bmatrix} 1 & 2 & 1 & 1 & 0 & 0 \\ 0 & 1 & 1 & -1 & 1 & 0 \\ 0 & -2 & 0 & -1 & 0 & 1 \end{bmatrix}.$$

Add (2) times row_2 to row_3

$$E_{32}(2) \, E_{31}(-1) \, E_{21}(-1) \, [A \ I] = \begin{bmatrix} 1 & 2 & 1 & 1 & 0 & 0 \\ 0 & 1 & 1 & -1 & 1 & 0 \\ 0 & 0 & 2 & -3 & 2 & 1 \end{bmatrix}.$$

Stage 2. Divide row_3 by 2 and then subtract it from row_2 and row_1

$$E_{13}(-1) \, E_{23}(-1) \, E_3(1/2) \, E_{32}(2) \, E_{31}(-1) \, E_{21}(-1) \, [A \ I] =$$

$$\begin{bmatrix} 1 & 2 & 0 & 5/2 & -1 & -1/2 \\ 0 & 1 & 0 & 1/2 & 0 & -1/2 \\ 0 & 0 & 1 & -3/2 & 1 & 1/2 \end{bmatrix}.$$

Finally, subtract (2) times row_2 from row_1

$$E_{12}(-2) \, E_{13}(-1) \, E_{23}(-1) \, E_3(1/2) \, E_{32}(2) \, E_{31}(-1) \, E_{21}(-1) \, [A \ I] =$$

$$\begin{bmatrix} 1 & 0 & 0 & 3/2 & -1 & 1/2 \\ 0 & 1 & 0 & 1/2 & 0 & -1/2 \\ 0 & 0 & 1 & -3/2 & 1 & 1/2 \end{bmatrix}.$$

One can easily verify the inverse is given by the right side of the transformed augmented matrix

$$A^{-1} = \begin{bmatrix} 3/2 & -1 & 1/2 \\ 1/2 & 0 & -1/2 \\ -3/2 & 1 & 1/2 \end{bmatrix}.$$

The solution of $A\mathbf{x} = \mathbf{d} = \begin{bmatrix} 1 & 2 & 3 \end{bmatrix}^T$ is easily computed by using the inverse matrix

$$\mathbf{x} = A^{-1}\mathbf{d} = \begin{bmatrix} 3/2 & -1 & 1/2 \\ 1/2 & 0 & -1/2 \\ -3/2 & 1 & 1/2 \end{bmatrix} \begin{bmatrix} 1 \\ 2 \\ 3 \end{bmatrix} = \begin{bmatrix} 1 \\ -1 \\ 2 \end{bmatrix}.$$

Example 3.5.2. Use the Gauss-Jordan method to find the inverse of

$$A = \begin{bmatrix} 0 & 5 & 6 \\ 1 & 3 & 1 \\ 2 & 1 & 1 \end{bmatrix}.$$

The augmented matrix is

$$[A \ I] = \begin{bmatrix} 0 & 5 & 6 & 1 & 0 & 0 \\ 1 & 3 & 1 & 0 & 1 & 0 \\ 2 & 1 & 1 & 0 & 0 & 1 \end{bmatrix}.$$

Stage 1. Because the 11-component is zero, row_1 and row_2 (or row_3) must be interchanged

$$P_{12}\,[A \ I] = \begin{bmatrix} 1 & 3 & 1 & 0 & 1 & 0 \\ 0 & 5 & 6 & 1 & 0 & 0 \\ 2 & 1 & 1 & 0 & 0 & 1 \end{bmatrix}$$

Subtract (2) times row_1 from row_3

$$E_{31}(-2)\,P_{12}\,[A \ I] = \begin{bmatrix} 1 & 3 & 1 & 0 & 1 & 0 \\ 0 & 5 & 6 & 1 & 0 & 0 \\ 0 & -5 & -1 & 0 & -2 & 1 \end{bmatrix}.$$

Add row_2 to row_3

$$E_{32}(1)\,E_{31}(-2)\,P_{12}\,[A \ I] = \begin{bmatrix} 1 & 3 & 1 & 0 & 1 & 0 \\ 0 & 5 & 6 & 1 & 0 & 0 \\ 0 & 0 & 5 & 1 & -2 & 1 \end{bmatrix}.$$

Stage 2. Divide row_2 and row_3 by 2

$$E_3(1/2)\,E_2(1/2)\,E_{32}(1)\,E_{31}(-2)\,P_{12}\,[A \ I]\,[A \ I] =$$

$$\begin{bmatrix} 1 & 3 & 1 & 0 & 1 & 0 \\ 0 & 1 & 6/5 & 1/5 & 0 & 0 \\ 0 & 0 & 1 & 1/5 & -2/5 & 1/5 \end{bmatrix}.$$

In order to transform the left side to the identity, subtract $(6/5)$ times row_3 from row_2, subtract row_3 from row_1 and last subtract (3) times row_2 from row_1

$$E_{12}(-3)\ E_{13}(-1)\ E_{23}(-6/5)\ E_3(1/2)\ E_2(1/2)\ E_{32}(1)\ E_{31}(-2)\ P_{12}\ [A\ I]\ =$$

$$\begin{bmatrix} 1 & 0 & 0 & -2/25 & -1/25 & 13/25 \\ 0 & 1 & 0 & -1/25 & 12/25 & -6/25 \\ 0 & 0 & 1 & 5/25 & -10/25 & 5/25 \end{bmatrix}.$$

The inverse is the right side of the transformed augmented matrix

$$A^{-1} = (1/25) \begin{bmatrix} -2 & -1 & 13 \\ -1 & 12 & -6 \\ 5 & -10 & 5 \end{bmatrix}.$$

The solution of $Ax = \mathbf{d} = [6\ 2\ 3]^T$ is computed by using the inverse matrix

$$\mathbf{x} = A^{-1}\mathbf{d} = (1/25) \begin{bmatrix} -2 & -1 & 13 \\ -1 & 12 & -6 \\ 5 & -10 & 5 \end{bmatrix} \begin{bmatrix} 6 \\ 2 \\ 3 \end{bmatrix} = \begin{bmatrix} 1 \\ 0 \\ 1 \end{bmatrix}.$$

Example 3.5.3. Consider the steady state heat wire problem in Sections 3.3 and 3.4 where there were five segments and current generating heat within the wire. The algebraic problem for the temperature in each interior segment is

$$\begin{bmatrix} 2 & -1 & 0 & 0 \\ -1 & 2 & -1 & 0 \\ 0 & -1 & 2 & -1 \\ 0 & 0 & -1 & 2 \end{bmatrix} \begin{bmatrix} u_1 \\ u_2 \\ u_3 \\ u_3 \end{bmatrix} = \begin{bmatrix} 470 \\ 400 \\ 400 \\ 470 \end{bmatrix}.$$

The augmented matrix for finding the inverse matrix is

$$[A\ I] = \begin{bmatrix} 2 & -1 & 0 & 0 & 1 & 0 & 0 & 0 \\ -1 & 2 & -1 & 0 & 0 & 1 & 0 & 0 \\ 0 & -1 & 2 & -1 & 0 & 0 & 1 & 0 \\ 0 & 0 & -1 & 2 & 0 & 0 & 0 & 1 \end{bmatrix}.$$

Stage 1. Obtain zeros in the subdiagonal by using

$$E_{43}(3/4)\ E_{32}(2/3)\ E_{21}(1/2)\ [A\ I] =$$

$$\begin{bmatrix} 2 & -1 & 0 & 0 & 1 & 0 & 0 & 0 \\ 0 & 3/2 & -1 & 0 & 1/2 & 1 & 0 & 0 \\ 0 & 0 & 4/3 & -1 & 1/3 & 2/3 & 1 & 0 \\ 0 & 0 & 0 & 5/4 & 1/4 & 2/4 & 3/4 & 1 \end{bmatrix}.$$

Stage 2. Multiply rows 1-4 by 1/2, 2/3, 3/4 and 4/5, respectively, to get for
$D \equiv E_4(4/5)\, E_3(3/4)\, E_2(2/3)\, E_1(1/2)$

$$D\, E_{43}(3/4)\, E_{32}(2/3)\, E_{21}(1/2)\, [A \ I] =$$

$$
\begin{bmatrix}
1 & -1/2 & 0 & 0 & 1/2 & 0 & 0 & 0 \\
0 & 1 & -2/3 & 0 & 1/3 & 2/3 & 0 & 0 \\
0 & 0 & 1 & -3/4 & 1/4 & 2/4 & 3/4 & 0 \\
0 & 0 & 0 & 1 & 1/5 & 2/5 & 3/5 & 4/5
\end{bmatrix}.
$$

In order to obtain zeros in the upper triangular part of the left side, use elementary matrices

$$E_{12}(1/2)\, E_{23}(2/3)\, E_{34}(3/4)\, D\, E_{43}(3/4)\, E_{32}(2/3)\, E_{21}(1/2)\, [A \ I] =$$

$$
\begin{bmatrix}
1 & 0 & 0 & 0 & 8/10 & 6/10 & 4/10 & 2/10 \\
0 & 1 & 0 & 0 & 6/10 & 12/10 & 8/10 & 4/10 \\
0 & 0 & 1 & 0 & 4/10 & 8/10 & 12/10 & 6/10 \\
0 & 0 & 0 & 1 & 2/10 & 4/10 & 6/10 & 8/10
\end{bmatrix}.
$$

So, the inverse matrix is

$$
A^{-1} = (1/10)
\begin{bmatrix}
8 & 6 & 4 & 2 \\
6 & 12 & 8 & 4 \\
4 & 8 & 12 & 6 \\
2 & 4 & 6 & 8
\end{bmatrix},
$$

and the solution of the above algebraic system is

$$
\mathbf{u} = A^{-1}\mathbf{d} = (1/10)
\begin{bmatrix}
8 & 6 & 4 & 2 \\
6 & 12 & 8 & 4 \\
4 & 8 & 12 & 6 \\
2 & 4 & 6 & 8
\end{bmatrix}
\begin{bmatrix}
470 \\
400 \\
400 \\
470
\end{bmatrix}
=
\begin{bmatrix}
870 \\
1270 \\
1270 \\
870
\end{bmatrix}.
$$

3.5.3 Properties of Inverse Matrices

Not all matrices have inverses! Any 2×2 with zero determinant cannot have
an inverse. For example, if

$$
A = \begin{bmatrix} 1 & 1 \\ 2 & 2 \end{bmatrix},
$$

then one cannot find the first column of the inverse matrix because of the
following contradiction

$$
\begin{array}{rcl}
A\mathbf{b}_1 & = & \mathbf{e}_1 \\
\begin{bmatrix} 1 & 1 \\ 2 & 2 \end{bmatrix}
\begin{bmatrix} b_{11} \\ b_{21} \end{bmatrix}
& = &
\begin{bmatrix} 1 \\ 0 \end{bmatrix} \\
b_{11} + b_{21} & = & 1 \text{ and } 2b_{11} + 2b_{21} = 0!
\end{array}
$$

In the last section of this chapter determinants of $n \times n$ matrices will be defined, and any matrix with a non-zero determinant will be shown to have an inverse matrix.

One very important property of inverse matrices is the solution of $A\mathbf{x} = \mathbf{d}$ is $A^{-1}\mathbf{d}$. Other properties are summarized in the following theorem.

Theorem 3.5.1 *(Properties of Inverse Matrices) Let A, A_1 and A_2 be $n \times n$ matrices that have inverses. Then the following statements hold.*

1.   *If $A^{-1} = B$, then A times column j of B is e_j.*
2.   *$A^{-1}d$ is the unique solution of $A\mathbf{x} = \mathbf{d}$.*
3.   *$(A_1 A_2)^{-1} = A_2^{-1} A_1^{-1}$ (note the reverse order).*
4.   *There is only one inverse matrix.*
5.   *The following are equivalent:*
     (i).   A has an inverse,
     (ii). the determinant of A is not zero and
     (iii). $A\mathbf{x} = \mathbf{0}$ implies $\mathbf{x} = \mathbf{0}$.

The proof of $(A_1 A_2)^{-1} = A_2^{-1} A_1^{-1}$ follows from the associative property of matrix products

$$
\begin{aligned}
(A_1 A_2)^{-1}(A_1 A_2) &= (A_2^{-1} A_1^{-1})(A_1 A_2) \\
&= A_2^{-1}(A_1^{-1}(A_1 A_2)) \\
&= A_2^{-1}((A_1^{-1} A_1) A_2) \\
&= A_2^{-1}((I) A_2) = I.
\end{aligned}
$$

The proof of the fifth property will be given in Section 3.7.

Example 3.5.4.   This illustrates the third property in the above theorem. Find the inverse matrix of $A = A_1 A_2$ where

$$
A = \begin{bmatrix} 4 & 0 & 0 \\ -4 & 2 & 0 \\ 0 & 0 & 1 \end{bmatrix} = \begin{bmatrix} 4 & 0 & 0 \\ 0 & 2 & 0 \\ 0 & 0 & 1 \end{bmatrix} \begin{bmatrix} 1 & 0 & 0 \\ -2 & 1 & 0 \\ 0 & 0 & 1 \end{bmatrix} = A_1 A_2.
$$

A_1 is a diagonal matrix and A_2 is an elementary matrix with inverses

$$
A_1^{-1} = \begin{bmatrix} 1/4 & 0 & 0 \\ 0 & 1/2 & 0 \\ 0 & 0 & 1 \end{bmatrix} \text{ and } A_2^{-1} = \begin{bmatrix} 1 & 0 & 0 \\ 2 & 1 & 0 \\ 0 & 0 & 1 \end{bmatrix}.
$$

The inverse of A is

$$
A^{-1} = A_2^{-1} A_1^{-1} = \begin{bmatrix} 1 & 0 & 0 \\ 2 & 1 & 0 \\ 0 & 0 & 1 \end{bmatrix} \begin{bmatrix} 1/4 & 0 & 0 \\ 0 & 1/2 & 0 \\ 0 & 0 & 1 \end{bmatrix} = \begin{bmatrix} 1/4 & 0 & 0 \\ 2/4 & 1/2 & 0 \\ 0 & 0 & 1 \end{bmatrix}.
$$

In the Gauss elimination method the augmented matrix was transformed by a product of elementary matrices, which have inverses. Therefore, the product of the elementary matrices, E, has an inverse and

$$E[A \ \mathbf{d}] = [U \ \widehat{\mathbf{d}}].$$

If all the diagonal components of U are not zero, the solutions of $A\mathbf{x} = \mathbf{d}$ and $U\mathbf{x} = EA\mathbf{x} = E\mathbf{d} = \widehat{\mathbf{d}}$ are the same.

3.5.4 Inverse Matrices and MATLAB

The by-hand computation of inverse matrices can be an extremely tedious process. Even using computing tools, the number of calculations is of order $2n^3/3$ arithmetic operations for an $n \times n$ matrix! The MATLAB command inv(A) will compute the inverse of a matrix A. Some illustrations of this and row operations are given in the MATLAB code inv_mat.m. Example 3.5.2 done using MATLAB is

```
>> A = [0 5 6;1 3 1;2 1 1]
   A =
           0 5 6
           1 3 1
           2 1 1
>> B = inv(A)                    % computes the inverse matrix
   B =
           -0.0800 -0.0400  0.5200
           -0.0400  0.4800 -0.2400
            0.2000 -0.4000  0.2000
>> A*B
       ans =
            1.0000 0.0000 0.0000
           -0.0000 1.0000 -0.0000
            0.0000 0.0000 1.0000
>> d = [6 2 3]'
   d =
       6
       2
       3
>> solution1 = inv(A)*d     % uses inverse matrix
   solution1 =
       1.0000
      -0.0000
       1.0000
>> solutions2 = A\d          % uses Gauss elimination
   solutions2 =
       1
```

$$0$$
$$1$$

In the previous section an application to a six-bar truss generated a 6×6 matrix. The inverse of the matrix can be found using the MATLAB code support.m, and the interested reader should examine this application and exercises 10 and 11.

3.5.5   Exercises

1.   Let $A = \begin{bmatrix} 3 & -1 \\ 1 & 3 \end{bmatrix}$.

(a).   Use Cramer's rule to find the inverse of A.
(b).   Use Gauss-Jordan method to find the inverse.
(c).   Find the inverse of

$$\begin{bmatrix} 3 & -1 & 0 & 0 \\ 1 & 3 & 0 & 0 \\ 0 & 0 & 3 & 0 \\ 0 & 0 & 0 & 4 \end{bmatrix}.$$

2.   Let $A = \begin{bmatrix} 1 & -1 & 1 \\ 1 & 0 & -3 \\ 0 & -2 & -3 \end{bmatrix}$.

(a).   Use the Gauss-Jordan method to find the inverse matrix.
(b).   Use the inverse to solve $A\mathbf{x} = \mathbf{d} = [0 \ \ 10 \ \ 20]^T$.

3.   Let $A = \begin{bmatrix} 3 & -1 & 0 \\ -1 & 3 & -1 \\ 0 & -1 & 3 \end{bmatrix}$.

(a).   Use the Gauss-Jordan method to find the inverse matrix.
(b).   Use the inverse to solve $A\mathbf{x} = \mathbf{d} = [1 \ \ 2 \ \ 3]^T$.

4.   Let $A = \begin{bmatrix} 0 & 2 & 1 \\ 1 & 3 & 5 \\ 2 & 2 & 0 \end{bmatrix}$.

(a).   Use the Gauss-Jordan method to find the inverse matrix.
(b).   Use the inverse to solve $A\mathbf{x} = \mathbf{d} = [4 \ \ 3 \ \ 5]^T$.

5.   Verify the row operations used in Example 3.5.3.

6.   Let $A = \begin{bmatrix} 1 & 2 & 0 & 1 \\ 2 & 1 & 0 & 0 \\ 0 & 0 & 2 & 0 \\ 1 & 2 & 0 & 2 \end{bmatrix}$.

(a).   Use the Gauss-Jordan method to find the inverse matrix.
(b).   Use the inverse to solve $A\mathbf{x} = \mathbf{d} = [2 \ \ 1 \ \ 4 \ \ -1]^T$.

7.   Let $A_1 = \begin{bmatrix} 1 & 0 & 0 \\ 0 & 1 & 0 \\ -4 & 0 & 1 \end{bmatrix}$ and $A_2 = \begin{bmatrix} 1 & 2 & 0 \\ 0 & 1 & 0 \\ 0 & 0 & 1 \end{bmatrix}$.

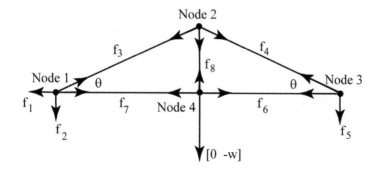

Figure 3.5.1: Five-bar Truss

(a).   Find the inverses of both matrices.

(b).   Find the inverses of $A_1 A_2$ and $A_2 A_1$.

8.   Prove: if A has an inverse, then there is only one inverse matrix.

9.   Use MATLAB to confirm any by-hand calculations done in exercises 2-4 or 6.

10.   Consider the six-bar truss in Section 3.4, see Figure 3.4.1.

(a).   Use the MATLAB command inv(A) to find the inverse of the 6×6 matrix.

(b).   Find the forces for $\theta = \pi/6, \pi/4$ and $2\pi/6$.

11.   Consider the five-bar truss in Figure 3.5.1, see the MATLAB code bridge.m

(a).   Formulate an algebraic system for the eight forces.

(b).   Use the MATLAB commands A\d and inv(A) to solve this when $w = -10000$ and variable $\theta = \pi/6, \pi/4$ and $2\pi/6$.

3.6   *LU* Factorization

If the given matrix can be written as a product of a lower and upper triangular matrices, then one may be able to solve $A\mathbf{x} = \mathbf{d}$. Suppose $A = LU$ and both L and U have non-zero diagonal components. By the associative property

$$A\mathbf{x} = (LU)\mathbf{x} = L(U\mathbf{x}) = \mathbf{d}.$$

So, solve by forward substitution $L\mathbf{y} = \mathbf{d}$ and then solve by backward substitution $U\mathbf{x} = \mathbf{y}$. An example was given in Section 3.3

$$\begin{bmatrix} 2 & -1 & 0 \\ -1 & 2 & -1 \\ 0 & -1 & 2 \end{bmatrix} \begin{bmatrix} x_1 \\ x_2 \\ x_3 \end{bmatrix} = \begin{bmatrix} 200 \\ 0 \\ 70 \end{bmatrix}.$$

By using the tridiagonal structure of the matrix we were able to find factors

$$
\begin{bmatrix} 2 & -1 & 0 \\ -1 & 2 & -1 \\ 0 & -1 & 2 \end{bmatrix} = \begin{bmatrix} 1 & 0 & 0 \\ -1/2 & 1 & 0 \\ 0 & -2/3 & 1 \end{bmatrix} \begin{bmatrix} 2 & -1 & 0 \\ 0 & 3/2 & -1 \\ 0 & 0 & 4/3 \end{bmatrix}
$$

$$
A \quad = \quad LU.
$$

The objective of this section is to show how one can use row operations and elementary matrices to possibly find the LU factors for more general matrices. Block versions of the LU factors will be described and applied to a three-loop circuit problem.

3.6.1   Attributes of LU Factorization

Before we present the formal definition, the following three 2×2 matrices reveal some important facts. First, not all matrices have LU factors, but the interchange of some rows may allow us to find the factors

$$
A \quad = \quad \begin{bmatrix} 0 & 1 \\ 1 & 1 \end{bmatrix}
$$

$$
P_{12}A \quad = \quad \begin{bmatrix} 1 & 1 \\ 0 & 1 \end{bmatrix} = \begin{bmatrix} 1 & 0 \\ 0 & 1 \end{bmatrix}\begin{bmatrix} 1 & 1 \\ 0 & 1 \end{bmatrix} = LU.
$$

Second, one can use row operations and elementary matrices to find LU factors

$$
A \quad = \quad \begin{bmatrix} 2 & -1 \\ -1 & 2 \end{bmatrix}
$$

$$
E_{21}(1/2)\, A \quad = \quad \begin{bmatrix} 2 & -1 \\ 0 & 3/2 \end{bmatrix} = U.
$$

Now let L be the inverse of the elementary matrix $E_{21}(1/2)$ giving

$$
E_{21}(1/2)^{-1}\,(E_{21}(1/2)\, A) \quad = \quad E_{21}(1/2)^{-1}\begin{bmatrix} 2 & -1 \\ 0 & 3/2 \end{bmatrix} = LU
$$

$$
A \quad = \quad \begin{bmatrix} 1 & 0 \\ -1/2 & 1 \end{bmatrix}\begin{bmatrix} 2 & -1 \\ 0 & 3/2 \end{bmatrix} = LU.
$$

Third, the lower and upper matrices may not have an inverse

$$
A \quad = \quad \begin{bmatrix} 1 & -1 \\ -1 & 1 \end{bmatrix}
$$

$$
E_{21}(1)\, A \quad = \quad \begin{bmatrix} 1 & -1 \\ 0 & 0 \end{bmatrix} = U
$$

$$
A \quad = \quad \begin{bmatrix} 1 & 0 \\ -1 & 1 \end{bmatrix}\begin{bmatrix} 1 & -1 \\ 0 & 0 \end{bmatrix} = LU.
$$

Definition 3.6.1. An $n \times n$ matrix A has an *LU factorization* if and only if $A = LU$ where L is a lower triangular $n \times n$ matrix (ij-components are zero for $j > i$), U is an upper triangular $n \times n$ matrix (ij-components are zero for $j < i$) and both L and U have inverse matrices (their ii-components are not zero).

The following theorem can be proved by using mathematical induction on the dimension of the matrix.

Theorem 3.6.1 *(LU Factorization) If the $n \times n$ matrix A has an inverse, then for some permutation of the rows $PA = LU$. Moreover, $A\mathbf{x} = \mathbf{d}$ has a unique solution, which can be found via*

$$PA\mathbf{x} = P\mathbf{d}$$
$$L(U\mathbf{x}) = P\mathbf{d}.$$

Definition 3.6.2. The *LU factorization method* uses row operations and inverses of the elementary matrices.

Stage 1. Use row operations to transform the matrix from A to U.

Stage 2. Let E be the product of the elementary matrices such that $EA = U$. Define $L = E^{-1}$ and note $A = E^{-1}(EA) = E^{-1}U = LU$.

Example 3.6.1. This example does not require the interchange of rows to find the factorization of the 3×3 matrix

$$A = \begin{bmatrix} 2 & -1 & 0 \\ -1 & 2 & -1 \\ 0 & -1 & 2 \end{bmatrix}.$$

Stage 1. Two elementary matrices are required to transform A to upper triangular form

$$EA = E_{32}(2/3)\, E_{21}(1/2)\, A = \begin{bmatrix} 2 & -1 & 0 \\ 0 & 3/2 & -1 \\ 0 & 0 & 4/3 \end{bmatrix}.$$

Stage 2. Multiply both sides from the left by $E_{32}(2/3)^{-1} = E_{32}(-2/3)$, then multiply from the left by $E_{21}(1/2)^{-1} = E_{21}(-1/2)$

$$\begin{aligned} A &= E_{21}(-1/2)\, E_{32}(-2/3) \begin{bmatrix} 2 & -1 & 0 \\ 0 & 3/2 & -1 \\ 0 & 0 & 4/3 \end{bmatrix} \\ &= \begin{bmatrix} 1 & 0 & 0 \\ -1/2 & 1 & 0 \\ 0 & -2/3 & 1 \end{bmatrix} \begin{bmatrix} 2 & -1 & 0 \\ 0 & 3/2 & -1 \\ 0 & 0 & 4/3 \end{bmatrix} = LU. \end{aligned}$$

Notice the inverse of E is easy to compute as are the matrix products

$$\begin{aligned} E^{-1} &= (E_{32}(2/3)\, E_{21}(1/2))^{-1} \\ &= E_{21}(1/2)^{-1}\, E_{32}(2/3)^{-1} \\ &= E_{21}(-1/2)\, E_{32}(-2/3). \end{aligned}$$

Example 3.6.2. This example does require the interchange of rows to find the factorization of the 3×3 matrix

$$A = \begin{bmatrix} 0 & 5 & 6 \\ 1 & 3 & 1 \\ 2 & 1 & 1 \end{bmatrix}.$$

Stage 1. Interchange rows one and two (or three)

$$P_{12}\, A = \begin{bmatrix} 0 & 1 & 0 \\ 1 & 0 & 0 \\ 0 & 0 & 1 \end{bmatrix} \begin{bmatrix} 0 & 5 & 6 \\ 1 & 3 & 1 \\ 2 & 1 & 1 \end{bmatrix} = \begin{bmatrix} 1 & 3 & 1 \\ 0 & 5 & 6 \\ 2 & 1 & 1 \end{bmatrix}.$$

Two elementary matrices are required to transform $P_{12}\, A$ to upper triangular form

$$E_{32}(1)\, E_{31}(-2)\, P_{12}\, A = \begin{bmatrix} 1 & 3 & 1 \\ 0 & 5 & 6 \\ 0 & 0 & 5 \end{bmatrix}.$$

Stage 2. Multiply both sides from the left by $E_{32}(1)^{-1} = E_{32}(-1)$, and then multiply from the left by $E_{31}(-2)^{-1} = E_{31}(2)$

$$\begin{aligned} P_{12}\, A &= E_{31}(2)\, E_{32}(-1) \begin{bmatrix} 1 & 3 & 1 \\ 0 & 5 & 6 \\ 0 & 0 & 5 \end{bmatrix} \\ &= \begin{bmatrix} 1 & 0 & 0 \\ 0 & 1 & 0 \\ 2 & -1 & 1 \end{bmatrix} \begin{bmatrix} 1 & 3 & 1 \\ 0 & 5 & 6 \\ 0 & 0 & 5 \end{bmatrix} = LU. \end{aligned}$$

In order to solve the algebraic system $A\mathbf{x} = \mathbf{d} = [6\ \ 2\ \ 3]^T$, use this LU factorization in $P_{12}A\mathbf{x} = P_{12}\mathbf{d}$.

$$\begin{matrix} & L(U\mathbf{x}) & = & P_{12}\mathbf{d} \\ \begin{bmatrix} 1 & 0 & 0 \\ 0 & 1 & 0 \\ 2 & -1 & 1 \end{bmatrix} (\begin{bmatrix} 1 & 3 & 1 \\ 0 & 5 & 6 \\ 0 & 0 & 5 \end{bmatrix} \begin{bmatrix} x_1 \\ x_2 \\ x_3 \end{bmatrix}) & = & \begin{bmatrix} 2 \\ 6 \\ 3 \end{bmatrix}. \end{matrix}$$

The solution of $L\mathbf{y} = P_{12}\mathbf{d}$ is $\mathbf{y} = [2\ \ 6\ \ 5]$, and the solution of $U\mathbf{x} = \mathbf{y}$ is $\mathbf{x} = [1\ \ 0\ \ 1]$.

3.6.2   Block LU Factors

Often it is useful to represent an $n \times n$ matrix A as a 2×2 block matrix where $n = n_1 + n_2$, A_{11} is $n_1 \times n_1$, A_{22} is $n_2 \times n_2$, A_{12} is $n_1 \times n_2$ and A_{21} is $n_2 \times n_1$

$$A = \begin{bmatrix} A_{11} & A_{12} \\ A_{21} & A_{22} \end{bmatrix}.$$

An algebraic system $Ax = d$ may be written with $\mathbf{x}^T = [\mathbf{x}_1^T \ \mathbf{x}_2^T]$, $\mathbf{d}^T = [\mathbf{d}_1^T \ \mathbf{d}_2^T]$ where $\mathbf{x}_1$ and $\mathbf{d}_1$ are $n_1 \times 1$ column vectors, $\mathbf{x}_2$ and $\mathbf{d}_2$ are $n_2 \times 1$ column vectors and

$$Ax = d$$

$$\begin{bmatrix} A_{11} & A_{12} \\ A_{21} & A_{22} \end{bmatrix} \begin{bmatrix} \mathbf{x}_1 \\ \mathbf{x}_2 \end{bmatrix} = \begin{bmatrix} \mathbf{d}_1 \\ \mathbf{d}_2 \end{bmatrix}. \tag{3.6.1}$$

An example is for $n = 5 = 2 + 3 = n_1 + n_2$ with A_{11} is 2×2 and A_{22} is a 3×3 diagonal matrix

$$\begin{bmatrix} 2 & -1 & 1 & 0 & 1 \\ -1 & 2 & 0 & 1 & 0 \\ 1 & 0 & 2 & 0 & 0 \\ 0 & 1 & 0 & 2 & 0 \\ 1 & 0 & 0 & 0 & 2 \end{bmatrix}.$$

In many applications the blocks are parts of models for physical applications such as velocity and pressure or voltages and currents. Moreover, one may have some previous knowledge about the square blocks and their inverse matrices.

Block matrix products can be defined as long as number of columns in the left submatrix is equal to the number of rows in the right submatrix. Block elementary matrices can be used to transform the block matrix into a block upper triangular matrix. Here, we must assume the 11-block has an inverse A_{11}^{-1}. When I_1 is $n_1 \times n_1$, I_2 is $n_2 \times n_2$, $\widehat{A}_{22} \equiv A_{22} - A_{21} A_{11}^{-1} A_{12}$, $\widehat{\mathbf{d}}_2 \equiv \mathbf{d}_2 - A_{21} A_{11}^{-1} \mathbf{d}_1$, Z is an $n_1 \times n_2$ zero matrix, then

$$\begin{bmatrix} I_1 & Z \\ -A_{21}A_{11}^{-1} & I_2 \end{bmatrix} \begin{bmatrix} A_{11} & A_{12} \\ A_{21} & A_{22} \end{bmatrix} \begin{bmatrix} \mathbf{x}_1 \\ \mathbf{x}_2 \end{bmatrix} = \begin{bmatrix} I_1 & Z \\ -A_{21}A_{11}^{-1} & I_2 \end{bmatrix} \begin{bmatrix} \mathbf{d}_1 \\ \mathbf{d}_2 \end{bmatrix}$$

$$\begin{bmatrix} A_{11} & A_{12} \\ Z^T & \widehat{A}_{22} \end{bmatrix} \begin{bmatrix} \mathbf{x}_1 \\ \mathbf{x}_2 \end{bmatrix} = \begin{bmatrix} \mathbf{d}_1 \\ \widehat{\mathbf{d}}_2 \end{bmatrix}.$$

The matrix $\widehat{A}_{22}$ is called the *Schur complement* of A_{11} in A. If $\widehat{A}_{22}$ also has an inverse, then one can solve for $\mathbf{x}_2$ and then $\mathbf{x}_1$

$$\widehat{A}_{22}\mathbf{x}_2 = \widehat{\mathbf{d}}_2 \text{ and} \tag{3.6.2}$$

$$A_{11}\mathbf{x}_1 = \mathbf{d}_1 - A_{12}\mathbf{x}_2. \tag{3.6.3}$$

An alternative view point is to use block *LU* factors of A. Multiplying A by the block elementary matrix

$$\begin{bmatrix} I_1 & Z \\ -A_{21}A_{11}^{-1} & I_2 \end{bmatrix} \begin{bmatrix} A_{11} & A_{12} \\ A_{21} & A_{22} \end{bmatrix} = \begin{bmatrix} A_{11} & A_{12} \\ Z^T & \widehat{A}_{22} \end{bmatrix}.$$

The block elementary matrix has an inverse, which is very easy to find

$$\begin{bmatrix} I_1 & Z \\ -A_{21}A_{11}^{-1} & I_2 \end{bmatrix}^{-1} = \begin{bmatrix} I_1 & Z \\ A_{21}A_{11}^{-1} & I_2 \end{bmatrix}.$$

Multiply both sides by this inverse to get a block LU factorization of A

$$\begin{bmatrix} A_{11} & A_{12} \\ A_{21} & A_{22} \end{bmatrix} = \begin{bmatrix} I_1 & Z \\ A_{21}A_{11}^{-1} & I_2 \end{bmatrix} \begin{bmatrix} A_{11} & A_{12} \\ Z^T & \widehat{A}_{22} \end{bmatrix}. \qquad (3.6.4)$$

Both the block lower and block upper triangular matrices will have inverses provided the diagonal blocks have inverses.

Example 3.6.3. Consider the above 5×5 matrix with $n_1 = 2$, $n_1 = 3$,

$$A_{11} = \begin{bmatrix} 2 & -1 \\ -1 & 2 \end{bmatrix}, A_{12} = \begin{bmatrix} 1 & 0 & 1 \\ 0 & 1 & 0 \end{bmatrix}$$

$$A_{21} = \begin{bmatrix} 1 & 0 \\ 0 & 1 \\ 1 & 0 \end{bmatrix} \text{ and } A_{22} = \begin{bmatrix} 2 & 0 & 0 \\ 0 & 2 & 0 \\ 0 & 0 & 2 \end{bmatrix}.$$

$\widehat{A}_{22} \equiv A_{22} - A_{21}A_{11}^{-1}A_{12}$ is easy to compute and has an inverse

$$\begin{aligned} \widehat{A}_{22} &\equiv \begin{bmatrix} 2 & 0 & 0 \\ 0 & 2 & 0 \\ 0 & 0 & 2 \end{bmatrix} - \begin{bmatrix} 1 & 0 \\ 0 & 1 \\ 1 & 0 \end{bmatrix} \begin{bmatrix} 2/3 & 1/3 \\ 1/3 & 2/3 \end{bmatrix} \begin{bmatrix} 1 & 0 & 1 \\ 0 & 1 & 0 \end{bmatrix} \\ &= 1/3 \begin{bmatrix} 4 & -1 & -2 \\ -1 & 4 & -1 \\ -2 & -1 & 4 \end{bmatrix} \text{ and} \end{aligned}$$

$$\widehat{A}_{22}^{-1} = 1/4 \begin{bmatrix} 5 & 2 & 3 \\ 2 & 4 & 2 \\ 3 & 2 & 5 \end{bmatrix}.$$

The solution of $A\mathbf{x} = \mathbf{d} = [1\ 2\ 3\ 4\ 5]^T$ can be computed by using equations (3.6.2) and (3.6.3)

$$\begin{aligned} \mathbf{x}_2 &= \widehat{A}_{22}^{-1}\widehat{\mathbf{d}}_2 = \widehat{A}_{22}^{-1}(\mathbf{d}_2 - A_{21}A_{11}^{-1}\mathbf{d}_1) \\ &= 1/4 \begin{bmatrix} 5 & 2 & 3 \\ 2 & 4 & 2 \\ 3 & 2 & 5 \end{bmatrix} \left(\begin{bmatrix} 3 \\ 4 \\ 5 \end{bmatrix} - \begin{bmatrix} 1 & 0 \\ 0 & 1 \\ 1 & 0 \end{bmatrix} \begin{bmatrix} 2/3 & 1/3 \\ 1/3 & 2/3 \end{bmatrix} \begin{bmatrix} 1 \\ 2 \end{bmatrix} \right) \\ &= \begin{bmatrix} 6 \\ 5 \\ 7 \end{bmatrix} \end{aligned}$$

and

$$\begin{aligned} \mathbf{x}_1 &= A_{11}^{-1}(\mathbf{d}_1 - A_{12}\mathbf{x}_2) \\ &= \begin{bmatrix} 2/3 & 1/3 \\ 1/3 & 2/3 \end{bmatrix} \left(\begin{bmatrix} 1 \\ 2 \end{bmatrix} - \begin{bmatrix} 1 & 0 & 1 \\ 0 & 1 & 0 \end{bmatrix} \begin{bmatrix} 6 \\ 5 \\ 7 \end{bmatrix} \right) \\ &= \begin{bmatrix} -9 \\ -6 \end{bmatrix}. \end{aligned}$$

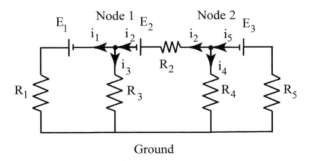

Figure 3.6.1: Three-loop Circuit

The block approach to solving systems is applicable to many important problems, but one point of concern is whether or not the $\widehat{A}_{22}$ matrix has an inverse. If A_{11} has an inverse, then

$$\begin{bmatrix} I_1 & Z \\ -A_{21}A_{11}^{-1} & I_2 \end{bmatrix}\begin{bmatrix} A_{11} & A_{12} \\ A_{21} & A_{22} \end{bmatrix} = \begin{bmatrix} A_{11} & A_{12} \\ Z^T & \widehat{A}_{22} \end{bmatrix} \text{ and}$$

$$\begin{bmatrix} A_{11} & A_{12} \\ Z^T & \widehat{A}_{22} \end{bmatrix}\begin{bmatrix} I_1 & -A_{11}^{-1}A_{12} \\ Z^T & I_2 \end{bmatrix} = \begin{bmatrix} A_{11} & Z \\ Z^T & \widehat{A}_{22} \end{bmatrix}.$$

If A and A_{11} have inverses, then the left side is a product of three matrices with inverses. Consequently, the product on the left side must have an inverse, and the 2×2 block diagonal matrix on the right side must have an inverse. This means the 22-block $\widehat{A}_{22}$ has an inverse. The following theorem summarizes the above results.

Theorem 3.6.2 *(Block LU Factors) Consider the 2×2 block system in equation (3.6.1). If A and A_{11} have inverses, then $\widehat{A}_{22}$ has an inverse. Moreover, the solution of (3.6.1) can be found by solving equations (3.6.2) and (3.6.3).*

3.6.3   Application to Three-loop Circuit

In Subsection 3.1.5 a two-loop circuit was considered. By using the Kirchhoff voltage and current laws an algebraic system for the three unknown currents was formulated and solved by using determinants. In circuits with more loops there will be a larger number of unknown currents and potentials. This requires more sophisticated solution methods such as Gauss elimination or block factorizations. In this section a three-loop circuit will be studied, see Figure 3.6.1. There will be five unknown currents and two unknown potentials at nodes one and two.

In order to further understand the potential within a circuit, consider a single loop circuit with a battery and two resistors as illustrated in the top half of Figure 3.6.2. We choose point a to be the ground potential set equal to zero.

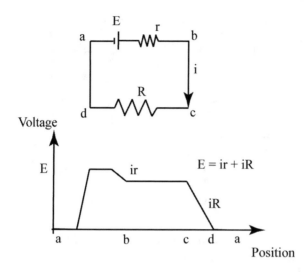

Figure 3.6.2: Potential in a Single-loop Circuit

The battery increases the potential from zero to E. Within the battery there is a small resistance r so that by Ohm's law there is a drop in potential of ir from E to $E - ir$, which is depicted by the segment from points a to b. The segment from b to c has very small resistance and is neglected. The larger resistance R between points c and d has a potential drop equal to iR from $E - ir$ to $E - ir - iR = 0$.

The derivation of the five voltage equations and two current equations at nodes one and two assumes the potentials at nodes one and two are given by x_1 and x_2 and the potential at the bottoms nodes is set to zero. The voltage drops across resistors R_k for $k = 1, \cdots, 5$ are plus or minus $i_k R_k$, which depends on the direction of the current.

Voltage from node one to ground:

$$x_1 + E_1 - i_1 R_1 = 0 \text{ and}$$
$$x_1 - i_3 R_3 = 0.$$

Voltage from node one to node two:

$$x_1 + E_2 - (-i_2)R_2 = x_2.$$

Voltage from node two to ground:

$$x_2 + E_3 - (-i_5)R_5 = 0 \text{ and}$$
$$x_2 - i_4 R_4 = 0.$$

At each node the sum of the currents must be zero:

$$-i_1 + i_2 - i_3 = 0 \text{ and}$$
$$-i_2 - i_4 + i_5 = 0.$$

The matrix version of these seven equations is

$$
\begin{bmatrix}
R_1 & 0 & 0 & 0 & 0 & -1 & 0 \\
0 & R_2 & 0 & 0 & 0 & 1 & -1 \\
0 & 0 & R_3 & 0 & 0 & -1 & 0 \\
0 & 0 & 0 & R_4 & 0 & 0 & -1 \\
0 & 0 & 0 & 0 & R_5 & 0 & 1 \\
-1 & 1 & -1 & 0 & 0 & 0 & 0 \\
0 & -1 & 0 & -1 & 1 & 0 & 0
\end{bmatrix}
\begin{bmatrix}
i_1 \\ i_2 \\ i_3 \\ i_4 \\ i_5 \\ x_1 \\ x_2
\end{bmatrix}
=
\begin{bmatrix}
E_1 \\ -E_2 \\ 0 \\ 0 \\ -E_3 \\ 0 \\ 0
\end{bmatrix}.
$$

One can view the coefficient matrix as either a 7×7 matrix or as 2×2 block matrix where the 11-block is a 5×5 diagonal matrix with the resistances on the diagonal. The block *LU* factorization method reduces to solving a 2×2 matrix equation $\hat{A}_{22}\mathbf{x}_2 = \hat{\mathbf{d}}_2$ where the unknown vector has the two potentials, and solving a 5×5 diagonal matrix equation $A_{11}\mathbf{x}_1 = \mathbf{d}_1 - A_{12}\mathbf{x}_2$. This reduction in the computational complexity illustrates the merits of using appropriate block structure.

Example 3.6.4. The MATLAB code circuit3.m can be used to solve this circuit problem. It uses three methods: Gauss elimination command A\d, the inverse matrix command inv(A)*d and block *LU* factorization as in equation (3.6.4). The following resistor and battery values were used: $R_1 = 1$, $R_2 = 0.1$, $R_3 = 3$, $R_4 = 2$, $R_5 = 4$ and $E_1 = 10$, $E_2 = 20$, $E_3 = 30$. The code generated the solution

$$
\begin{bmatrix}
i_1 \\ i_2 \\ i_3 \\ i_4 \\ i_5 \\ x_1 \\ x_2
\end{bmatrix}
=
\begin{bmatrix}
-5.2290 \\
-10.3053 \\
-5.0763 \\
1.8702 \\
-8.4351 \\
-15.2290 \\
3.7405
\end{bmatrix}.
$$

The negative currents indicate the current flow is in an opposite direction as indicated in Figure 3.6.1. The potentials are relative to the bottom nodes.

3.6.4   LU Factorization Using MATLAB

The MATLAB command lu(A) can be used to factor an $n \times n$ matrix A. The *LU* factor of Example 3.6.1 is

```
>> A = [2 -1 0;-1 2 -1;0 -1 2]
       A =
            2 -1  0
           -1  2 -1
            0 -1  2
>> [L U] = lu(A)
       L =
            1.0000 0.0000 0.0000
```

$$
\begin{array}{ccc}
-0.5000 & 1.0000 & 0.0000 \\
0.0000 & -0.6667 & 1.0000
\end{array}
$$

$U =$

$$
\begin{array}{ccc}
2.0000 & -1.0000 & 0.0000 \\
0.0000 & 1.5000 & -1.0000 \\
0.0000 & 0.0000 & 1.3333
\end{array}
$$

The *LU* factor of Example 5.6.2 is a little more confusing because rows one and three are interchanged and not rows one and two

>> [L U P] = lu(A)
L =

$$
\begin{array}{ccc}
1.0000 & 0.0000 & 0.0000 \\
0.0000 & 1.0000 & 0.0000 \\
0.5000 & 0.5000 & 1.0000
\end{array}
$$

·U =

$$
\begin{array}{ccc}
2.0000 & 1.0000 & 1.0000 \\
0.0000 & 5.0000 & 6.0000 \\
0.0000 & 0.0000 & -2.5000
\end{array}
$$

P =

$$
\begin{array}{ccc}
0 & 0 & 1 \\
1 & 0 & 0 \\
0 & 1 & 0
\end{array}
$$

3.6.5 Exercises

1. Let A be the 4×4 matrix

$$
A = \begin{bmatrix}
2 & -1 & 0 & 0 \\
-1 & 2 & -1 & 0 \\
0 & -1 & 2 & -1 \\
0 & 0 & -1 & 2
\end{bmatrix}.
$$

(a).   Use row operations and elementary matrices to find the *LU* factors.
(b).   Use these to solve $Ax = d = [470\ \ 400\ \ 400\ \ 470]^T$.

2. Let A be the 3×3 matrix

$$
A = \begin{bmatrix}
1 & 2 & 1 \\
1 & 3 & 6 \\
1 & 0 & 1
\end{bmatrix}.
$$

(a).   Use row operations and elementary matrices to find the *LU* factors.
(b).   Use these to solve $Ax = d = [1\ \ 2\ \ 3]^T$.

3. Let A be the 3×3 matrix in Example 3.6.2. Interchange row one and three, that is, use P_{13} and not P_{12}.

(a).   Use row operations and elementary matrices to find the *LU* factors.
(b).   Use these to solve $Ax = d = [6\ \ 3\ \ 2]^T$.

4.   Let A be the 3×3 matrix

$$A = \begin{bmatrix} 1 & 2 & 3 \\ 1 & 2 & 6 \\ 2 & 3 & 9 \end{bmatrix}.$$

    (a).   Use row operations and elementary matrices to find the LU factors.

    (b).   Use these to solve $A\mathbf{x} = \mathbf{d} = [6 \ 3 \ 2]^T$.

5.   Let A be the 4×4 matrix in exercise 1.

    (a).   View this as a block 2×2 matrix where $n = 4 = 2 + 2 = n_1 + n_2$. Find the four blocks and the Schur complement matrix $\widehat{A}_{22}$.

    (b).   Use equations (3.6.2) and (3.6.3) to solve $A\mathbf{x} = \mathbf{d} = [470 \ \ 400 \ \ 400 \ \ 470]^T$.

6.   Let A be the 5×5 matrix

$$\begin{bmatrix} 2 & 0 & 1 & 0 & 1 \\ 0 & 3 & 0 & 2 & 0 \\ 1 & 0 & 3 & -1 & 0 \\ 0 & 1 & -1 & 3 & -1 \\ 1 & 1 & 0 & -1 & 3 \end{bmatrix}.$$

    (a).   View this as a block 2×2 matrix where $n = 5 = 2 + 3 = n_1 + n_2$. Find the four blocks and the Schur complement matrix $\widehat{A}_{22}$.

    (b).   Use equations (3.6.2) and (3.6.3) to solve $A\mathbf{x} = \mathbf{d} = [5 \ 6 \ 7 \ 8 \ 9]^T$.

7.   Use MATLAB to confirm any calculations in exercises 1-6.

8.   Consider the 7×7 matrix in Example 3.6.4. By-hand calculations and Gauss elimination solve this problem.

9.   Consider the 7×7 matrix in Example 3.6.4. By-hand calculations and LU factorization solve this problem.

10.   Consider the 7×7 matrix in Example 3.6.4. By-hand calculations and block LU factorization solve this problem.

11.   Consider the 7×7 matrix in Example 3.6.4.

    (a).   Use the MATLAB code circuit3.m to confirm any calculations in exercises 8-10.

    (b).   Use this code to experiment with different resistors and batteries.

    (c).   What happens when R_2 is near or is zero?

3.7   Determinants and Cramer's Rule

In Sections 1.4 and 2.2 determinants of 2×2 and 3×3 matrices were introduced as natural ways to compute areas and volumes between vectors, which are columns in the matrix. Also, the unique solution of algebraic systems with two or three unknowns is characterized by the determinant of the 2×2 or 3×3 coefficient matrix not being zero. In this section these topics will be reviewed and extended to $n \times n$ matrices. Basic properties of determinants will be presented, inverse matrix computation via determinants will be formulated and an application to steady state three-tanking mixing will illustrate the use of Cramer's rule.

3.7.1   Determinants for 2×2 and 3×3 Matrices

Consider a 2×2 matrix with column vectors **a** and **b**

$$A = \begin{bmatrix} a_{11} & a_{12} \\ a_{21} & a_{22} \end{bmatrix} = [\mathbf{a} \ \mathbf{b}].$$

The area of the parallelogram formed by the vectors, which are not parallel, is plus or minus the determinant of A

$$\det(A) \equiv a_{11}a_{22} - a_{12}a_{21}.$$

As indicated in Section 2.4 determinants can be used to solve 2×2 algebraic systems $A\mathbf{x} = \mathbf{d}$ by Cramer's rule when $\det(A) \neq 0$

$$x_1 = \det\left(\begin{bmatrix} d_1 & a_{12} \\ d_2 & a_{22} \end{bmatrix}\right)/\det(A) \text{ and } x_2 = \det\left(\begin{bmatrix} a_{11} & d_1 \\ a_{21} & d_2 \end{bmatrix}\right)/\det(A).$$

Notice the $\det(A)$ does change if the rows and columns are interchanged

$$\det\left(\begin{bmatrix} a_{11} & a_{21} \\ a_{12} & a_{22} \end{bmatrix}\right) = \det(A^T) = \det(A).$$

And, $\det(A)$ changes sign if the rows are interchanged

$$\det\left(\begin{bmatrix} a_{21} & a_{22} \\ a_{11} & a_{12} \end{bmatrix}\right) = -\det(A).$$

The extension of determinants to 3×3 matrices can be done in an inductive way. Consider a 3×3 matrix with column vectors **a**, **b** and **c**

$$A = \begin{bmatrix} a_{11} & a_{12} & a_{13} \\ a_{21} & a_{22} & a_{23} \\ a_{31} & a_{32} & a_{33} \end{bmatrix} = [\mathbf{a} \ \mathbf{b} \ \mathbf{c}].$$

The volume of the parallelepiped formed by the three vectors is plus or minus the dot product of **a** with $\mathbf{b} \times \mathbf{c}$, which can be written as the determinant of A

$$\begin{aligned} \det(A) \ &\equiv \ a_{11}(a_{22}a_{33} - a_{23}a_{32}) \\ &\quad -a_{12}(a_{21}a_{33} - a_{23}a_{31}) \\ &\quad +a_{13}(a_{21}a_{32} - a_{22}a_{31}) \\ &= \ a_{11}(a_{22}a_{33} - a_{23}a_{32}) \\ &\quad -a_{21}(a_{12}a_{33} - a_{13}a_{32}) \\ &\quad +a_{31}(a_{12}a_{23} - a_{13}a_{22}) \\ &= \ \mathbf{a} \bullet (\mathbf{b} \times \mathbf{c}). \end{aligned}$$

Here we have interchanged rows and columns and the determinant does not change. That is, $\det(A) = \det(A^T)$ where A^T is called the *transpose of A* and is defined as

$$A^T \equiv \begin{bmatrix} a_{11} & a_{21} & a_{31} \\ a_{12} & a_{22} & a_{32} \\ a_{13} & a_{23} & a_{33} \end{bmatrix} = \begin{bmatrix} \mathbf{a}^T \\ \mathbf{b}^T \\ \mathbf{c}^T \end{bmatrix}.$$

Cramer's rule, for $\det(A) \neq 0$, allows one to easily solve the 3×3 algebraic systems $A\mathbf{x} = \mathbf{d}$

$$\begin{aligned} x_1 &= \det([\mathbf{d} \ \ \mathbf{b} \ \ \mathbf{c}])/\det(A), \\ x_2 &= \det([\mathbf{a} \ \ \mathbf{d} \ \ \mathbf{c}])/\det(A) \text{ and} \\ x_2 &= \det([\mathbf{a} \ \ \mathbf{b} \ \ \mathbf{d}])/\det(A). \end{aligned}$$

The justification of the formula for the first unknown follows from the properties of the cross product and the column version of a matrix-vector product

$$\begin{aligned} A\mathbf{x} &= \mathbf{d} \\ x_1\mathbf{a} + x_2\mathbf{b} + x_3\mathbf{c} &= \mathbf{d} \\ x_1\mathbf{a} &= \mathbf{d} - x_2\mathbf{b} - x_3\mathbf{c}. \end{aligned}$$

Recall the cross product $\mathbf{b} \times \mathbf{c}$ is perpendicular to both $\mathbf{b}$ and $\mathbf{c}$. Use this fact and take the dot product of both sides with $\mathbf{b} \times \mathbf{c}$

$$\begin{aligned} (x_1\mathbf{a}) \bullet (\mathbf{b} \times \mathbf{c}) &= (\mathbf{d} - x_2\mathbf{b} - x_3\mathbf{c}) \bullet (\mathbf{b} \times \mathbf{c}) \\ (\mathbf{a} \bullet (\mathbf{b} \times \mathbf{c}))x_1 &= \mathbf{d} \bullet (\mathbf{b} \times \mathbf{c}) - \mathbf{b} \bullet (\mathbf{b} \times \mathbf{c})x_2 - \mathbf{c} \bullet (\mathbf{b} \times \mathbf{c})x_3 \\ \det(A)x_1 &= \det([\mathbf{d} \ \ \mathbf{b} \ \ \mathbf{c}]) - 0x_2 - 0x_3. \end{aligned}$$

The derivations of the equations for the other two unknowns are similar. Furthermore, these concepts generalize to algebraic systems with n unknowns.

3.7.2 Determinant of an $n \times n$ Matrix

The determinant of an $n \times n$ matrix is defined inductively from determinant of $(n-1) \times (n-1)$ matrices in much the same way as for 3×3 matrices. Let M_{ij} be the determinant of the $(n-1) \times (n-1)$ matrix formed by deleting row i and column j of A. The determinant of A may be computed by expanding about any row of column. Here we use the first row of the matrix.

Definition 3.7.1. The row expansion of a *determinant* of an $n \times n$ matrix A is along the first row

$$\det(A) \equiv a_{11}(-1)^{1+1}M_{11} + a_{12}(-1)^{1+2}M_{12} + \cdots + a_{1n}(-1)^{1+n}M_{1n}.$$

The coefficients of a_{ij}, $(-1)^{i+j}M_{ij}$, are called *cofactors*.

For $n = 3$ it is easy to expand the 2×2 determinants to reveal the following seven rules:

1. The determinant of the transpose matrix is $\det(A^T) = \det(A)$.
2. The determinant of the matrix formed by an interchange of
    two rows is $\det(P_{ij}A) = -\det(A)$.
3. The determinant of the matrix formed by adding a
    multiple of a row j to row i is unchanged
    $\det(E_{ij}(a)A) = \det(A)$.
4. The determinant of the matrix formed by multiplying a row by
    c is $\det(E_i(c)A) = c\det(A)$.
5. $\det(AB) = \det(A)\det(B)$.
6. $\det(A) = a_{i1}(-1)^{i+1}M_{i1} + a_{i2}(-1)^{i+2}M_{i2} + a_{i3}(-1)^{i+3}M_{i3}$
    (row i expansion).
7. $\det(A) = a_{1j}(-1)^{1+j}M_{1j} + a_{2j}(-1)^{2+j}M_{2j} + a_{3j}(-1)^{3+j}M_{3j}$
    (column j expansion).

These seven rules extend to $n \times n$ matrices. The following example illustrates
how they may be used to reduce the computational burden.

Example 3.7.1. Consider the matrix in Sections 3.1 and 3.4 (Example 3.4.3)
from the two-loop circuit. First, the determinant is evaluated by expanding
down column one, which has only two non-zero components,

$$
\begin{aligned}
\det(A) &= \det\left(\begin{bmatrix} 1 & -1 & 1 \\ 1 & 0 & -3 \\ 0 & -2 & -3 \end{bmatrix}\right) \\
&= (-1)^2(1)\det\left(\begin{bmatrix} 0 & -3 \\ -2 & -3 \end{bmatrix}\right) + (-1)^{2+1}(1)\det\left(\begin{bmatrix} -1 & 1 \\ -2 & -3 \end{bmatrix}\right) \\
&= -6 - 5 = -11.
\end{aligned}
$$

One can also use the following elementary row operations to transform the
matrix to upper triangular form

$$
E_{32}(2)\, E_{11}(-1)\, A = U = \begin{bmatrix} 1 & -1 & 1 \\ 0 & 1 & -4 \\ 0 & 0 & -11 \end{bmatrix}.
$$

Apply rule three twice to get $\det(E_{32}(2)(E_{11}(-1)\,A)) = \det(E_{11}(-1)\,A) = \det(A) = \det(U)$. The determinant of an upper triangular matrix can be expanded
by the left column so that the determinant is the product of the diagonal com-
ponents. Therefore, $\det(A) = \det(U) = (1)(1)(-11) = -11$.

Example 3.7.2. Consider the 4×4 matrix for the heat conduction problem in
Sections 3.3 and 3.4 (Example 3.4.4) and expand it by column one using rule

one

$$
\det(A) = \det\left(\begin{bmatrix} 2 & -1 & 0 & 0 \\ -1 & 2 & -1 & 0 \\ 0 & -1 & 2 & -1 \\ 0 & 0 & -1 & 2 \end{bmatrix}\right)
$$

$$
= 2\det\left(\begin{bmatrix} 2 & -1 & 0 \\ -1 & 2 & -1 \\ 0 & -1 & 2 \end{bmatrix}\right) - (-1)\det\left(\begin{bmatrix} -1 & 0 & 0 \\ -1 & 2 & -1 \\ 0 & -1 & 2 \end{bmatrix}\right)
$$

$$
= 2(4) - 3 = 5.
$$

This determinant may also be computed by elementary matrix transformation to obtain an upper triangular form

$$
E_{43}(3/4)\, E_{32}(2/3)\, E_{21}(1/2)\, A = U = \begin{bmatrix} 2 & -1 & 0 & 0 \\ 0 & 3/2 & -1 & 0 \\ 0 & 0 & 4/3 & -1 \\ 0 & 0 & 0 & 5/4 \end{bmatrix}.
$$

Use the third rule three times and expand the upper triangular matrix by the left column to get

$$
\det(E_{43}(3/4)\, E_{32}(2/3)\, E_{21}(1/2)\, A) = \det(A) = \det(U) = 1(3/2)(4/3)(5/4) = 5.
$$

3.7.3 Cramer's Rule and Inverses

In order to generalize Cramer's rule from 3×3 to $n \times n$ matrices, we must generalize perpendicular three dimensional vectors to n-dimensional vectors. Two n-dimensional column vectors $\mathbf{a} = [a_1 \cdots a_n]^T$ and $\mathbf{b} = [b_1 \cdots b_n]^T$ are called *perpendicular or orthogonal* if and only if the dot product is zero, that is, $\mathbf{a}^T\mathbf{b} = \mathbf{a}\bullet\mathbf{b} \equiv a_1 b_1 + \cdots + a_n b_n = 0$. Let A be an $n \times n$ matrix with components a_{ij}, and denote its column j by $\mathbf{a}_{:j}$ so that $A = [\mathbf{a}_{:1} \cdots \mathbf{a}_{:n}]$. The matrix-vector product $A\mathbf{x}$ can be written as a linear combination of the column vectors, see Theorem 3.2.1,

$$
A\mathbf{x} = \mathbf{d}
$$
$$
x_1\mathbf{a}_{:1} + \cdots + x_n\mathbf{a}_{:n} = \mathbf{d}.
$$

In order to find the first unknown x_1, we must solve

$$
x_1\mathbf{a}_{:1} = \mathbf{d} - (x_2\mathbf{a}_{:2} + \cdots + x_n\mathbf{a}_{:n}).
$$

In the case $n = 3$, we computed the dot product of both sides with the cross product of the second and third columns. Since the cross product of two vectors is perpendicular to both the vectors, the dot product on the right side simplified to the $\det([\mathbf{d} \ \ \mathbf{a}_{:2} \ \ \mathbf{a}_{:3}])$ and, consequently, $x_1 = \det([\mathbf{d} \ \ \mathbf{a}_{:2} \ \ \mathbf{a}_{:3}])/\det(A)$.

The generalization of the cross product to higher dimensions requires an n-dimensional vector that will be perpendicular to $n - 1$ given n-dimensional vectors. Let the $n - 1$ given column vectors $j = 2, \cdots, n$ in the matrix A be $\mathbf{a}_{:2}, \cdots, \mathbf{a}_{:n}$. We now claim the column vector $\mathbf{c}$ of cofactors with $j = 1$ will be perpendicular to each of these vectors

$$\mathbf{c} = \begin{bmatrix} (-1)^{1+1}M_{11} \\ (-1)^{2+1}M_{21} \\ \vdots \\ (-1)^{n+1}M_{n1} \end{bmatrix}.$$

To see how this is established, consider the reduced notational burden of the case $n = 4$. We must find x_1 in

$$x_1\mathbf{a}_{:1} = \mathbf{d} - (x_2\mathbf{a}_{:2} + x_3\mathbf{a}_{:3} + x_4\mathbf{a}_{:4}).$$

Compute the dot product of both sides with $\mathbf{c}$

$$\begin{aligned} (x_1\mathbf{a}_{:1}) \bullet \mathbf{c} &= (\mathbf{d} - (x_2\mathbf{a}_{:2} + x_3\mathbf{a}_{:3} + x_4\mathbf{a}_{:4})) \bullet \mathbf{c} \qquad (3.7.1) \\ (\mathbf{a}_{:1} \bullet \mathbf{c})x_1 &= \mathbf{d} \bullet \mathbf{c} - (\mathbf{a}_{:2} \bullet \mathbf{c})x_2 - (\mathbf{a}_{:3} \bullet \mathbf{c})x_3 - (\mathbf{a}_{:4} \bullet \mathbf{c})x_4. \end{aligned}$$

All five of the dot products can be expressed as determinants.

$$\begin{aligned} \mathbf{a}_{:1} \bullet \mathbf{c} &= a_{11}(-1)^{1+1}M_{11} + a_{21}(-1)^{2+1}M_{21} + \\ &\quad a_{31}(-1)^{3+1}M_{31} + a_{41}(-1)^{4+1}M_{41} \\ &= \det(A) \quad \text{(expansion by column one) and similarly} \\ \mathbf{d} \bullet \mathbf{c} &= \det([\mathbf{d} \ \ \mathbf{a}_{:2} \ \ \mathbf{a}_{:3} \ \ \mathbf{a}_{:4}]). \end{aligned}$$

The other three determinants are also expanded by column one and are equal to zero because they have two identical columns.

$$\begin{aligned} (\mathbf{a}_{:2} \bullet \mathbf{c}) &= a_{12}(-1)^{1+1}M_{11} + a_{22}(-1)^{2+1}M_{21} + \\ &\quad a_{32}(-1)^{3+1}M_{31} + a_{42}(-1)^{4+1}M_{41} \\ &= \det(\begin{bmatrix} a_{12} & a_{12} & a_{13} & a_{14} \\ a_{22} & a_{22} & a_{23} & a_{24} \\ a_{32} & a_{32} & a_{33} & a_{34} \\ a_{42} & a_{42} & a_{43} & a_{44} \end{bmatrix}) = \det([\mathbf{a}_{:2} \ \ \mathbf{a}_{:2} \ \ \mathbf{a}_{:3} \ \ \mathbf{a}_{:4}]) = 0, \\ (\mathbf{a}_{:3} \bullet \mathbf{c}) &= \det([\mathbf{a}_{:3} \ \ \mathbf{a}_{:2} \ \ \mathbf{a}_{:3} \ \ \mathbf{a}_{:4}]) = 0 \text{ and} \\ (\mathbf{a}_{:4} \bullet \mathbf{c}) &= \det([\mathbf{a}_{:4} \ \ \mathbf{a}_{:2} \ \ \mathbf{a}_{:3} \ \ \mathbf{a}_{:4}]) = 0. \end{aligned}$$

Then equation (3.7.1) becomes

$$\det(A)x_1 = \det([\mathbf{d} \ \ \mathbf{a}_{:2} \ \ \mathbf{a}_{:3} \ \ \mathbf{a}_{:4}]).$$

By using other columns of cofactors one can solve for the other unknowns.

Theorem 3.7.1 *(Cramer's Rule) Let A be an $n \times n$ matrix, and let A_i be an $n \times n$ matrix with the same columns of A except column i is replaced by column $\mathbf{d}$. If $\det(A) \neq 0$, then the solution of $A\mathbf{x} = \mathbf{d}$ is given by*

$$x_i = \det(A_i)/\det(A).$$

Cramer's rule is useful for smaller dimensional matrices, special calculations where only a few of the unknown components are required and some theoretical considerations. However, it does allow one to give an explicit formula for the inverse matrix

$$
\begin{aligned}
AA^{-1} &= I \\
A[\mathbf{b}_1 \cdots \mathbf{b}_n] &= [\mathbf{e}_1 \cdots \mathbf{e}_n] \text{ or equivalently} \\
A\mathbf{b}_1 &= \mathbf{e}_1, \cdots, A\mathbf{b}_n = \mathbf{e}_n.
\end{aligned}
$$

In order to find the first column $\mathbf{b}_1$ of the inverse matrix, apply Cramer's rule with $\mathbf{d} = \mathbf{e}_1$

$$
\begin{aligned}
b_{i1} &= \det(A_i)/\det(A) \text{ where} \\
b_{11} &= \det([\mathbf{e}_1 \ \mathbf{a}_{:2} \cdots \mathbf{a}_{:n}])/\det(A) = (-1)^{1+1} M_{11}/\det(A) \\
b_{21} &= \det([\mathbf{a}_{:1} \ \mathbf{e}_1 \cdots \mathbf{a}_{:n}])/\det(A) = (-1)^{1+2} M_{12}/\det(A) \\
&\vdots \\
b_{n1} &= \det([\mathbf{a}_{:1} \ \mathbf{a}_{:2} \cdots \mathbf{e}_1])/\det(A) = (-1)^{1+n} M_{1n}/\det(A).
\end{aligned}
$$

The other columns of the inverse matrix are similar.

Theorem 3.7.2 *(Inverse Matrix via Determinants) Let A be an $n \times n$ matrix with non-zero determinant. Then the inverse matrix exists and column j in the inverse matrix is*

$$
\mathbf{b}_j = \begin{bmatrix} b_{1j} \\ b_{2j} \\ \vdots \\ b_{nj} \end{bmatrix} = \begin{bmatrix} (-1)^{j+1} M_{j1}/\det(A) \\ (-1)^{j+2} M_{j2}/\det(A) \\ \vdots \\ (-1)^{j+n} M_{jn}/\det(A) \end{bmatrix}.
$$

Example 3.7.3. Use determinants to find the inverse of

$$
A = \begin{bmatrix} 1 & 2 & 1 \\ 1 & 3 & 2 \\ 1 & 0 & 1 \end{bmatrix}.
$$

Since $\det(A) = 2 \neq 0$, the inverse matrix exists. The first column in the inverse matrix $A^{-1} = B = [\mathbf{b}_1 \ \mathbf{b}_2 \ \mathbf{b}_3]$ is

$$
\mathbf{b}_1 = \begin{bmatrix} b_{11} \\ b_{21} \\ b_{31} \end{bmatrix} = \begin{bmatrix} (-1)^{1+1} M_{11}/2 \\ (-1)^{1+2} M_{12}/2 \\ (-1)^{1+3} M_{13}/2 \end{bmatrix} = \begin{bmatrix} \det(\begin{bmatrix} 3 & 2 \\ 0 & 1 \end{bmatrix})/2 \\ -\det(\begin{bmatrix} 1 & 2 \\ 1 & 1 \end{bmatrix})/2 \\ \det(\begin{bmatrix} 1 & 3 \\ 1 & 0 \end{bmatrix})/2 \end{bmatrix} = \begin{bmatrix} 3/2 \\ 2/2 \\ -3/2 \end{bmatrix}.
$$

The other two columns in the inverse matrix may be computed in a similar fashion to yield

$$A^{-1} = (1/2) \begin{bmatrix} 3 & -2 & 1 \\ 1 & 0 & -1 \\ -3 & 2 & 1 \end{bmatrix}.$$

If any column of an $n \times n$ matrix is a linear combination of the other columns, then the determinant of the matrix must be zero. Equivalently, if the determinant of the matrix is not zero, then no column of the matrix A is a linear combination of the other columns (the columns are *linearly independent,* see Section 5.1 for a more general discussion)

$$A\mathbf{x} = x_1 \mathbf{a}_{:1} + \cdots + x_n \mathbf{a}_{:n} = \mathbf{0} \text{ implies } \mathbf{x} = \mathbf{0}.$$

The bold font $\mathbf{0}$ represents a column vector of zeros. In the cases $n = 2$ or $n = 3$, this condition requires the vectors not be parallel for $n = 2$, and not be in the same plane for $n = 3$. Each column of such matrices must have at least one non-zero component so that row operations can be done to transform the matrix to upper triangular form. This eventually leads to the following theorem, which gives a characterization of matrices that have inverses.

Theorem 3.7.3 *(Inverse Matrix Equivalence) Let A be an $n \times n$ matrix. The following are equivalent:*

(i).   *A has an inverse,*
(ii). *$A\mathbf{x} = \mathbf{0}$ implies $\mathbf{x} = \mathbf{0}$ and*
(iii). *$\det(A) \neq 0$.*

3.7.4   Application to Three-tank Mixing

Mixing chemicals occurs in many industrial and biological systems and include hot tubs, lead poisoning and brine solutions. Here we consider three well-stirred tanks each with 24 liters (L) with flow rates (L/min) such that the tanks remain full, see Figure 3.7.1. The amounts of a chemical, say salt, are given by the unknown functions $x(t)$, $y(t)$ and $z(t)$ (Kg) so that the concentrations are $x(t)/24$, $y(t)/24$ and $z(t)/24$ (Kg/L). The flow rates in and out of the tanks are the numbers (L/min) adjacent to the horizontal arrows in the figure. Also, the chemical is directly added to the tanks as indicated by the rate numbers (Kg/min) above the vertical arrows in the figure.

The model is derived by equating the rate of change for the amount to the rate of change incoming minus the rate of change outgoing. The flow rate (L/min) times the concentration (Kg/L) is the rate of change in or out of a tank from an adjacent tank. For the tank on the left, pure water is flowing in at a rate equal to 6 (L/min) and the mixture from the center tank is flowing in at a rate equal to 4 (L/min) with a concentration equal to $y/24$ (Kg/L). The rate of change for the amount incoming is $1 + 4(y/24)$ and the rate of change of amount outgoing is $10(x/24)$. This gives the differential equation for the left

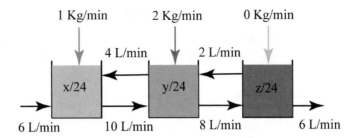

Figure 3.7.1: Three-tank Mixing

tank, and the differential equations for the center and right tanks are similar, generating a system of three differential equations

$$\begin{cases} x' = \frac{dx}{dt} = (1 + 4(y/24)) - 10(x/24) \\ y' = \frac{dy}{dt} = (2 + 10(x/24) + 2(z/24)) - (4(y/24) + 8(y/24)) \\ z' = \frac{dz}{dt} = 8(y/24) - (2(z/24) + 6(z/24)). \end{cases}$$

The solution of the system of differential equations is more complicated than the system for the projectile problem in Subsection 2.3.3. The time dependent solution will be discussed in more detail in chapters six and seven.

Presently, consider the steady state solution, which requires all amounts to be independent of time so that their time derivatives are zero. The above system of differential equations becomes an algebraic system

$$\begin{cases} 0 = (1 + 4(y/24)) - 10(x/24) \\ 0 = (2 + 10(x/24) + 2(z/24)) - (4(y/24) + 8(y/24)) \\ 0 = 8(y/24) - (2(z/24) + 6(z/24)). \end{cases}$$

The equivalent vector equation is

$$\begin{bmatrix} -10/24 & 4/24 & 0 \\ 10/24 & -12/24 & 2/24 \\ 0 & 8/24 & -8/24 \end{bmatrix} \begin{bmatrix} x \\ y \\ z \end{bmatrix} = \begin{bmatrix} -1 \\ -2 \\ 0 \end{bmatrix}.$$

This has a solution because the determinant of the matrix is not zero

$$\det(\begin{bmatrix} -10/24 & 4/24 & 0 \\ 10/24 & -12/24 & 2/24 \\ 0 & 8/24 & -8/24 \end{bmatrix}) = (-10/24) \begin{bmatrix} -12/24 & 2/24 \\ 8/24 & -8/24 \end{bmatrix}$$

$$-(2/24) \begin{bmatrix} 10/24 & 2/24 \\ 0 & -8/24 \end{bmatrix}$$

$$= -5/144.$$

Cramer's rule can easily be used to find the steady state solutions $x = 7.2$, $y = 12.0$ and $z = 12.0$. In chapter seven we will show that the time dependent solutions will approach the steady state solutions as time goes to infinity.

3.7.5   Determinants Using MATLAB

Return to Example 3.7.2 where the matrix is 4×4 and we want to solve $A\mathbf{u} = \mathbf{d} = [470 \ 400 \ 400 \ 470]^T$ using Cramer's rule. The first two unknowns are as follows.

```
>> A =[2 -1  0  0;
       -1  2 -1  0;
        0 -1  2 -1;
        0  0 -1  2];
>> d = [470 400 400 470]';
>> det(A)
   ans =
        5
>> A1 = [d A(:,2) A(:,3) A(:,4)]
   A1 =
        470 -1  0  0
        400  2 -1  0
        400 -1  2 -1
        470  0 -1  2
>> u1 = det(A1)/5
   u1 =
        870
>> A2 = [A(:,1) d A(:,3) A(:,4)]
   A2 =
        2 470  0  0
       -1 400 -1  0
        0 400  2 -1
        0 470 -1  2
>> u2 = det(A2)/5
   u2 =
     1270.
```

Reconsider the application to the three-tank mixing where the right tank now has the chemical being directly inserted from the top at a rate equal to 3 (Kg/min). The differential equation for the right tank in Figure 3.7.1 becomes

$$z' = \frac{dz}{dt} = (3 + 8(y/24)) - (2(z/24) + 6(z/24)).$$

The steady state solution must satisfy

$$\begin{bmatrix} -10/24 & 4/24 & 0 \\ 10/24 & -12/24 & 2/24 \\ 0 & 8/24 & -8/24 \end{bmatrix} \begin{bmatrix} x \\ y \\ z \end{bmatrix} = \begin{bmatrix} -1 \\ -2 \\ -3 \end{bmatrix}.$$

Multiplying both sides by 24 and use Cramer's rule to find the solution.

```
>> A = [-10  4  0;
```

```
                    10 -12   2;
                     0   8  -8];
        >> det(A)
            ans =
                       -480
        >> d = [-24 -48 -72]';
        >> A\d                                 % uses Gauss elimination
            ans =
                     8.4000
                    15.0000
                    24.0000
        >> A1 = [d A(:,2) A(:,3)];             % uses Cramer's rule
        >> A2 = [A(:,1) d A(:,3)];
        >> A3 = [A(:,1) A(:,2) d];
        >> x = det(A1)/-480
            x =
                     8.4000
        >> y = det(A2)/-480
            y =
                       15
        >> z = det(A3)/-480
            z =
                       24.
```

3.7.6   Exercises

1.   Consider Example 3.7.1 and verify
   (a). $\det(A^T) = \det(A)$ and
   (b). $\det(P_{23}A) = -\det(A)$, that is, interchange rows 2 and 3.
2.   Consider Example 3.7.1 and verify
   (a). $\det(E_{32}(10)\,A) = \det(A)$, that is, multiply row 2 by 10 and add to row 3 and
   (b). $\det(E_2(4)A) = 4\det(A)$, that is, multiply row 2 by 4.
3.   Consider Example 3.7.1 and compute the determinant
   (a).   by expanding row $i = 3$ and
   (b).   by expanding column $j = 2$.
4.   Let A be the 4×4 matrix

$$A = \begin{bmatrix} 3 & -1 & 0 & 0 \\ -1 & 3 & -1 & 0 \\ 0 & -1 & 3 & -1 \\ 0 & 0 & -1 & 3 \end{bmatrix}.$$

   (a).   Find the determinant by expanding row one.
   (b).   Use elementary matrices to transform the matrix to upper triangular form and then compute the determinant.

5.   Let A be the 4×4 matrix

$$A = \begin{bmatrix} 1 & 2 & 1 & 0 \\ 0 & 1 & 0 & 0 \\ 8 & 16 & 11 & 0 \\ 1 & 2 & 3 & 7 \end{bmatrix}.$$

(a).   Find the determinant by expanding row two.
(b).   Use elementary matrices to transform the matrix to upper triangular form and then compute the determinant.
6.   In Example 3.7.3 use determinants to compute the other two columns of the inverse matrix.
7.   In Example 3.7.1 use determinants to compute inverse matrix.
8.   Consider the derivation of Cramer's rule for $n = 3$ (see the end of Subsection 3.7.1) and justify the equations for x_2 and x_3.
9.   Consider the derivation of Cramer's rule for $n = 4$ (see Subsection 3.7.3) and justify the equation for x_4.
10.   Use Cramer's rule to compute the solution to the three-tank mixing problem.
11.   Consider the three-tank mixing problem where the chemical is now directly inserted from the top at rates 2, 3 and 4 (Kg/min) for the left, center and right tanks, respectively. Assume the flow rates between the 24 liter tanks remain as in Figure 3.7.1.
(a).   Find the three differential equations.
(b).   Find the matrix equation for the steady state solution.
(c).   Solve the matrix equation.
12.   Use MATLAB to confirm any calculations done in exercises 4-7.
13.   Use MATLAB to confirm any calculations done in exercises 10 or 11.
14.   Use MATLAB and determinants to compute the other two solutions of $A\mathbf{u} = \mathbf{d} = [470 \ 400 \ 400 \ 470]^T$ where A is from Example 3.7.2.

Chapter 4

Ax = d: Least Squares Solution

This chapter contains methods for "solving" matrix equations of the form $A\mathbf{x} = \mathbf{d}$ where A is a matrix with more rows (equations) than columns (variables). The normal equations will be developed and used to solve these over determined algebraic systems. Applications include data fitting such as sales predictions, radioactive decay, world population models, flow rate data and parameter identification. By-hand calculations will be done as well as implementations in MATLAB for larger and nonlinear problems.

4.1 Curve Fitting to Data

This section is a continuation of the price prediction problem introduced in Subsection 2.4.4 where a person is trying to predict the price of a new product such as high definition televisions based on past observations of the price as a function of time. There were three observations made at months 1, 2 and 3 with prices $2000, $1950 and $1910. In an attempt to "model" this data by a line $y = mt + c$, the slope m and the intercept c were to be found from the three algebraic equations in (2.4.5) or in matrix form

$$
\begin{bmatrix} 1 & 1 \\ 2 & 1 \\ 3 & 1 \end{bmatrix} \begin{bmatrix} m \\ c \end{bmatrix} = \begin{bmatrix} 2000 \\ 1950 \\ 1910 \end{bmatrix}
$$

$$
A\mathbf{x} = \mathbf{d}.
$$

In this case A in 3×2, $\mathbf{x} = [m \quad c]^T$ is 2×1 and there are more equations than unknowns. So, it is a good possibility that there is no single straight line containing the three data points. This was viewed geometrically where we chose m and c so that the point in the plane given by the linear combination of the two column vectors of A would be closest to the point $\mathbf{d} = [2000 \ 1950 \ 1910]^T$.

Another way given by the residual vector $\mathbf{r} \equiv \mathbf{d} - A\mathbf{x}$ to describe this is to choose m and c so that the least squares function $\mathbf{r}^T\mathbf{r}$ is a minimum.

The objective is to extend both the geometric and residual approaches to additional data points (rows in A) and to additional model parameters (columns in A) so that more complicated and accurate data modeling can be done. This section will focus on more data points with two model parameters. Applications will be given to sales prediction, radioactive decay and population models.

4.1.1　The Least Squares Problem

Consider $A = [\mathbf{a}\ \ \mathbf{b}]$ with two $n \times 1$ columns where n may be larger than 3. The possible solution of the system

$$
\begin{aligned}
A\mathbf{x} &= \mathbf{d} \\
[\mathbf{a}\ \ \mathbf{b}]\begin{bmatrix} m \\ c \end{bmatrix} &= \mathbf{d} \\
m\mathbf{a} + c\mathbf{b} &= \mathbf{d}
\end{aligned}
$$

is the m and c that forces the residual vector $\mathbf{r} \equiv \mathbf{d} - (m\mathbf{a}+c\mathbf{b}) = [r_1\ \ r_2 \cdots r_n]$ to be as "close" as possible to the zero vector. If the sum of the squares of the residual vector's components is a minimum, then this is called the *least squares solution*. The objective is to find m and c that minimize

$$
\mathbf{r}^T\mathbf{r} = r_1^2 + r_2^2 + \cdots + r_n^2.
$$

The *least squares function* of m and c can be explicitly computed as

$$
\begin{aligned}
F(m,c) &\equiv \mathbf{r}^T\mathbf{r} = (\mathbf{d} - (m\mathbf{a} + c\mathbf{b}))^T(\mathbf{d} - (m\mathbf{a} + c\mathbf{b})) \\
&= \mathbf{d}^T\mathbf{d} - 2\mathbf{d}^T\mathbf{a}m - 2\mathbf{d}^T\mathbf{b}c + \\
&\quad 2\mathbf{a}^T\mathbf{b}mc + \mathbf{a}^T\mathbf{a}m^2 + \mathbf{b}^T\mathbf{b}c^2.
\end{aligned} \tag{4.1.1}
$$

This problem will be solved three ways: graphically, calculus and normal equations (in Section 4.2).

For the above high definition television data $\mathbf{a} = [1\ \ 2\ \ 3]^T$, $\mathbf{b} = [1\ \ 1\ \ 1]^T$ and $\mathbf{d} = [2000\ \ 1950\ \ 1910]^T$ and gives

$$
F(m,c) = 11450600 - 2(11630)m - 2(5860)c + 2(6)mc + 14m^2 + 3c^2.
$$

In Subsection 2.4.4 a geometric argument required the minimum to be the solution of the following algebraic system

$$
\begin{cases}
\mathbf{a}^T\mathbf{r} = \mathbf{a}^T(\mathbf{d} - (m\mathbf{a} + c\mathbf{b})) = \mathbf{a}^T\mathbf{d} - (\mathbf{a}^T\mathbf{a}m + \mathbf{a}^T\mathbf{b}c) = 0 \\
\mathbf{b}^T\mathbf{r} = \mathbf{b}^T(\mathbf{d} - (m\mathbf{a} + c\mathbf{b})) = \mathbf{b}^T\mathbf{d} - (\mathbf{b}^T\mathbf{a}m + \mathbf{b}^T\mathbf{b}c) = 0.
\end{cases} \tag{4.1.2}
$$

This is a special case of the normal equations, and as we shall see they give a solution to the least squares problem even if the number of data points is larger than $n = 3$.

Table 4.1.1: Computer Sales Data

Months	Computers Sold
1	78
2	85
3	90
4	95
5	104
6	113

Example 4.1.1. Suppose a new computer company has had increased sales and would like to make a prediction of future sales based on past sales. During the past six months sales have been 78, 85, 90, 95, 104 and 113 computers, see Table 4.1.1.

A first look at this data suggests an increase of about 8 computers each month. A more precise way to view this is to determine the slope and intercept so that $y = mt + c$ is "closest" to the data points as is depicted in the left graph in Figure 4.1.1, which will be discussed in detail in the next section. Each of the data points can be inserted into the linear equation and gives a system with six equations

$$\begin{cases} m1 + c = 78 \\ m2 + c = 85 \\ m3 + c = 90 \\ m4 + c = 96 \\ m5 + c = 104 \\ m6 + c = 113. \end{cases}$$

The matrix form is

$$\begin{bmatrix} 1 & 1 \\ 2 & 1 \\ 3 & 1 \\ 4 & 1 \\ 5 & 1 \\ 6 & 1 \end{bmatrix} \begin{bmatrix} m \\ c \end{bmatrix} = \begin{bmatrix} 78 \\ 85 \\ 90 \\ 96 \\ 104 \\ 113 \end{bmatrix}.$$

The column vectors now have six components, but the least squares function $F(m, c)$ is a function of two variables. The least squares solution will be determined at the end of this section and will give the straight line indicated in Figure 4.1.1.

Example 4.1.2. Estimates of the world population over the past hundred years are in Table 4.1.2, see [9]. Careful inspection of the populations reveals larger increases in the populations as time progresses. So, an attempt to model this data by a straight line may not yield very accurate predictions about the world population. One possible model, among many, is to use a quadratic function

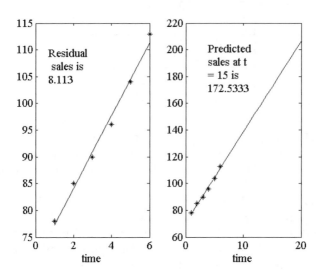

Figure 4.1.1: Sales Data

Table 4.1.2: World Population Data

Year	World Pop. (billions)
1900	1.65
1910	1.75
1920	1.86
1930	2.07
1940	2.30
1950	2.52
1960	3.02
1970	3.70
1980	4.44
1990	5.27
2000	6.06

of time where the three coefficients are chosen so that the quadratic function is increasing and concave up for times near the data. Let $y = at^2 + bt + c$ where time starts in 1900, $t = 0$, and ends in 2000, $t = 100$. There are $n = 11$ equations giving the algebraic system

$$\begin{cases} a0^2 + b0 + 1c = 1.65 \\ a10^2 + b10 + 1c = 1.75 \\ a20^2 + b20 + 1c = 1.86 \\ \vdots \\ a100^2 + b100 + c = 6.06. \end{cases}$$

The equivalent matrix equation uses an 11×3 matrix A and has the form

$$\begin{bmatrix} 0 & 0 & 1 \\ 10^2 & 10 & 1 \\ 20^2 & 20 & 1 \\ \vdots & \vdots & \vdots \\ 100^2 & 100 & 1 \end{bmatrix} \begin{bmatrix} a \\ b \\ c \end{bmatrix} = \begin{bmatrix} 1.65 \\ 1.75 \\ 1.86 \\ \vdots \\ 6.06 \end{bmatrix}.$$

In this case the least square function $F(a, b, c)$ depends on the data and the three unknown coefficients a, b and c. The normal equations as described in the Section 4.2 will be used to solve this system.

4.1.2   Least Squares Solution and Graphing

Consider the least squares problem for fitting data to a straight line with n data points so that the matrix is $n \times 2$ and $A = [\mathbf{a} \ \mathbf{b}]$ with least squares function $F(m,c)$ as given in equation (4.1.1). Since the least squares function depends on two variables, we may graph the value $z = F(m, c)$ on the z-axis with the independent variables $x = m$ and $y = c$. Then upon doing the graphing for appropriate choices of m and c, the minimum of the least squares function can be estimated and the corresponding values of m and c approximated.

The MATLAB code ls_graph.m has been used to generate the 3D graph on the left in Figure 4.1.2 and the contour graph on the right of the same figure for the least squares function $F(m, c)$ in Example 4.1.1. Some trial and error was required to locate the minimum of the least squares function, which is approximately $F(m, c) \approx 10$ where $(m, c) \approx (6.6, 71)$. MATLAB commands mesh() and contour() were used to generate the left and right graphs in the Figure 4.1.2, respectively. This method is not very accurate and not applicable when more than two parameters are to be found. But, it does give geometric interpretation and can be useful when examining the sensitivity of the least squares solution to variations in the data. In Figure 4.1.2 notice the solution is located in a long thin "valley" of the graph. This suggests small changes in the data might cause significant changes in the solution!

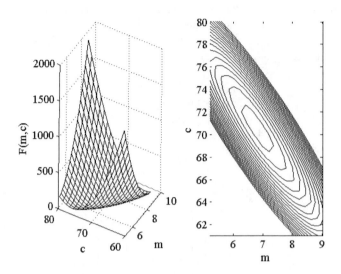

Figure 4.1.2: Least Squares Function for Sales Data

4.1.3   Least Squares Solution and Calculus

The least square function $F(m,c)$ in equation (4.1.1) is a quadratic function of
two variables. By fixing one of the variables, it becomes a function of the other
variable. The left graph in Figure 4.1.2 is formed from the trace curves in both
the c and m directions. The minimum of the least squares function $F(m,c)$
must be located at the intersection of trace curves in the m and c directions so
that both trace curves are at a minimum. This allows one to use calculus of a
single variable, but applied twice: once to the trace curve in the m direction
and once to the trace curve in the c direction.

Fix $c = c_0$ and consider $F(m, c_0)$ as a function of m. Define $f(m)$

$$f(m) \equiv F(m, c_0) = \mathbf{d}^T\mathbf{d} - 2\mathbf{d}^T\mathbf{a}m - 2\mathbf{d}^T\mathbf{b}c_0 + 2\mathbf{a}^T\mathbf{b}mc_0 + \mathbf{a}^T\mathbf{a}m^2 + \mathbf{b}^T\mathbf{b}c_0^2.$$

Compute the first derivative with respect to m and set it equal to zero

$$\frac{df}{dm} = 0 = 0 - 2\mathbf{d}^T\mathbf{a} - 0 + 2\mathbf{a}^T\mathbf{b}c_0 + \mathbf{a}^T\mathbf{a}2m + 0.$$

This derivative is also called the *partial derivative of $F(m,c)$ with respect to the
m variable* and is written as $\frac{\partial F}{\partial m}$.

Fix $m = m_0$ and consider $F(m, c)$ as a function of c. Define the $g(c)$

$$g(c) \equiv F(m_0, c) = \mathbf{d}^T\mathbf{d} - 2\mathbf{d}^T\mathbf{a}m_0 - 2\mathbf{d}^T\mathbf{b}c + 2\mathbf{a}^T\mathbf{b}m_0c + \mathbf{a}^T\mathbf{a}m_0^2 + \mathbf{b}^T\mathbf{b}c^2.$$

Compute the first derivative with respect to c and set it equal to zero

$$\frac{dg}{dc} = 0 = 0 - 0 - 2\mathbf{d}^T\mathbf{b} + 2\mathbf{a}^T\mathbf{b}m_0 + \mathbf{b}^T\mathbf{b}2c + 0.$$

This derivative is also called the *partial derivative of $F(m,c)$ with respect to the c variable* and is written as $\frac{\partial F}{\partial c}$.

These two equations can be written in vector form

$$\begin{bmatrix} \mathbf{a}^T\mathbf{a} & \mathbf{a}^T\mathbf{b} \\ \mathbf{b}^T\mathbf{a} & \mathbf{b}^T\mathbf{b} \end{bmatrix} \begin{bmatrix} m_0 \\ c_0 \end{bmatrix} = \begin{bmatrix} \mathbf{a}^T\mathbf{d} \\ \mathbf{b}^T\mathbf{d} \end{bmatrix}.$$

Note they are the same as in equations (4.1.2), but now the length of the vectors is not restricted to $n = 3$! The matrix equation can also be written as

$$A^T A\mathbf{x} = A^T\mathbf{d} \text{ where}$$
$$A = [\mathbf{a} \ \mathbf{b}], \ A^T = \begin{bmatrix} \mathbf{a}^T \\ \mathbf{b}^T \end{bmatrix} \text{ and } \mathbf{x} = \begin{bmatrix} m_0 \\ c_0 \end{bmatrix}.$$

This is called the *normal equations* corresponding to the least squares problem $A\mathbf{x} = \mathbf{d}$.

Return to the sales prediction problem in Example 4.1.1 where

$$\mathbf{a} = \begin{bmatrix} 1 & 2 & 3 & 4 & 5 & 6 \end{bmatrix}^T$$
$$\mathbf{b} = \begin{bmatrix} 1 & 1 & 1 & 1 & 1 & 1 \end{bmatrix}^T \text{ and}$$
$$\mathbf{d} = \begin{bmatrix} 78 & 85 & 90 & 96 & 104 & 113 \end{bmatrix}^T,$$

giving $\mathbf{a}^T\mathbf{a} = 91$, $\mathbf{a}^T\mathbf{b} = 21 = \mathbf{b}^T\mathbf{a}$, $\mathbf{b}^T\mathbf{b} = 6$, $\mathbf{a}^T\mathbf{d} = 2100$ and $\mathbf{b}^T\mathbf{d} = 566$. The normal equation is

$$\begin{bmatrix} 91 & 21 \\ 21 & 6 \end{bmatrix} \begin{bmatrix} m_0 \\ c_0 \end{bmatrix} = \begin{bmatrix} 2100 \\ 566 \end{bmatrix},$$

and the solution is easily computed to be $m_0 = 34/5 = 6.8000$ and $c_0 = 1058/15 \approx 70.5333$ with $F(m_0, c_0) \approx 8.1333$. These numbers are consistent with the estimates given by the graphical approach in Figure 4.1.2. Now use the equation $y = m_0 t + c_0$ to predict any future sales. For example, at the month $t = 15$ the predicted sales is

$$y(15) = m_0 15 + c_0 = 6.8000(15) + 70.5333 \approx 172.$$

The data, straight line model $y = m_0 t + c_0$ and prediction are illustrated in Figure 4.1.1, which was generated by the MATLAB code market.m.

4.1.4 Application to Radioactive Decay

The model for the amount of radioactive material, $u(t)$, is

$$\frac{du}{dt} = -ru \text{ where}$$
$$r \text{ is the decay rate and } u(0) \text{ is the initial amount.}$$

Table 4.1.3: Radioactive Decay Data

Times	Amounts
10	101
11	95
12	90
13	86.5
14	82
15	79
16	78.2

The solution of this simple differential equation is $u(t) = u(0)e^{-rt}$. If both r and $u(0)$ are known, then it is easy to make predictions about the amount of radioactive material by simply evaluating this function at the desired time.

Often these constants are not known! By making observations at different times one can formulate a least squares problem for r and $u(0)$. In order to convert this to a linear model, compute the natural log of both sides of the model to get

$$\ln(u(t)) = \ln(u(0)e^{-rt}) = \ln(u(0)) + \ln(e^{-rt}) = \ln(u(0)) + (-r)t.$$

Use the linear model $y = mt + c$ with $y = \ln(u(t))$, $m = -r$ and $c = \ln(u(0))$.

Consider following data taken at times $\mathbf{t} = \begin{bmatrix} 10 & 11 & 12 & 13 & 14 & 15 & 16 \end{bmatrix}^T$ and measured amounts $\mathbf{u} = \begin{bmatrix} 101 & 95 & 90 & 86.5 & 82 & 79 & 78.2 \end{bmatrix}^T$, see Table 4.1.3. There $n = 7$ algebraic equations give the system

$$\left\{ \begin{array}{c} m10 + c = \ln(101) \\ m11 + c = \ln(95) \\ \vdots \\ m16 + c = \ln(78.2). \end{array} \right.$$

The equivalent 7×2 matrix equation is

$$\begin{bmatrix} 10 & 1 \\ 11 & 1 \\ \vdots & \vdots \\ 16 & 1 \end{bmatrix} \begin{bmatrix} m \\ c \end{bmatrix} = \begin{bmatrix} \ln(101) \\ \ln(95) \\ \vdots \\ \ln(78.2) \end{bmatrix}.$$

The normal equations are solved, see the MATLAB code decay.m, to give $m = -0.0439$ and $c = 5.0372$. This implies $r = -m = 0.0439$ and $u(0) = e^{\ln(u(0))} = e^c = e^{5.0372} = 154.0352$ giving

$$u(t) = u(0)e^{-rt} = 154.0352e^{-0.0439t}.$$

The predicted amount at time $t = 60$ is 11.5172. The data and exponential model are illustrated in Figure 4.1.3

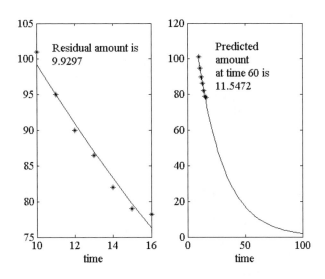

Figure 4.1.3: Radioactive Decay

4.1.5 Least Squares Computations Using MATLAB

Consider Example 4.1.1 with the sales data in Table 4.1.1.

```
>> t = [1 2 3 4 5 6]';
>> sales = [ 78 85 90 96 104 113 ]';
>> a = t;
>> b = ones(6,1);
>> d = sales;
>> A=[a b];              % 6x2 matrix
>> a'*a                  % compute the components
        ans =
            91
>> a'*b
        ans =
            21
>> b'*a
        ans =
            21
>> b'*b
        ans =
            6
>> a'*d
        ans =
            2100
```

```
>> b'*d
        ans =
              566
>> A'*A                   % matrix product for normal equations
        ans =
              91 21
              21 6
>> A'*d
        ans =
              2100
              566
>> lssol=(A'*A)\A'*d      % solution via Gauss elimination
        lssol =
              6.8000
              70.5333
>> lssol2 = A\d           % better method...uses QR factors
        lssol2 =
              6.8000
              70.5333
```

The reader should examine the MATLAB codes market.m and decay.m for additional least squares computations.

4.1.6   Exercises

1.   Consider the least squares solution of the system

$$\begin{cases} 1m + c = 1 \\ 2m + c = 3 \\ 3m + c = 7. \end{cases}$$

(a).   Find the least squares function in equation (4.1.1).
(b).   Solve the system in equation (4.1.2).
(c).   Compute $y = mt + c$ at time $t = 5$.

2.   Consider the price of high definition television after four months of price watching

Months	HDTV Prices
1	2000
2	1950
3	1910
4	1875

(a).   Find the least squares function in equation (4.1.1).
(b).   Solve the system in equation (4.1.2).
(c).   Compute $y = mt + c$ at time $t = 10$ months.

3.   Consider the computer sales problem in Example 4.1.1 after two additional months of sales: month 7 there were 120 computers sold and month 8 there were 131 computers sold. Consider all 8 data points, see Table 4.1.1.
    (a).   Find the least squares function in equation (4.1.1).
    (b).   Solve the system in equation (4.1.2).
    (c).   Compute $y = mt + c$ at time $t = 15$ months.

4.   Consider tuition, fees and books (lumped as costs) at University Expensive. These have risen over the past four years and one would like to predict what they will be four years from now. Suppose

Years	Costs
1	11,300
2	11,900
3	12,700
4	13,400

    (a).   Find the least squares function in equation (4.1.1).
    (b).   Solve the system in equation (4.1.2).
    (c).   Compute $y = mt + c$ at time $t = 8$ years.

5.   Consider the radioactive decay problem, see Table 4.1.3. Suppose there was a data entry error at time $t = 13$ where the measured amount should have been 85.6 and not 86.5.
    (a).   Find the least squares solution.
    (b).   Predict the amount at time $t = 60$.

6.   Consider a population such as fish in a large lake, which for a limited time interval, can be modeled by $P' = (b - d)P$ where $P(t)$ is the population size at time t, b is the birth rate and d is the death rate. The solution of this differential equation is

$$P(t) = P(0)e^{(b-d)t}.$$

These rates and the initial population are usually not known and must be determined from observations. Use the radioactive decay model as a template and the following population data

Years	Population Estimates
1	1000
2	1300
3	1750
4	2400

    (a).   Solve the system in equation (4.1.2).
    (b).   Compute $P(t)$ at time $t = 8$ years.

7.   Use MATLAB to confirm any calculations done in exercises 1-6.

8.   Use the MATLAB codes market.m and ls_graph.m to examine sensitively of the least squares solution when there is a 10% error or uncertainty in the data for Example 4.1.1.

9.   Use the MATLAB codes decay.m to examine sensitively of the least squares solution when there is a 5% error or uncertainty in the data for the radioactive decay problem.

4.2   Normal Equations

The normal equations will be developed for the general least squares solution of $Ax = d$ where A is $m \times n$ and $m > n$ (more rows or equations than columns or unknowns). In order to achieve this, the transpose of a matrix and its properties will be studied. The normal equations $A^T Ax = A^T d$ will be shown to solve the least squares problem, provided $A^T A$ has an inverse. These concepts will be related to the calculus and geometric approaches that were discussed in previous sections. An application to population predictions, based on past population data, will be discussed.

4.2.1   Transpose Matrix

The transpose of a single $m \times 1$ column vector $\mathbf{a}$ is a $1 \times m$ row vector, which we have been denoting by $\mathbf{a}^T = [a_1 \ a_2 \cdots a_m]$. Likewise, the transpose of a $1 \times m$ row vector will be $m \times 1$ column vector. As in the past, we will continue to assume lower case bold font means a column vector. The transpose operation may be applied to matrices.

Definition 4.2.1. Let $A = [\mathbf{a}_1 \ \mathbf{a}_2 \cdots \mathbf{a}_n]$ be an $m \times n$ matrix whose j column $m \times 1$ vectors are $\mathbf{a}_j$. The *transpose of A, A^T*, is an $n \times m$ matrix whose j row vectors are the transpose of the j column vectors of A, $\mathbf{a}_j^T$,

$$A^T \equiv \begin{bmatrix} \mathbf{a}_1^T \\ \mathbf{a}_2^T \\ \vdots \\ \mathbf{a}_n^T \end{bmatrix}.$$

Another way of defining the transpose is $A^T \equiv [a_{ji}]$ where $A = [a_{ij}]$, $1 \le i \le m$ and $1 \le j \le n$.

Example 4.2.1. Consider the following matrices

$$\mathbf{x} = \begin{bmatrix} 3 \\ 4 \end{bmatrix}, \ A = \begin{bmatrix} 2 & 7 \\ 8 & -1 \end{bmatrix} \text{ and } B = \begin{bmatrix} 1 & 3 & 4 \\ 7 & 8 & 9 \end{bmatrix}.$$

The transpose of these are

$$\mathbf{x}^T = \begin{bmatrix} 3 & 4 \end{bmatrix}, \ A^T = \begin{bmatrix} 2 & 8 \\ 7 & -1 \end{bmatrix} \text{ and } B^T = \begin{bmatrix} 1 & 7 \\ 3 & 8 \\ 4 & 9 \end{bmatrix}.$$

Note the following identity

$$
\begin{aligned}
(A\mathbf{x})^T &= (\begin{bmatrix} 2 & 7 \\ 8 & -1 \end{bmatrix} \begin{bmatrix} 3 \\ 4 \end{bmatrix})^T \\
&= \begin{bmatrix} 34 \\ 20 \end{bmatrix}^T = \begin{bmatrix} 34 & 20 \end{bmatrix} \\
&= \begin{bmatrix} 3 & 4 \end{bmatrix} \begin{bmatrix} 2 & 8 \\ 7 & -1 \end{bmatrix} \\
&= \mathbf{x}^T A^T.
\end{aligned}
$$

The transpose of a matrix-vector product may also be extended to a matrix-matrix product

$$
AB = \begin{bmatrix} 2 & 7 \\ 8 & -1 \end{bmatrix} \begin{bmatrix} 1 & 3 & 4 \\ 7 & 8 & 9 \end{bmatrix} = \begin{bmatrix} 51 & 62 & 71 \\ 1 & 16 & 23 \end{bmatrix}
$$

$$
\begin{aligned}
(AB)^T &= \begin{bmatrix} 51 & 1 \\ 62 & 16 \\ 71 & 23 \end{bmatrix} = \begin{bmatrix} 1 & 7 \\ 3 & 8 \\ 4 & 9 \end{bmatrix} \begin{bmatrix} 2 & 8 \\ 7 & -1 \end{bmatrix} \\
&= B^T A^T.
\end{aligned}
$$

Theorem 4.2.1 *(Properties of Transpose)* *Let A and B be $m \times n$ matrices, C be an $n \times p$ matrix, $\mathbf{x}$ and $\mathbf{y}$ be $n \times 1$ column vectors and s be a real number.*

1. *$(sA)^T = s(A^T)$ and $(A+B)^T = A^T + B^T$.*
2. *$\mathbf{x}^T\mathbf{y} = \mathbf{y}^T\mathbf{x}$, $(A\mathbf{x})^T = \mathbf{x}^T A^T$ and $(AC)^T = C^T A^T$.*
3. *If A is $m \times m$ and has an inverse, then $(A^{-1})^T = (A^T)^{-1}$.*

The proof of $(A\mathbf{x})^T = \mathbf{x}^T A^T$ follows from the column version of a matrix-vector product, see Theorem 3.2.1,

$$
\begin{aligned}
(A\mathbf{x})^T &= (x_1\mathbf{a}_1 + x_2\mathbf{a}_2 + \cdots + x_n\mathbf{a}_n)^T \\
&= (x_1\mathbf{a}_1)^T + (x_2\mathbf{a}_2)^T + \cdots + (x_n\mathbf{a}_n)^T \\
&= x_1\mathbf{a}_1^T + x_2\mathbf{a}_2^T + \cdots + x_n\mathbf{a}_n^T \\
&= \begin{bmatrix} x_1 & x_2 & \cdots & x_n \end{bmatrix} \begin{bmatrix} \mathbf{a}_1^T \\ \mathbf{a}_2^T \\ \vdots \\ \mathbf{a}_m^T \end{bmatrix} \\
&= \mathbf{x}^T A^T.
\end{aligned}
$$

The proofs of the other properties are also direct consequences of the definitions.

4.2.2 Normal Equations and Least Squares

If the exact solution of $A\mathbf{x} = \mathbf{d}$ cannot be found, then perhaps one can find the $\mathbf{x}$ such that the residual $\mathbf{r}(\mathbf{x}) \equiv \mathbf{d} - A\mathbf{x}$ is as "small" as possible. As in the previous section we shall define "small" in terms of the least squares function $F(\mathbf{x}) \equiv \mathbf{r}(\mathbf{x})^T \mathbf{r}(\mathbf{x})$, which is a function from n-dimensional space into the real numbers, that is, $F : \mathbb{R}^n \longrightarrow \mathbb{R}$.

Definition 4.2.2. Consider $A\mathbf{x} = \mathbf{d}$ where A is $m \times n$ and $m > n$. The *least squares solution* is an $n \times 1$ column vectors $\mathbf{x}$ such that the least squares function evaluated at this $\mathbf{x}$ is less than or equal to all other evaluations of the least squares function

$$F(\mathbf{x}) \leq F(\mathbf{y}) \text{ for all } n \times 1 \text{ column vectors } \mathbf{y}.$$

In the previous section with two variables it was not too complicated to solve the minimization problem. For larger number of columns $n > 2$, it is useful to take advantage of the properties of the transpose matrix to derive the normal equations. Let $\mathbf{y} = \mathbf{x} + (\mathbf{y} - \mathbf{x})$ where $\mathbf{x}$ is the least squares solution. Then

$$
\begin{aligned}
\mathbf{r}(\mathbf{y}) &= \mathbf{d} - A\mathbf{y} \\
&= \mathbf{d} - A(\mathbf{x} + (\mathbf{y} - \mathbf{x})) \\
&= \mathbf{d} - A\mathbf{x} - A(\mathbf{y} - \mathbf{x}) \\
&= \mathbf{r}(\mathbf{x}) - A(\mathbf{y} - \mathbf{x}).
\end{aligned}
$$

Put this into the least squares function and use $(A(\mathbf{y} - \mathbf{x}))^T (A(\mathbf{y} - \mathbf{x})) \geq 0$

$$
\begin{aligned}
F(\mathbf{y}) &= \mathbf{r}(\mathbf{y})^T \mathbf{r}(\mathbf{y}) \\
&= (\mathbf{r}(\mathbf{x}) - A(\mathbf{y} - \mathbf{x}))^T (\mathbf{r}(\mathbf{x}) - A(\mathbf{y} - \mathbf{x})) \\
&= (\mathbf{r}(\mathbf{x})^T - (A(\mathbf{y} - \mathbf{x}))^T)(\mathbf{r}(\mathbf{x}) - A(\mathbf{y} - \mathbf{x})) \\
&= \mathbf{r}(\mathbf{x})^T \mathbf{r}(\mathbf{x}) - 2(A(\mathbf{y} - \mathbf{x}))^T \mathbf{r}(\mathbf{x}) + (A(\mathbf{y} - \mathbf{x}))^T (A(\mathbf{y} - \mathbf{x})) \\
&\geq F(\mathbf{x}) - 2(\mathbf{y} - \mathbf{x})^T A^T \mathbf{r}(\mathbf{x}) + 0. \quad\quad (4.2.1)
\end{aligned}
$$

If $A^T \mathbf{r}(\mathbf{x}) = \mathbf{0}$, then for all $\mathbf{y}$ $F(\mathbf{y}) \geq F(\mathbf{x})$. The vector equation $A^T \mathbf{r}(\mathbf{x}) = \mathbf{0}$ can be written as

$$
\begin{aligned}
A^T \mathbf{r}(\mathbf{x}) &= \mathbf{0} \\
A^T (\mathbf{d} - A\mathbf{x}) &= \mathbf{0} \\
A^T A\mathbf{x} &= A^T \mathbf{d}.
\end{aligned}
$$

If the $n \times n$ matrix $A^T A$ has an inverse, then the solution of this equation will also be the solution of the least squares problem.

Definition 4.2.3. Consider $A\mathbf{x} = \mathbf{d}$ where A is $m \times n$ and $m > n$. The *normal equation* uses the $n \times n$ matrix $A^T A$ and is

$$A^T A\mathbf{x} = A^T \mathbf{d}.$$

In order to ensure the $n \times n$ matrix $A^T A$ has an inverse, see Theorem 3.7.3, we must show $A^T A\mathbf{z} = \mathbf{0}$ implies $\mathbf{z} = \mathbf{0}$. This leads to important concept of linearly independent vectors, which will be more completely studied in Section 5.1.

Definition 4.2.4. Consider n $m \times 1$ column vectors. They are called *linearly independent* if and only if no column vector is a linear combination of the other column vectors. This is equivalent to $A\mathbf{z} = \mathbf{0}$ implies $\mathbf{z} = \mathbf{0}$ where A is the $m \times n$ matrix formed by the column vectors.

Example 4.2.2. Consider the HDTV price data problem in Subsection 2.4.2 where $x_1 = m$, $x_2 = c$ and

$$\begin{bmatrix} 1 & 1 \\ 2 & 1 \\ 3 & 1 \end{bmatrix} \begin{bmatrix} x_1 \\ x_2 \end{bmatrix} = \begin{bmatrix} 2000 \\ 1950 \\ 1910 \end{bmatrix}.$$

The normal equations are

$$\begin{bmatrix} 1 & 2 & 3 \\ 1 & 1 & 1 \end{bmatrix} \begin{bmatrix} 1 & 1 \\ 2 & 1 \\ 3 & 1 \end{bmatrix} \begin{bmatrix} x_1 \\ x_2 \end{bmatrix} = \begin{bmatrix} 1 & 2 & 3 \\ 1 & 1 & 1 \end{bmatrix} \begin{bmatrix} 2000 \\ 1950 \\ 1910 \end{bmatrix}$$

$$\begin{bmatrix} 14 & 6 \\ 6 & 2 \end{bmatrix} \begin{bmatrix} x_1 \\ x_2 \end{bmatrix} = \begin{bmatrix} 11630 \\ 5860 \end{bmatrix},$$

and the solution is $x_1 = -45.00$ and $x_2 \approx 1863.33$. The column vectors in the 3×2 matrix are linearly independent because $A\mathbf{z} = \mathbf{0}$ implies $\mathbf{z} = \mathbf{0}$ as follows from

$$A\mathbf{z} = \mathbf{0}$$

$$\begin{bmatrix} 1 & 1 \\ 2 & 1 \\ 3 & 1 \end{bmatrix} \begin{bmatrix} z_1 \\ z_2 \end{bmatrix} = \begin{bmatrix} 0 \\ 0 \\ 0 \end{bmatrix} \quad \text{or equivalently}$$

$$z_1 + z_2 = 0$$
$$2z_1 + z_2 = 0 \text{ and}$$
$$3z_1 + z_2 = 0.$$

The only way the three equations can hold is if $z_1 = z_2 = 0$, that is, $\mathbf{z} = \mathbf{0}$.

In order to show $A^T A$ has an inverse, suppose $A^T A\mathbf{z} = \mathbf{0}$ and show $\mathbf{z} = \mathbf{0}$. Use the associative and the transpose properties

$$\mathbf{z}^T (A^T A\mathbf{z}) = \mathbf{z}^T \mathbf{0}$$
$$(\mathbf{z}^T A^T) A\mathbf{z} = 0$$
$$(A\mathbf{z})^T (A\mathbf{z}) = 0.$$

This means $A\mathbf{z}$ must be the zero vector. Since $A\mathbf{z}$ is a linear combination of the columns of A, it is sufficient to require that no column of A can be written as a linear combination of the other columns. This proves the next theorem.

Theorem 4.2.2 *(Normal Equation Solution) Consider $A\mathbf{x} = \mathbf{d}$ where A is $m \times n$ and $m > n$. If the columns of A are linearly independent, then $A^T A$ has an inverse and the normal equation has a unique solution, which is also the least squares solution of $A\mathbf{x} = \mathbf{d}$.*

One difficulty in solving the normal equations is that they are often "ill-conditioned" problems. This means small variations in the data or numerical approximations can result in large deviations from the exact solution. For such problems, other methods that use the QR factorization of the matrix should be used, see [5]. The MATLAB command A\d is an implementation of these more robust methods.

4.2.3   Normal Equations and Calculus

The minimization of the least squares function $F(\mathbf{x}) = \mathbf{r}(\mathbf{x})^T \mathbf{r}(\mathbf{x})$ has n independent variables in the vector $\mathbf{x}$. Because of the special structure of the least squares function, this can be simplified to a function of a single real variable t. Replace $\mathbf{x}$ by $\mathbf{x}+t\mathbf{y}$ where $\mathbf{x}$ is the least square solution and define

$$f(t) \equiv \mathbf{r}(\mathbf{x} + t\mathbf{y})^T \mathbf{r}(\mathbf{x}+t\mathbf{y}).$$

This function must have a minimum at $t = 0$ and is a quadratic function of t (again use the properties of the transpose)

$$f(t) = \mathbf{r}(\mathbf{x})^T \mathbf{r}(\mathbf{x}) - 2t(A\mathbf{y})^T \mathbf{r}(\mathbf{x}) + t^2 (A\mathbf{y})^T (A\mathbf{y}). \qquad (4.2.2)$$

The first and second derivatives are

$$\frac{df}{dt}(t) = 0 - 2(A\mathbf{y})^T \mathbf{r}(\mathbf{x}) + 2t(A\mathbf{y})^T (A\mathbf{y}) \text{ and } \frac{d^2 f}{dt^2}(t) = 2(A\mathbf{y})^T (A\mathbf{y}).$$

If the columns of A are linearly independent and $\mathbf{y} \neq \mathbf{0}$, then $A\mathbf{y} \neq \mathbf{0}$. At $t = 0$ the first derivative must be zero and the second derivative must be positive

$$\frac{df}{dt}(0) = 0 = -2(A\mathbf{y})^T \mathbf{r}(\mathbf{x}).$$

Choose $\mathbf{y}$ to be the j unit column vector so that $A\mathbf{y} = A\mathbf{e}_j = \mathbf{a}_{:j}$ is the j column vector of A. Thus, for all columns of A

$$\mathbf{a}_{:j}^T \mathbf{r}(\mathbf{x}) = 0,$$

which means the normal equations must hold because all rows of the transpose matrix times the residual vector are zero.

In Subsection 2.4.4 a geometric interpretation was given for the special case $m = 3$ and $n = 2$. The least squares solution is the point on the plane, given by the linear combination of the two column vectors, that is closest to the point associated with the vector $\mathbf{d}$. The column vectors must be perpendicular to the residual vector, and these two resulting equations are just the normal equations.

When $m > 3$ or $n > 2$, one can generalize this approach. The plane will now be replaced by the *range of the matrix*, which is the linear combination of n $m \times 1$ column vectors and can be written as $A\mathbf{y}$. The concept of *distance* between two vectors $\mathbf{d}$ and $A\mathbf{y}$ in m-dimensional space will be

$$((\mathbf{d} - A\mathbf{y})^T (\mathbf{d} - A\mathbf{y}))^{1/2}$$

and, hence, the distance squared between $\mathbf{d}$ and $A\mathbf{y}$ is $\mathbf{r}(\mathbf{y})^T \mathbf{r}(\mathbf{y})$. The least squares solution is the point in the range of A that is closest to $\mathbf{d}$. Define *perpendicular or orthogonal* vectors $\mathbf{a}$ and $\mathbf{r}$ in m-dimensional space to mean $\mathbf{a}^T \mathbf{r} = 0$. The normal equations now simply require the columns of A to be orthogonal to the residual vector.

4.2.4 Application to World Population Modeling

Consider the world population data in Table 4.1.2, which was initially presented in Example 4.1 2 of the previous section. There were 11 data points from 1900 to 2000 in decade increments, and examination of the data indicated an increasing and concave upward population estimates. The objective is to find a quadratic curve $y = at^2 + bt + c$ so that the data will be "close" to this curve. For a limited time interval this function can be used to make predictions about the world population based on just the past observations. This is perhaps one of the most primitive population models and ignores changing birth and death rates and variable population densities.

This model is the least square problem with $m = 11$, $n = 3$ and $x_1 = a$, $x_2 = b$ and $x_3 = c$. The 11×3 matrix and 11 equations that correspond to the data (where time equals zero is 1900) is derived from the equations

$$at_i^2 + bt_i + c = \text{population estimates at times } t_i.$$

The $m = 11$ equations can be written in vector form

$$
\begin{bmatrix}
0 & 0 & 1 \\
10^2 & 10 & 1 \\
20^2 & 20 & 1 \\
30^2 & 30 & 1 \\
40^2 & 40 & 1 \\
50^2 & 50 & 1 \\
60^2 & 60 & 1 \\
70^2 & 70 & 1 \\
80^2 & 80 & 1 \\
90^2 & 90 & 1 \\
100^2 & 100 & 1
\end{bmatrix}
\begin{bmatrix}
a \\
b \\
c
\end{bmatrix}
=
\begin{bmatrix}
1.65 \\
1.75 \\
1.86 \\
2.07 \\
2.30 \\
2.52 \\
3.02 \\
3.70 \\
4.44 \\
5.27 \\
6.06
\end{bmatrix}.
$$

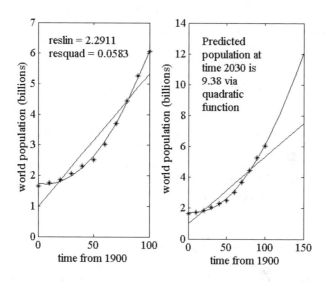

Figure 4.2.1:   World Population Prediction

The normal equation requires one to solve the 3×3 algebraic system

$$A^T A\mathbf{x} \;=\; A^T\mathbf{d}$$

$$10^4 \begin{bmatrix} 25333.00 & 302.50 & 3.85 \\ 302.50 & 3.85 & .0550 \\ 3.85 & .0550 & .0011 \end{bmatrix} \begin{bmatrix} a \\ b \\ c \end{bmatrix} = 10^4 \begin{bmatrix} 17.347 \\ .221 \\ .003 \end{bmatrix}$$

whose solution is $a = 0.0005$, $b = -0.0075$ and $c = 1.7393$.

Figure 4.2.1 illustrates the data and a linear and quadratic least squares approximation, which was done using the MATLAB code pop.m. The left graph in Figure 4.2.1 clearly indicates the data are more quadratic than linear. Once the two least squares problems have been solved, the two least squares functions are evaluated and are $reslin = 2.2911$ and $resquad = .0583$, which confirms our visual observations. The right graph in Figure 4.2.1 extends the time axis from 100 (year 2000) to 150 (year 2050) so that any predictions can be made. For example, the predicted population at $t = 130$ (year 2030) based on the quadratic least squares function is

$$\begin{aligned} y \;&=\; at^2 + bt + c \\ &=\; (0.0005)130^2 + (-0.0075)\,130 + 1.7393 \approx 9.38. \end{aligned}$$

4.2.5   Least Squares Using MATLAB

The following is the MATLAB code pop.m used to generate Figure 4.2.1. It uses the MATLAB command A\d to accurately solve the least squares problem for the linear and quadratic fit to the world population data.

MATLAB code pop.m

```
% This code finds the linear and quadratic least squares curve
% fit to the world population data from 1900 to 2000.
% Time data where 1900 starts at 0 in increments of 10.
t = [0 10 20 30 40 50 60 70 80 90 100]';
% Population data in billions.
popw = [1.65 1.75 1.86 2.07 2.30 2.52 3.02 3.70 4.44 5.27 6.06]';
one = ones(11,1);
Alin = [t one];
lin= Alin\popw          % Solves normal equation for linear fit.
poplin = lin(1)*t + lin(2);
Aquad = [t.^2 t one];
quad = Aquad\popw       % Solves normal equation for quadratic fit.
popquad = quad(1)*t.^2 + quad(2)*t + quad(3);
reslin = (popw - poplin)'*(popw - poplin)
resquad = (popw - popquad)'*(popw - popquad)
subplot(1,2,1)
plot(t,popw,'*', t,poplin, t,popquad)
newt = 0:10:150';
poplin = lin(1)*newt + lin(2);
popquad = quad(1)*newt.^2 + quad(2)*newt + quad(3);
subplot(1,2,2)
plot(t,popw,'*', newt,poplin, newt,popquad)
```

The quadratic function can be evaluated at $t = 130$ by executing quad(1)*130^2 + quad(2)*130 + quad(3) where the 3×1 vector quad is the solution of the normal equations given by quad = Aquad\popw. The less robust method for solving the normal equations is the Gauss elimination method that is implemented by the MATLAB command (Aquad'*Aquad)\(Aquad'*popw).

4.2.6   Exercises

1.   Consider the following matrices

$$\mathbf{x} = \begin{bmatrix} 3 \\ 4 \end{bmatrix}, \ A = \begin{bmatrix} 2 & 0 & 1 \\ 7 & -1 & 2 \end{bmatrix} \text{ and } B = \begin{bmatrix} 1 & 3 & 4 \\ 1 & 0 & 5 \\ 3 & -1 & 2 \end{bmatrix}.$$

(a).   Find $\mathbf{x}^T$, A^T and B^T.

(b).   Verify $(AB)^T = B^T A^T$.

2.   Let $A = \begin{bmatrix} 1 & 2 & 1 \\ 1 & 3 & 2 \\ 1 & 0 & 1 \end{bmatrix}$.

(a).   Find A^T and A^{-1}.

(b).   Verify $(A^T)^{-1} = (A^{-1})^T$.

3.   Prove: $(A^T)^{-1} = (A^{-1})^T$ when the $n \times n$ matrix A has an inverse.

4.   Use the properties of the transpose.

   (a).   Justify each step in the derivation of the inequality in line (4.2.1).

   (b).   Derive equation (4.2.2).

5.   Consider the four data points in the HDTV price problem and the resulting linear least squares problem with $x_1 = m$ and $x_2 = c$

$$\begin{bmatrix} 1 & 1 \\ 2 & 1 \\ 3 & 1 \\ 4 & 1 \end{bmatrix} \begin{bmatrix} x_1 \\ x_2 \end{bmatrix} = \begin{bmatrix} 2000 \\ 1950 \\ 1910 \\ 1875 \end{bmatrix}.$$

   (a).   Find the normal equations.

   (b).   Show the columns in the 4×2 matrix are linearly independent.

   (c).   Solve the normal equations.

   (d).   Use the above to predict the HDTV price at time $t = 8$.

6.   Consider the first four data points in the world population problem and the resulting quadratic least squares problem with $x_1 = a$, $x_2 = b$ and $x_3 = c$

$$\begin{bmatrix} 0 & 0 & 1 \\ 100 & 10 & 1 \\ 400 & 20 & 1 \\ 900 & 30 & 1 \end{bmatrix} \begin{bmatrix} x_1 \\ x_2 \\ x_3 \end{bmatrix} = \begin{bmatrix} 1.65 \\ 1.75 \\ 1.86 \\ 2.07 \end{bmatrix}.$$

   (a).   Find the normal equations.

   (b).   Show the columns in the 4×3 matrix are linearly independent.

   (c).   Solve the normal equations.

   (d).   Use the above to predict the world population at time $t = 130$, and compare with the prediction based on 11 data points.

7.   Consider the world population problem with 11 data points.

   (a).   Find the normal equations for the linear least squares problem.

   (b).   Confirm the normal equations for the quadratic least squares problem.

   (c).   Compare the predicted world population at time $t = 120$ for the linear and quadratic models.

8.   Use MATLAB to confirm any calculations in exercises 1, 2, 5, 6 or 7.

9.   Use the MATLAB code pop.m.

   (a).   Confirm the calculations and predict the population at $t = 160$.

   (b).   Use the Gauss elimination method, via (A'*A)\(A'*d), to solve the normal equations for both the linear and quadratic models.

   (c).   Suppose the population estimates for the last two data points were too low and should have been 5% higher. Solve the new normal equations and make a new prediction of the population at $t = 160$.

Table 4.3.1: Multlinear Data

i	x_i	y_i	z_i (dependent)
1	1	1	100
2	2	1	98
3	1	2	95
4	2	2	94

4.3 Multilinear Data Fitting

Data may be dependent on more than one variable. In this section the focus will be on multilinear data approximation such as $z = f(x, y) = m_1x + m_2y + c$ where the parameters for the slopes in the x and y directions and the z intercept must be found so that the data z_i is "closest" to the surface points $f(x_i, y_i)$. Problems like this are easily formulated as least squares problems whose solution will be given by the normal equations. Applications will include price data for sales levels in several markets, real estate appraisal as function of several attributes, and identification of flow rates in "mixing tank" models.

4.3.1 Multilinear in Two Variables

Consider the following four data points in Table 4.3.1 where the measured data is z_i with $i = 1, \cdots, 4$ and may depend in a linear way on the independent variables x and y. The objective is to find m_1, m_2 and c so that $z_i \approx m_1x_i + m_2y_i + c$. This means the residuals $r_i = z_i - (m_1x_i + m_2y_i + c)$ should be as small as possible. From inspection of the data points it appears the two slopes should be negative and the intercept should be positive.

In order to formulate the problem in terms of the normal equations, one must list the possible equations in the system

$$\begin{cases} m_11 + m_21 + c = 100 \\ m_12 + m_21 + c = 98 \\ m_11 + m_22 + c = 95 \\ m_12 + m_22 + c = 94. \end{cases}$$

The vector version of this, where $\mathbf{x} = [m_1 \ m_2 \ c]^T$, $\mathbf{d} = [100 \ 98 \ 95 \ 94]^T$ and A is 4×3, is

$$\begin{bmatrix} 1 & 1 & 1 \\ 2 & 1 & 1 \\ 1 & 2 & 1 \\ 2 & 2 & 1 \end{bmatrix} \begin{bmatrix} m_1 \\ m_2 \\ c \end{bmatrix} = \begin{bmatrix} 100 \\ 98 \\ 95 \\ 94 \end{bmatrix}.$$

The least squares solution is determined by the normal equations

$$A^T A \mathbf{x} = A^T \mathbf{d}$$

$$\begin{bmatrix} 1 & 2 & 1 & 2 \\ 1 & 1 & 2 & 2 \\ 1 & 1 & 1 & 1 \end{bmatrix} \begin{bmatrix} 1 & 1 & 1 \\ 2 & 1 & 1 \\ 1 & 2 & 1 \\ 2 & 2 & 1 \end{bmatrix} \begin{bmatrix} m_1 \\ m_2 \\ c \end{bmatrix} = \begin{bmatrix} 1 & 2 & 1 & 2 \\ 1 & 1 & 2 & 2 \\ 1 & 1 & 1 & 1 \end{bmatrix} \begin{bmatrix} 100 \\ 98 \\ 95 \\ 94 \end{bmatrix}$$

$$\begin{bmatrix} 6 & 9 & 6 \\ 9 & 10 & 6 \\ 6 & 6 & 4 \end{bmatrix} \begin{bmatrix} m_1 \\ m_2 \\ c \end{bmatrix} = \begin{bmatrix} 579 \\ 576 \\ 387 \end{bmatrix}.$$

The solution is $m_1 = -1.50$, $m_2 = -4.50$ and $c = 105.75$. The residual vector is $\mathbf{r} = \mathbf{d} - A\mathbf{x} = [.25 \ -.25 \ -.25 \ .25]^T$ so that $\mathbf{r}^T \mathbf{r} = 0.25$. The predicted value of z for $(x, y) = (3, 2)$ is

$$m_1 3 + m_2 2 + c = (-1.50)3 + (-4.50)2 + 105.75 = 92.25.$$

Generally, there should be a much larger set of data points than four for the three unknowns.

4.3.2   Application to Price Data in Multiple Markets

Investors are always interested in the price per unit of an item. As the number of units sold increases, usually the price per unit decreases. The price per unit in one market will be different than the price per unit in another market. In the data in Table 4.3.1 you might consider the z values to be a price per unit in either country A or country B. Let x be the units sold in country A and y be the units sold in country B. The price function for country A $f(x, y) = m_1 x + m_2 y + c$ is a function of the units sold in the two markets. The two markets have some interaction so that the number of units sold in one market does have an impact on the price in the other market. A similar price function for the other country could also be modeled.

Consider three adjacent countries A, B and C with x, y and z units sold, respectively. Table 4.3.2 has 9 data points for the price per unit for country A given by p_i where $i = 1, \cdots, 9$. The objective is to find m_1, m_2, m_3 and c so that $p_i \approx m_1 x_i + m_2 y_i + m_3 z_i + c$. The corresponding equations give the system

$$\begin{cases} m_1 2 + m_2 3 + m_3 1 + c = 20,000 \\ m_1 3 + m_2 4 + m_3 2 + c = 19,900 \\ \quad \vdots \\ m_1 27 + m_2 13 + m_3 9 + c = 16,500. \end{cases}$$

Table 4.3.2: Price Data for Three Markets

i	x_i	y_i	z_i	p_i (price for country A)
1	2	3	1	20,000
2	3	4	2	19,900
3	5	4	3	19,300
4	8	5	3	19,100
5	12	6	4	18,900
6	15	8	6	18,400
7	19	10	7	17,800
8	23	11	7	17,200
9	27	13	9	16,500

The matrix equation is $A\mathbf{x} = \mathbf{d}$ where A is 9×4 and there four unknowns $\mathbf{x} = [m_1 \ m_2 \ m_3 \ c]^T$

$$\begin{bmatrix} 2 & 3 & 1 & 1 \\ 3 & 4 & 2 & 1 \\ \vdots & \vdots & \vdots & \vdots \\ 27 & 13 & 9 & 1 \end{bmatrix} \begin{bmatrix} m_1 \\ m_2 \\ m_3 \\ c \end{bmatrix} = \begin{bmatrix} 20,000 \\ 19,900 \\ \vdots \\ 16,500 \end{bmatrix}.$$

The normal equations reduce to a 4×4 matrix solve, which can easily be done using MATLAB command A\d.

```
>> [A d]
     2  3 1 1 20000
     3  4 2 1 19900
     5  4 3 1 19300
     8  5 3 1 19100
    12  6 4 1 18900
    15  8 6 1 18400
    19 10 7 1 17800
    23 11 7 1 17200
    27 13 9 1 16500
>> A'*A
    2090 1064 722 114
    1064  556 374  64
     722  374 254  42
     114   64  42   9
>> A'*d
    2031100
    1154600
     754500
     167100
>> A\d                 % (A'*A)\(A'*d) is a less robust solver
```

1.0e+004 *
-0.00934343434343
-0.00636363636364
-0.00474747474747
2.04242424242424

This means $m_1 \approx -93.43$, $m_2 \approx -63.63$, $m_3 \approx -47.47$ and $c \approx 20,424.24$. The predicted price in country A at a sales level $(x, y, z) = (40, 16, 12)$ is

$$m_1 40 + m_2 16 + m_3 12 + c \approx 15,098.98.$$

The accuracy of the model depends on the data and the choice of the linear approximation to the data. In this case the price data is fairly large and the least squares function $\mathbf{r}^T \mathbf{r}$ should be compared with $\mathbf{d}^T \mathbf{d}$ such as in $\mathbf{r}^T \mathbf{r} / \mathbf{d}^T \mathbf{d} = 6.03$ 10^{-5}.

4.3.3 Application to Real Estate Appraisal

Consider the appraisal of a home based on its location and five attributes: age, number of bathrooms, number of bedrooms, heated square feet and general condition. Several homes in this location have recently been sold for a known price and known values of the five attributes. The home being appraised will have known values of the five attributes, but not the price. The objective is to estimate the price of the home based on the recent sales of comparable homes.

Let the independent variables be x_1 (age), x_2 (bathrooms), x_3 (bedrooms), x_4 (square feet) and x_5 (condition). Suppose there are seven comparable homes in this location, and they have the following attributes and values v given in Table 4.3.3. The model will assume the value of the property depends linearly on the values of the attributes and has the form

$$v \approx m_1 x_1 + m_2 x_2 + m_3 x_3 + m_4 x_4 + m_5 x_5 + c.$$

The equations for the seven comparable homes yield the system

$$\begin{cases} m_1 10 + m_2 2.5 + m_3 3 + m_4 2000 + m_5 9 + c = 260 \\ m_1 12 + m_2 2.0 + m_3 5 + m_4 2500 + m_5 8 + c = 270 \\ \quad \vdots \\ m_1 10 + m_2 2.0 + m_3 4 + m_4 2400 + m_5 9 + c = 280. \end{cases}$$

The corresponding vector equation has a 7×6 matrix, five unknown slopes and an unknown intercept

$$\begin{bmatrix} 10 & 2.5 & 3 & 2000 & 9 & 1 \\ 12 & 2.0 & 5 & 2500 & 8 & 1 \\ \vdots & \vdots & \vdots & \vdots & \vdots & \vdots \\ 10 & 2.0 & 4 & 2400 & 9 & 1 \end{bmatrix} \begin{bmatrix} m_1 \\ m_2 \\ m_3 \\ m_4 \\ m_5 \\ c \end{bmatrix} = \begin{bmatrix} 260 \\ 270 \\ \vdots \\ 280 \end{bmatrix}.$$

Table 4.3.3: Home Appraisal Data

i (home)	x_{i1}	x_{i2}	x_{i3}	x_{i4}	x_{i5}	v_i (thousand)
1	10	2.5	3	2000	9	260
2	12	2	5	2500	8	270
3	15	2	4	1900	6	230
4	8	3	3	1600	9	250
5	11	2.5	4	2300	7	260
6	9	2	3	2000	8	250
7	10	2	4	2400	9	280

As in the price data application, the normal equation is easy to solve using MATLAB.

>> [A d]
        1.0e+003 *
        0.0100 0.0025 0.0030 2.0000 0.0090 0.0010 0.2600
        0.0120 0.0020 0.0050 2.5000 0.0080 0.0010 0.2700
        0.0150 0.0020 0.0040 1.9000 0.0060 0.0010 0.2300
        0.0080 0.0030 0.0030 1.6000 0.0090 0.0010 0.2500
        0.0110 0.0025 0.0040 2.3000 0.0070 0.0010 0.2600
        0.0090 0.0020 0.0030 2.0000 0.0080 0.0010 0.2500
        0.0100 0.0020 0.0040 2.4000 0.0090 0.0010 0.2800
>> lssol = A\d
    lssol =
        -0.5489   %   equals m_1
        6.7198   %   equals m_2
        0.8193   %   equals m_3
        0.0425   %   equals m_4
        7.7542   %   equals m_5
        93.3308   %   equals c
>> r = d - A*lssol
>> r'*r/(d'*d)
        5.7814e-005
>> lssol(1)*9 + lssol(2)*2.5 + lssol(3)*4 +
        lssol(4)*1900 + lssol(5)*9 + lssol(6)
        259.0114   %   equals the appraised value of the home

Assume the home being appraised has the following attributes: 9 years old, 2.5 bathrooms, 4 bedrooms, 1900 heated square feet and general condition equal to 9. Once the five slopes and intercept have been computed, the home can be appraised by evaluating

$$
\begin{aligned}
\text{appraised value} \;&=\; m_1 x_1 + m_2 x_2 + m_3 x_3 + m_4 x_4 + m_5 x_5 + c \\
&=\; m_1 9 + m_2 2.5 + m_3 4 + m_4 1900 + m_5 9 + c \\
&\approx\; 259.0.
\end{aligned}
$$

Notice the first slope, m_1, is a negative number, which indicates the value of the home decreases as the age increases. The other four attributes add to the value of the home and so their slopes should be positive.

4.3.4   Application to Flow Rate Identification

In Subsection 3.7.4 the three-tank mixing of a chemical was modeled provided one knew the flow rates between the tanks. The objective was to find the steady state concentrations of the chemical in each of the tanks. A related application is to find the concentration of a chemical such as a medicine or toxin in an animal. A simplistic model is to view the three "tanks" as blood, bones or tissue. The chemical can pass in and out of these, but the flow rates and volumes may not be directly found. The following is one possible way of estimating the coefficients based on past observations of the concentrations in the three "tanks."

Consider three tanks whose concentrations are $x(t)$, $y(t)$ and $z(t)$. Suppose the model for the first tank has the form

$$\frac{dx}{dt} = ax + by + cz.$$

This is a simplification because often the coefficients must be related such as by $a = -b-c$. The coefficients a, b and c are to be found from the past observations $x(t_i)$, $y(t_i)$ and $z(t_i)$. Let $x_i = x(t_i)$, $y_i = y(t_i)$, $z_i = z(t_i)$ and approximate the derivative by centered finite differences

$$\frac{dx}{dt}(t_i) \approx \frac{x_{i+1} - x_{i-1}}{t_{i+1} - t_{i-1}} \equiv d_i.$$

Then the above differential equation gives n algebraic equations where there were $n + 2$ observations

$$ax_i + by_i + cz_i = d_i.$$

For example, consider the 10 observations recorded in Table 4.3.4. The algebraic equations give the system

$$\begin{cases} a3 + b9 + c12 = 1.5 \\ a5 + b8 + c13 = 2.0 \\ \vdots \\ a15 + b1 + c29 = 1.5. \end{cases}$$

The corresponding 8×3 vector equation $A\mathbf{x} = \mathbf{d}$ where $\mathbf{x} = \begin{bmatrix} a & b & c \end{bmatrix}^T$ is

$$\begin{bmatrix} 3 & 9 & 12 \\ 5 & 8 & 13 \\ \vdots & \vdots & \vdots \\ 15 & 1 & 29 \end{bmatrix} \begin{bmatrix} a \\ b \\ c \end{bmatrix} = \begin{bmatrix} 1.5 \\ 2.0 \\ \vdots \\ 1.5 \end{bmatrix}.$$

The least squares solution is easy to find using the MATLAB command A\d.

Table 4.3.4: Three-tank Mixing Data

$\mathbf{t}_i$	$\mathbf{x}_i$	$\mathbf{y}_i$	$\mathbf{z}_i$	$\mathbf{d}_i = \frac{x_{i+1}-x_{i-1}}{t_{i+1}-t_{i-1}}$
1	2	10	11	
2	3	9	12	1.5
3	5	8	13	2.0
4	7	7	16	2.0
5	9	5	18	2.0
6	11	4	21	1.5
7	12	3	23	1.5
8	14	2	26	1.5
9	15	1	29	1.5
10	17	1	33	

```
>> [A d]
        3.0000 9.0000 12.0000 1.5000
        5.0000 8.0000 13.0000 2.0000
        7.0000 7.0000 16.0000 2.0000
        9.0000 5.0000 18.0000 2.0000
        11.0000 4.0000 21.0000 1.5000
        12.0000 3.0000 23.0000 1.5000
        14.0000 2.0000 26.0000 1.5000
        15.0000 1.0000 29.0000 1.5000
>> lssol = A\d
    lssol =
        0.2849          % equals a
        0.2361          % equals b
        -0.1099         % equals c
>> r = d - A*lssol;
>> r'*r
        0.2313
```

The coefficients in the other two differential equations

$$\frac{dy}{dt} = \widehat{a}x + \widehat{b}y + \widehat{c}z \text{ and}$$
$$\frac{dz}{dt} = \widetilde{a}x + \widetilde{b}y + \widetilde{c}z$$

may be computed in a similar way by replacing d_i with $\widehat{d}_i$ and $\widetilde{d}_i$, respectively,

$$\frac{dy}{dt}(t_i) \approx \frac{y_{i+1} - y_{i-1}}{t_{i+1} - t_{i-1}} \equiv \widehat{d}_i \text{ and}$$
$$\frac{dz}{dt}(t_i) \approx \frac{z_{i+1} - z_{i-1}}{t_{i+1} - t_{i-1}} \equiv \widetilde{d}_i.$$

4.3.5   Exercises

1.   Consider the multilinear two variable data in Table 4.3.1.
   (a).   By-hand calculations confirm the normal equations.
   (b).   Suppose the data for $x = 2$ and $y = 2$ is 105 and not 94. Is it appropriate to model the data by $m_1 x + m_2 y + c$?
   (c).   Do the calculations with 105 and compare the new residual vector with the old residual vector.

2.   Consider the multilinear two variable data

i	x_i	y_i	z_i (dependent)
1	1	1	100
2	2	1	98
3	1	2	95
4	2	2	94
5	3	1	95
6	1	3	89
7	3	2	91
8	2	3	86
9	3	3	84

   (a).   Find 9×3 matrix A.
   (b).   Use MATLAB to solve the least squares problem.
   (c).   Use the least squares solution to predict the value of $m_1 x + m_2 y + c$ when $x = 4$ and $y = 6$.

3.   Consider the price data application in three markets as given for country A in Table 4.3.2. Suppose country B has price data $\mathbf{p} = [25000 \ \ 24200 \ \ 23900 \ \ 23200 \ \ 22600 \ \ 22100 \ \ 21500 \ \ 20500 \ \ 20100]^T$ and the other data in the table remain the same.
   (a).   Find 11×4 matrix A.
   (b).   Use MATLAB to solve the least squares problem.
   (c).   Use the least squares solution to predict the unit price in country B from $m_1 x + m_2 y + m_3 z + c$ when $x = 40$, $y = 16$ and $z = 12$.

4.   Consider the real estate appraisal data in Table 4.3.3. Suppose value of the comparable homes is $\mathbf{v} = [260 \ \ 270 \ \ 230 \ \ 250 \ \ 260 \ \ 250 \ \ 295]^T$ and the other data in the table remain the same.
   (a).   Find 7×6 matrix A.
   (b).   Use MATLAB to solve the least squares problem.
   (c).   Use the least squares solution to predict value of the same home.

5.   Consider the three-tank data in Table 4.3.4. Find the coefficients in

$$\frac{dy}{dt} = \widehat{a}x + \widehat{b}y + \widehat{c}z.$$

   (a).   Find 8×3 matrix A and the 8×1 column vector $\widehat{\mathbf{d}}$.
   (b).   Use MATLAB to solve the least squares problem.

4.4 Parameter Identification

Models evolving from differential equations have parameters, which must be approximated. An example is radioactive decay as described in Section 4.1. The model leads to $u(0)e^{-rt}$ where the parameters are the initial amount $u(0)$ and the decay rate r. We were able to convert the exponential function into linear curve fit by using the natural logarithm. However, this does not always simplify all models! An alternative is to use a search algorithm to locate the minimum of the more complicated nonlinear least squares function. The MATLAB command fminsearch() is an implementation of the Nelder-Mead search algorithm. Applications to radioactive decay, logistics population and steady state heat conduction parameter identifications will be given.

4.4.1 Search Algorithm for Minimization

Minimization problems that occur in parameter identification are from a nonlinear least squares function

$$\begin{aligned} F(p_1, p_2, \cdots, p_n) &= (d_1 - u(t_1, p_1, p_2, \cdots, p_n))^2 + \cdots \\ &\quad + (d_m - u(t_m, p_1, p_2, \cdots, p_n))^2 \end{aligned}$$

where $u(t, p_1, p_2, \cdots, p_n)$ is a solution of a differential equation and n is the number of parameters and m is the number of data points. In the radioactive decay problem there are two parameters $p_1 = u(0)$ and $p_2 = r$, and the differential equation is

$$\frac{du}{dt} = -ru \text{ with solution } u(t, p_1, p_2) = p_1 e^{-p_2 t} = u(0)e^{-rt}.$$

The *Nelder-Mead simplex method* searches for the minimum of a real-valued function $F(\mathbf{p})$ of m variables. In the case of three variables a sketchy outline of the method is as follows.

Step 1. Pick four points $\mathbf{p}^0, \mathbf{p}^1, \mathbf{p}^2$ and $\mathbf{p}^3$ to form a tetrahedron in three dimensional space.
Step 2. Reorder the points so that $F(\mathbf{p}^0) \le F(\mathbf{p}^1) \le F(\mathbf{p}^2) \le F(\mathbf{p}^3)$.
Step 3. Choose a new fourth point, *new_*$\mathbf{p}^3$, by locating it on the line containing $\mathbf{p}^3$ and $(\mathbf{p}^0 + \mathbf{p}^1 + \mathbf{p}^2)/3$.
Step 4. Repeat steps two and three until values of $F(\mathbf{p})$ in step two are "nearly" the same.

In step two $(\mathbf{p}^0 + \mathbf{p}^1 + \mathbf{p}^2)/3$ is the center of the face (triangle) in the tetrahedron with vertices $\mathbf{p}^0, \mathbf{p}^1$ and $\mathbf{p}^2$ and

$$new\_\mathbf{p}^3 = \mathbf{p}^3 + \alpha((\mathbf{p}^0 + \mathbf{p}^1 + \mathbf{p}^2)/3 - \mathbf{p}^3).$$

If $\alpha = 0$, then *new_*$\mathbf{p}^3 = \mathbf{p}^3$. If $\alpha = 1$, then *new_*$\mathbf{p}^3 = (\mathbf{p}^0 + \mathbf{p}^1 + \mathbf{p}^2)/3$. The tricky part of the algorithm is a "good" choice of α, where the objective is to

choose α so that the new set of four points is "closer" to the minimum of $F(\mathbf{p})$.
We shall rely on the implementation given by MATLAB.

The MATLAB implementation of the Nelder-Mead method has the form
fminsearch('Function', guess_ vector). 'Function' is a function file and guess_
vector is an initial guess for the parameters of the function $F(\mathbf{p})$ such that it is
a minimum. For example, use fminsearch() to find the minimum of

$$F(p_1, p_2, p_3) = 100 + (p_1 - 2)^2 + (p_2 - 3)^2 + (p_3 - 4)^2.$$

The function file and execution of fminsearch() are

```
function Fnm = Fnm(p)
    Fnm = 100 + (p(1)-2)^2 + (p(2)-3)^2 + (p(3)-4)^2;
>> [p Fval] = fminsearch('Fnm', [10 10 10])
    p =
        1.9999 3.0000 4.0000
    Fval =
        100.0000.
```

4.4.2 Application to Radioactive Decay

Return to the radioactive decay data in Section 4.1 where at times $\mathbf{t} = [10$
$11 \ 12 \ 13 \ 14 \ 15 \ 16]$ the measured amounts of a radioactive substance were
$\mathbf{u} = [101 \ 95 \ 90 \ 86.2 \ 82 \ 79 \ 78.2]$. The exponential function with unknown
initial amount and decay rate must be fit to these seven data points. So, $m = 7$,
$n = 2$ and the nonlinear least squares function is

$$F(u(0), r) = (u_1 - u(0)e^{-rt_1})^2 + \cdots$$
$$+ (u_7 - u(0)e^{-rt_7})^2.$$

The following MATLAB implementation has two function files and a code file.
The function file rad_sol.m has the exponential function and the file ls_rad.m
has the above least squares function. The code file rad_fmin.m has the call to
the MATLAB command fminsearch(), and it also lists the two function files.

MATLAB Code rad_fmin.m
```
% function r = rad_sol(x1,x2,t)
% r = x1*exp(-x2*t);
%
%function lsrad = ls_rad(xx)
% t = 10:16;
% a = [101 95 90 86.5 82 79 78.2 ];
% lsrad = 0;
% for i = 1:7
% lsrad = lsrad + (a(i)-rad_sol(xx(1),xx(2),t(i)))^2;
% end
%
```

```
format long
[xx lsval] = fminsearch('ls_rad',[120 .03])
rad_sol(xx(1),xx(2),60)
for i = 0:100
        amount(i+1) = rad_sol(xx(1),xx(2),i);
end
plot(amount)
```

The two function files must be in your directory as well as the above code file rad_fmin.m. Execution of this code file gives

```
>> rad_fmin
xx =
        1.0e+002 *
        1.55946611533598 0.00044865208356
lsval =
        9.73565029005973
rad_sol(xx(1),xx(2),60)
        10.56557646847834.
```

This means the initial amount is 155.95 and the decay rate is 0.04486. The predicted amount at time equal to 60 is 10.5655. These values are very close to the output from using the linear fit to $\ln(u(t)) = \ln(u(0)) + (-r)t$, where the initial amount is 154.0352, decay rate is 0.0439 and amount at time 60 is 11.5172.

4.4.3 Application to United States Population Data

In Section 4.2 we fit a linear and quadratic function to world population data. In this section population data from the United States will be examined. Four models of the data will be considered: linear, quadratic, exponential and logistic curves. Variation in population data and the methods for solving the least squares problem will be discussed.

The population data in Table 4.4.1, see [10], has a number of inconsistencies in how and who were included. Before 1950 the populations in Hawaii and Alaska were not counted. There has been some variation in how resident and nonresident citizens were counted. Also, note the 2000 entry, which was 275.13 million before the 2000 census and 282.12 after the census. The four models for the population are

Model	Function	Differential Equation
linear	$mt + c$	none
quadratic	$at^2 + bt + c$	none
exponential	$u(0)e^{(b-d)t}$	$\frac{du}{dt} = (b-d)u$
logistic	$\frac{MC}{C+e^{-Mct}}$	$\frac{du}{dt} = c(M-u)u$

The least square problems for the first three can be solved by the normal equations. The differential equations for the population $u(t)$ are based on the birth

Table 4.4.1: US Population Data

Year	U.S. Pop. (millions)
1900	76.09
1910	92.41
1920	106.46
1930	123.08
1940	132.12
1950	152.27
1960	180.67
1970	205.05
1980	227.22
1990	249.46
2000	275.12
2000 (after census)	282.12

and death rates, b and d. For small time intervals they may be approximated by constants and in this case we get the exponential function. In the logistic model the birth and death rates are allowed to vary linearly with the population so that $b - d = c(M - u)$. One can view M as the maximum population, and c will determine how fast the maximum population will be attained.

The parameters M and c can be determined from a least squares approximation of the data derived by using centered finite differences to approximate the derivative

$$\frac{du}{dt}(t_i)\frac{1}{u(t_i)} = c(M - u(t_i))$$

$$\frac{u(t_{i+1}) - u(t_{i-1})}{t_{i+1} - t_{i-1}}\frac{1}{u(t_i)} = cM - cu(t_i).$$

An implementation is in the MATLAB code pop_us.m where the minimum of the least squares functions is computed $reslin = 825.37$, $resquad = 108.81$, $resexpo = 278.22$ and $reslogistic = 144.08$. The least squares function for the quadratic curve fit is the smallest, but the model for larger time intervals does not have any other way to track births, deaths and emigration.

Another way of determining the constants in the logistic model is to apply the nonlinear least squares approach to the logistic function

$$u(t) = \frac{MC}{C + e^{-Mct}}.$$

There are eleven data points and three parameters $p_1 = M$, $p_2 = C$ and $p_3 = c$ so that $m = 11$ and $n = 3$. The second parameter reflects the initial population because

$$u(0) = \frac{MC}{C + 1} \text{ giving } C = \frac{u(0)}{M - u(0)}.$$

The nonlinear least squares function is

$$F(M,C,c) = (u_1 - \frac{MC}{C + e^{-Mct_1}})^2 + \cdots$$
$$+ (u_{11} - \frac{MC}{C + e^{-Mct_{11}}})^2.$$

The following MATLAB implementation has two function files and a code file. The function file log_sol.m has the logistic function and the file ls_log.m has the above least squares function. The code file pop_us_fmin.m has the call to the MATLAB command fminsearch(), and it also lists the two function files.

MATLAB Code pop_us_fmin.m

```
% function r = log_sol(x1,x2,x3,t)
% r = x1*x2/(x2+exp(-x1*x3*t));
%
%function ls = ls_log(xx)
% t = 0:10:100;
% p = [76.09 92.41 106.46 123.08 132.12 152.27 180.67
%          205.05 227.22 249.46 275.13];
% ls = 0;
% for i = 1:11
% ls = ls + (p(i)-log_sol(xx(1),xx(2),xx(3),t(i)))^2;
% end
%
format long
[xx lsval] = fminsearch('ls_log',[500 .2 .0001])
log_sol(xx(1),xx(2),xx(3),150)
x2 = 282.12*exp(-xx(1)*xx(3)*100)/(xx(1)-282.12);
log_sol(xx(1),x2,xx(3),150)
t = 0:10:100;
p = [76.09 92.41 106.46 123.08 132.12 152.27 180.67
         205.05 227.22 249.46 275.13];
for i = 0:30
    newt(i+1) = i*10+1900;
    popus(i+1) = log_sol(xx(1),x2,xx(3),i*10);
end
plot(t+1900,p,'*',newt, popus)
```

The two function files must be in your directory as well as the above code file pop_us_fmin.m and execution of this code file gives

```
>> pop_us_fmin
    xx =
           1.0e+002 *
           5.90373049898959  0.00150901875126  0.00000029836503
        lsval =
```

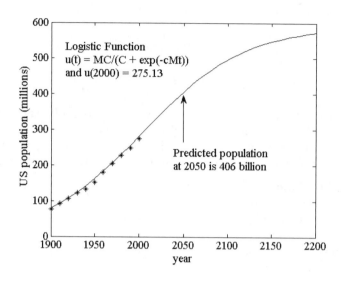

Figure 4.4.1:   US Population and Logistic Model

1.051582719560504e+002
% predicted population at 2050 using 273.13 at 2000
4.011070115341667e+002
% predicted population at 2050 using 282.12 to restart at 2000.
4.063488610657162e+002

The computed values for the parameters are $M = 590.37$, $C = 0.15092$ and $c = 0.00002983$. The least squares function evaluated at these parameters is equal to 105.04. Figure 4.4.1 is a graph of this logistic function and the data used to find above parameters.

The following table indicates different interpretations of the data depending on the data and the method used.

Method	2000 Data	M	C	c	2050 Pred.
pop_us.m	275.13	586	.1613	.00002982	404
pop_us.m	282.12	676	.1328	.00002489	422
pop_us_fmin.m	275.13	590	.1509	.00002983	406
pop_us_fmin.m	282.12	744	.1175	.00002208	433
United States?	additional	*	*	*	419

4.4.4   Application to Heat Conduction

Consider a long thin rod of length L and rectangular cross section $W \times H$. The left end is attached to a hot mass. Heat diffuses down the rod and is lost through the surface of the rod to a cooler surrounding region whose temperature

Table 4.4.2: Temperature Data

x_i	u_i (temperature)	x_i	u_i (temperature)
10	115.95	42	48.21
14	101.94	46	44.48
18	90.18	50	40.66
22	80.01	54	38.51
26	71.37	58	37.81
30	63.75	62	36.11
34	56.94	66	35.18
38	51.97		

is u_{sur}. Let the temperature be independent of time but be a function of x from the left end where $u(x)$ is the temperature. The steady state model is derived from the Fourier heat law. Let K be the thermal conductivity and c be the proportionality constant that controls heat loss from the rod to the surrounding region.

change is heat $=$ diffusion from left and right $+$ loss through the surface

$$0 = K(WH)\frac{d^2u}{dx^2} + c(2H + 2W)(u_{sur} - u)$$

$$0 = \frac{d^2u}{dx^2} + C(u_{sur} - u) \text{ where } C \equiv \frac{c(2H + 2W)}{K(WH)}.$$

The unknown parameters are C, $\frac{du}{dx}(0) = Q_1$ and $\frac{du}{dx}(L) = Q_2$.

The solution of the ordinary differential equation can be shown to have the form

$$u(x) = C_1 \sinh(C^{1/2}x) + C_2 \cosh(C^{1/2}x) + u_{sur}.$$

The constants are determined from the left and right boundary conditions $\frac{du}{dx}(0) = Q_1$ and $\frac{du}{dx}(L) = Q_2$ and are

$$C_1 = Q_1/C^{1/2} \text{ and}$$

$$C_2 = \frac{Q_2 - Q_1 \cosh(C^{1/2}L)}{C^{1/2} \sinh(C^{1/2}L)}.$$

This means the temperature $u(x, C, Q_1, Q_2)$ is a function of the three parameters as well as x.

The temperature data was taken in rod with length $L = 70$ and at points starting at $x = 10$ in increments of 4 until $x = 66$ is recorded in Table 4.4.2, which was taken at the Math Instructional and Research Laboratory [3]. The surrounding temperature was $u_{sur} = 21.47$. The nonlinear least squares function has $m = 15$ data points and $n = 3$ parameters and is

$$F(C, Q_1, Q_2) = (u_1 - u(x_1, C, Q_1, Q_2))^2 + \cdots$$
$$+ (u_{15} - u(x_{15}, C, Q_1, Q_2))^2.$$

Figure 4.4.2 was generated by the MATLAB code heat_fmin.m and illustrates the temperature data and the temperature function with the parameters

$$p_1 = C = 0.0018447, \ p_2 = Q_1 = -6.257308 \text{ and } p_3 = Q_2 = -0.00000769.$$

The value of the least squares function was 7.2725.

MATLAB Code heat_fmin.m

```
% function r = heat_sol(x1,x2,x3,t)
% r = 2147/100+x2/x1^(1/2)*sinh(x1^(1/2)*t)-
     (x2*cosh(70*x1^(1/2))-x3)/x1^(1/2)/sinh(70*x1^(1/2))
     *cosh(x1^(1/2)*t);
%
% function lsr1 = ls_heat(xx)
% x = [10 14 18 22 26 30 34 38 42 46 50 54 58 62 66];
% temp_data = [ 115.95 101.94 90.18 80.01 71.37 63.75 56.94
               51.97 48.21 44.48 40.66 38.51 37.81 36.11 35.18];
% lsr1 = 0;
% for i = 1:15
%     lsr1 = lsr1 + (temp(i)-sol1(xx(1),xx(2),xx(3),x(i)))^2;
% end
%
format long
[xx fctval] = fminsearch('ls_heat', [.001 -4.3 .0001])
for i = 1:57
    xnew(i) = i+9;
    temperature(i) = heat_sol(xx(1),xx(2),xx(3),xnew(i));
end
x = [10 14 18 22 26 30 34 38 42 46 50 54 58 62 66];
temp_data = [ 115.95 101.94 90.18 80.01 71.37 63.75 56.94
             51.97 48.21 44.48 40.66 38.51 37.81 36.11 35.18];
plot(x,temp_data,'*',xnew,temperature)
```

4.4.5   Exercises

1.   Use the MATLAB command fminsearch() to find the minimum of

$$F(p_1, p_2, p_3) = 100 + 2(p_1 - 2)^2 + 4(p_2 - 3)^2 + 10(p_3 - 4)^2.$$

2.   Use the MATLAB command fminsearch() to find some of the many minimum points of

$$F(p_1, p_2, p_3) = 100 \sin(\pi p_1) \sin(.5\pi p_2) \sin(.25\pi p_3).$$

3.   Use the MATLAB command fminsearch() to solve the radioactive decay problem in exercise 5 in Section 4.1.

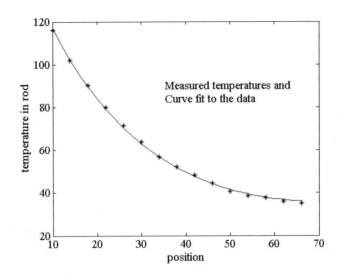

Figure 4.4.2:   Temperature Data and Curve Fit

4.   Use the MATLAB command fminsearch() to solve the population problem in exercise 6 in Section 4.1.

5.   Verify the computation for the United States population, which was done by the MATLAB code pop_us_fmin.m.

6.   Consider the world population data given in Example 4.1.2, see Table 4.1.2. Use a variation of the MATLAB code pop_us_fmin.m to find a logistic curve fit to this data, see pop_world_fmin.m.

7.   Consider the heat conduction model. Derive the equations for C_1 and C_2.

8.   Consider the heat conduction model. Suppose there was an error in collecting the temperature data, which should have been 5% less than recorded in Table 4.4.2. Find the new solution to the curve fitting problem.

Chapter 5

Ax = d: Multiple Solutions

This chapter contains methods for solving algebraic equations $A\mathbf{x} = \mathbf{d}$ where A is a matrix with more columns than rows, that is, more variables than equations. Row operations are used to find the basis of the nullspace solution set and a particular solution. Applications to circuits, trusses and fluid flow are studied in the context of equilibrium equations where the block Gauss elimination and nullspace methods are introduced.

5.1 Subspaces and Solutions in $\mathbb{R}^3$

In this section the foundation is prepared for the systematic representation of multiple solutions when there are m equations, n unknowns, n is larger than m and they are large numbers. First, we consider subspaces in $\mathbb{R}^n$ and the general form of the solutions of $A\mathbf{x} = \mathbf{d}$. Then we focus on the cases with three variables ($n = 3$) and one or two equations ($m = 1$ or 2). The bridging concepts of subspaces, spanning sets and basis will be illustrated. They will be used to represent solutions of $A\mathbf{x} = \mathbf{d}$, and an application to a two-loop circuit will be given.

5.1.1 Subspaces in $\mathbb{R}^n$

Let $\mathbf{a}$, $\mathbf{b}$ and $\mathbf{c}$ be vectors in $\mathbb{R}^n$ and let s and t be real numbers. Let $\mathbf{a} + \mathbf{b}$ and $s\mathbf{a}$ denote vector addition and scalar multiplication. Theorem 3.1.1 listed the following vector space properties for the set of all vectors in $\mathbb{R}^n$:

$\mathbf{a} + \mathbf{b} \in \mathbb{R}^n$, $\mathbf{a} + \mathbf{0} = \mathbf{a}$ where $\mathbf{0} \in \mathbb{R}^n$ has zeros as components, $\mathbf{a} + (-\mathbf{a}) = \mathbf{0}$,
$\mathbf{a} + \mathbf{b} = \mathbf{b} + \mathbf{a}$, $\mathbf{a} + (\mathbf{b} + \mathbf{c}) = (\mathbf{a} + \mathbf{b}) + \mathbf{c}$;
$s\mathbf{a} \in \mathbb{R}^n$, $1\mathbf{a} = \mathbf{a}$, $s(\mathbf{a} + \mathbf{b}) = s\mathbf{a} + s\mathbf{b}$, $(s + t)\mathbf{a} = s\mathbf{a} + t\mathbf{a}$ and $s(t\mathbf{a}) = (st)\mathbf{a}$.

Definition 5.1.1. A subset S of $\mathbb{R}^n$ is called a *vector subspace* of $\mathbb{R}^n$ if and only if for all $\mathbf{a}$, $\mathbf{b}$ and $\mathbf{c}$ in S

$\mathbf{a} + \mathbf{b} \in S$, $\mathbf{a} + \mathbf{0} = \mathbf{a}$ where $\mathbf{0} \in S$ has zeros as components, $\mathbf{a} + (-\mathbf{a}) = \mathbf{0}$

where $-\mathbf{a} \in S$, $\mathbf{a} + \mathbf{b} = \mathbf{b} + \mathbf{a}$, $\mathbf{a} + (\mathbf{b} + \mathbf{c}) = (\mathbf{a} + \mathbf{b}) + \mathbf{c}$;
$s\mathbf{a} \in S$, $1\mathbf{a} = \mathbf{a}$, $s(\mathbf{a} + \mathbf{b}) = s\mathbf{a} + s\mathbf{b}$, $(s + t)\mathbf{a} = s\mathbf{a} + t\mathbf{a}$ and $s(t\mathbf{a}) = (st)\mathbf{a}$.

The lines and planes that contain the origin are examples of subspaces in $\mathbb{R}^3$. The following characterization of a subspace in $\mathbb{R}^n$ easily follows from the vector space properties of $\mathbb{R}^n$.

Theorem 5.1.1 *(Vector Subspace) A subset S in $\mathbb{R}^n$ is subspace if and only if*

(i). *the scalar multiple of any vector in S must also be in S and*
(ii). *the sum of any two vectors in S is also in S.*

Example 5.1.1. In Section 2.3 a line in $\mathbb{R}^3$ was described by a given point and direction vector as a row vector $\vec{r} = \vec{r}_0 + t\vec{d}$ where the parameter t is a free real number. For the position of particle moving in a straight line t could represent time. It will be convenient to change to column vector notation so that a line in $\mathbb{R}^3$ is the set of vectors

$$\mathbf{x} = \mathbf{x}_0 + t\mathbf{d}$$

where $\mathbf{x}_0$ represents the given point in $\mathbb{R}^3$, t is a free real number and $\mathbf{d}$ is the given direction vector in $\mathbb{R}^3$. If the given point is the origin, then the line is called a *one-dimensional subspace* of $\mathbb{R}^3$ and

$$S = \{\mathbf{x} \in \mathbb{R}^3 : \ \mathbf{x} = t\mathbf{d} \text{ and } t \in \mathbb{R}\}.$$

The set of all points in a line, $\mathbf{x} = \mathbf{0} + t\mathbf{d} = t\mathbf{d}$, has the two properties in Theorem 5.1.1. Any scalar multiple of a point is in the line, which follows from $s\mathbf{x} = s(t\mathbf{d}) = (st)\mathbf{d}$ where s is a real number. The sum of any two points in the line is also in the line, which follows from $\mathbf{x} = t_1\mathbf{d}$ and $\mathbf{y} = t_2\mathbf{d}$ giving $\mathbf{x} + \mathbf{y} = t_1\mathbf{d} + t_2\mathbf{d} = (t_1 + t_2)\mathbf{d}$. If a line in $\mathbb{R}^3$ does not contain the origin, then it is called an *affine* subset of $\mathbb{R}^3$.

Example 5.1.2. Planes in $\mathbb{R}^3$ were described in Section 2.4 by a given point and two vectors that are not parallel as $\vec{r} = \vec{r}_0 + r\vec{a} + s\vec{b}$ where the two parameters r and s are free real numbers. The column vector notation for a plane in $\mathbb{R}^3$ is

$$\mathbf{x} = \mathbf{x}_0 + r\mathbf{a} + s\mathbf{b}$$

where the two non-parallel vectors are $\mathbf{a}$ and $\mathbf{b}$ with $\mathbf{a} \neq \mathbf{b}k$ for any real number k. If the given point is the origin, then the plane is called a *two-dimensional subspace* of $\mathbb{R}^3$ and

$$S = \{\mathbf{x} \in \mathbb{R}^3 : \ \mathbf{x} = r\mathbf{a} + s\mathbf{b} \text{ and } r, s \in \mathbb{R}\}.$$

Any point in such a plane is represented by $\mathbf{x} = \mathbf{0} + r\mathbf{a} + s\mathbf{b} = r\mathbf{a} + s\mathbf{b}$ and has the two properties in Theorem 5.1.1. Any scalar multiple of a point in the plane is also in the plane, which follows from $t(r\mathbf{a} + s\mathbf{b}) = (tr)\mathbf{a} + (ts)\mathbf{b}$ where t

is a real number. Let $\mathbf{x} = r_1\mathbf{a} + s_1\mathbf{b}$ and $\mathbf{y} = r_2\mathbf{a} + s_2\mathbf{b}$ be in a plane containing the origin. Then the sum is clearly in this plane because

$$\mathbf{x} + \mathbf{y} = (r_1\mathbf{a} + s_1\mathbf{b}) + (r_2\mathbf{a} + s_2\mathbf{b}) = (r_1 + r_2)\mathbf{a} + (s_1 + s_2)\mathbf{b}.$$

If the given point is not the origin, then the plane is also called an *affine* subset of $\mathbb{R}^3$.

Example 5.1.3. Any linear combination of vectors is also a subspace such as the linear combination of the two vector in $\mathbb{R}^4$. Let

$$\mathbf{a} = [-2/6 \ -1/3 \ 1 \ 0]^T \text{ and } \mathbf{b} = [-1/6 \ -8/3 \ 0 \ 1]^T$$

be given vectors in $\mathbb{R}^4$. The set of all linear combination is a subspace of $\mathbb{R}^4$

$$S = \{\mathbf{x} \in \mathbb{R}^4 : \mathbf{x} = r\mathbf{a} + s\mathbf{b} \text{ and } r, s \in \mathbb{R}\}.$$

The two vectors $\mathbf{a}$ and $\mathbf{b}$ will be derived in Example 5.2.3 from the set of vectors in $\mathbb{R}^4$ such that

$$A\mathbf{x} = \begin{bmatrix} 2 & 1 & 1 & 3 \\ 2 & 4 & 2 & 11 \\ 0 & 3 & 1 & 8 \end{bmatrix} \mathbf{x} = \begin{bmatrix} 0 \\ 0 \\ 0 \\ 0 \end{bmatrix}.$$

Definitions 5.1.2. Let A be an $m \times n$ matrix. The *nullspace* of A is *subspace* $null(A) \subset \mathbb{R}^n$

$$null(A) \equiv \{\mathbf{x} \in \mathbb{R}^n : A\mathbf{x} = \mathbf{0}\}.$$

In other words, the nullspace of A is the set of homogeneous solutions. The *range space* of A is subspace $range(A) \subset \mathbb{R}^m$

$$range(A) \equiv \{\mathbf{y} \in \mathbb{R}^m : A\mathbf{x} = \mathbf{y} \text{ for some } \mathbf{x} \in \mathbb{R}^n\}.$$

In other words, the range of A is the set of all linear combinations of its column vectors.

Example 5.1.4. Let

$$A = \begin{bmatrix} 2 & 1 & 1 & 3 \\ 2 & 4 & 2 & 11 \\ 0 & 3 & 1 & 8 \end{bmatrix}.$$

The range space of A is the linear combination of the four column vectors in A. By subtracting row one from row two and then row two from row three we have

$$EA = E_{32}(-1) \ E_{21}(-1) \ A = \begin{bmatrix} 2 & 1 & 1 & 3 \\ 0 & 3 & 1 & 8 \\ 0 & 0 & 0 & 0 \end{bmatrix}.$$

Because the third and fourth columns can be written as a linear combination of the first two columns, the range space of EA may be described by the linear combination of the first two column vectors in EA.

Definition 5.1.3. A set of vectors $\mathbf{y}_1, \cdots, \mathbf{y}_k$ in $\mathbb{R}^m$ is called *linearly indepen-dent* if and only if

$$c_1\mathbf{y}_1 + \cdots + c_k\mathbf{y}_k = \mathbf{0} \text{ implies } c_1 = c_2 = \cdots = c_k = 0.$$

This may be stated as $Y\mathbf{c} = \mathbf{0}$ implies $\mathbf{c} = \mathbf{0}$ where Y is the $m \times k$ matrix of column vectors and $\mathbf{c}$ and $\mathbf{0}$ are $k \times 1$ column vectors.

Example 5.1.1 (continued). In a line $\mathbb{R}^3$ containing the origin with one non-zero vector $\mathbf{y}_1 = \mathbf{d}$ is an example of linearly independent set.

Example 5.1.2 (continued). In a plane $\mathbb{R}^3$ containing the origin can be described by a linear combination of two vectors $\mathbf{y}_1 = \mathbf{a}$ and $\mathbf{y}_2 = \mathbf{b}$ that are not parallel. Thus, this is an example of a subspace in $\mathbb{R}^3$

Example 5.1.3 (continued). Here a subspace in $\mathbb{R}^4$ is defined by a linear combination of two vectors in $\mathbb{R}^4$ where $\mathbf{y}_1 = \mathbf{a} = [-2/6 \ \ -1/3 \ \ 1 \ \ 0]^T$ and $\mathbf{y}_2 = \mathbf{b} = [-1/6 \ \ -8/3 \ \ 0 \ \ 1]^T$. They are linearly independent because the third and fourth equations in

$$c_1 \begin{bmatrix} -2/6 \\ -1/3 \\ 1 \\ 0 \end{bmatrix} + c_2 \begin{bmatrix} -1/6 \\ -8/3 \\ 0 \\ 1 \end{bmatrix} = \begin{bmatrix} 0 \\ 0 \\ 0 \\ 0 \end{bmatrix}$$

imply $c_1 = 0$ and $c_2 = 0$.

Example 5.1.4 (continued). The first two column vectors in the new matrix EA are linearly independent because

$$c_1 \begin{bmatrix} 2 \\ 0 \\ 0 \end{bmatrix} + c_2 \begin{bmatrix} 1 \\ 3 \\ 0 \end{bmatrix} = \begin{bmatrix} 0 \\ 0 \\ 0 \end{bmatrix}$$

implies $c_1 2 + c_2 1 = 0$ and $c_1 0 + c_2 3 = 0$ and, hence, $c_1 = 0$ and $c_2 = 0$.

An important application is the case where one has $k = m$ linearly inde-pendent vectors in $\mathbb{R}^m$. Let A be the $m \times m$ matrix formed by the columns of these vectors

$$A = [\mathbf{y}_1 \ \ \mathbf{y}_2 \cdots \mathbf{y}_m].$$

$A\mathbf{c} = \mathbf{0}$ implies $\mathbf{c} = \mathbf{0}$ is equivalent to linearly independent vectors, and in this case it is characterized by $\det(A) \neq 0$, see Theorem 3.7.3.

Another significant case is from the least squares problem as described in Section 4.2 where the $m \times k$ coefficient matrix A has more rows than columns $m > k$. If the columns are linearly independent, then the $k \times k$ matrix $A^T A$ in the normal equation has an inverse, see Theorem 4.2.2.

Definition 5.1.4. Let $S \subset \mathbb{R}^m$ be a subspace. A set of vectors $\mathbf{y}_1, \cdots, \mathbf{y}_k$ in S is a *basis* of S if it has following two properties:

        (i). any vector in S is a linear combination of the basis vectors and

        (ii). the basis vectors are linearly independent.

The *dimension* of the subspace is k where k is the number of vectors in the basis.

In $\mathbb{R}^2$ and $\mathbb{R}^3$ the Cartesian unit vectors are the classical examples of a basis. In Examples 5.1.3 and 5.2.3 (in the next section) the two vectors in $\mathbb{R}^4$ will be shown to be a basis for a nullspace of the 3×4 matrix. In Example 5.1.4 the range space of EA has a basis given by $\mathbf{y}_1 = [2 \ 0 \ 0]^T$ and $\mathbf{y}_2 = [1 \ 3 \ 0]^T$ and has dimension equal to two.

The set of all solutions $A\mathbf{x} = \mathbf{d}$ can be broken into the *homogeneous* and *particular* parts $\mathbf{x} = \mathbf{x}_h + \mathbf{x}_p$. The homogeneous part must satisfy $A\mathbf{x}_h = \mathbf{0}$, and the particular part must satisfy $A\mathbf{x}_p = \mathbf{d}$. For any $m \times n$ matrix where $m < n$ the general solution of $A\mathbf{x} = \mathbf{d}$ is $\mathbf{x} = \mathbf{x}_h + \mathbf{x}_p$ because

$$A\mathbf{x} = A(\mathbf{x}_h + \mathbf{x}_p) = A\mathbf{x}_h + A\mathbf{x}_p = \mathbf{0} + \mathbf{d} = \mathbf{d}.$$

Thus, if one can find all homogeneous solutions and at least one particular solution, then the general form of all solutions is known. The homogeneous solutions are elements of the nullspace. If the nullspace has a basis given by $\mathbf{y}_1, \cdots, \mathbf{y}_k$, then the general homogeneous solution is a linear combination of these vectors

$$\mathbf{x}_h = c_1\mathbf{y}_1 + \cdots + c_k\mathbf{y}_k.$$

If $\mathbf{d}$ is in the range space of A, then there is a particular solution $A\mathbf{x}_p = \mathbf{d}$.

Theorem 5.1.2 *(General Solution of $A\mathbf{x} = \mathbf{d}$) If A is an $m \times n$ matrix where $m < n$, $\mathbf{y}_1, \cdots, \mathbf{y}_k$ is a basis for the nullspace of A and $\mathbf{x}_p$ is a particular solution of $A\mathbf{x}_p = \mathbf{d}$, then the general solution of $A\mathbf{x} = \mathbf{d}$ is*

$$\mathbf{x} = c_1\mathbf{y}_1 + \cdots + c_k\mathbf{y}_k + \mathbf{x}_p.$$

Moreover, if E is an $m \times m$ matrix with an inverse matrix, then the solution sets of $A\mathbf{x} = \mathbf{d}$ and $EA\mathbf{x} = E\mathbf{d}$ are equal. In this case, the algebraic systems are often called equivalent.

5.1.2 Solving One Equation in $\mathbb{R}^3$

Let a plane containing the origin in $\mathbb{R}^3$ be represented by all linear combinations of two non-parallel vectors $\mathbf{y}_1 = \mathbf{a}$ and $\mathbf{y}_2 = \mathbf{b}$. This is significant because for each vector in such a plane there is only one linear combination of $\mathbf{a}$ and $\mathbf{b}$ that represents of the vector. In order to see this, let $\mathbf{x} = r_1\mathbf{a} + s_1\mathbf{b}$ and $\mathbf{x} = r_2\mathbf{a} + s_2\mathbf{b}$ giving

$$\mathbf{0} = \mathbf{x} - \mathbf{x} = (r_1\mathbf{a} + s_1\mathbf{b}) - (r_2\mathbf{a} + s_2\mathbf{b}) = (r_1 - r_2)\mathbf{a} + (s_1 - s_2)\mathbf{b}.$$

The linear independence implies $r_1 - r_2 = 0$ and $s_1 - s_2 = 0$. The following example shows there may be more that one basis for a particular two-dimensional subspace, which in Example 5.1.6 is a nullspace of a 1×3 matrix $A = [2\ \ 2\ \ -2]$.

Example 5.1.5. Consider the two-dimensional subspace in $\mathbb{R}^3$ given by the linear combinations of two vectors $\mathbf{a} = [1\ \ 2\ \ 3]^T$ and $\mathbf{b} = [1\ \ 0\ \ 1]^T$. This is a plane containing the origin whose vector equation is $\mathbf{n} \bullet (\mathbf{x} - \mathbf{0}) = 0$ where $\mathbf{n} = \mathbf{a} \times \mathbf{b} = [2\ \ 2\ \ -2]^T$. This gives the traditional algebraic equation

$$2x_1 + 2x_2 - 2x_3 = 0.$$

Two other vectors that satisfy this equation and are not parallel are $\widehat{\mathbf{a}} = [1\ \ -1\ \ 0]^T$ and $\widehat{\mathbf{b}} = [0\ \ 1\ \ 1]^T$. They also describe the original plane because their vector equation is equivalent to the above algebraic equation. That is, $\widehat{\mathbf{n}} \bullet (\mathbf{x} - \mathbf{0}) = 0$ where $\widehat{\mathbf{n}} = \widehat{\mathbf{a}} \times \widehat{\mathbf{b}} = [-1\ \ -1\ \ 1]^T$ and so $\widehat{\mathbf{n}} = (-1/2)\mathbf{n}$ and $-x_1 - x_2 + x_3 = 0$. One can explicitly show $\widehat{\mathbf{a}}$ is a linear combination of $\mathbf{a}$ and $\mathbf{b}$

$$r\mathbf{a} + s\mathbf{b} \quad = \quad \widehat{\mathbf{a}}$$

$$r \begin{bmatrix} 1 \\ 2 \\ 3 \end{bmatrix} + s \begin{bmatrix} 1 \\ 0 \\ 1 \end{bmatrix} = \begin{bmatrix} 1 \\ -1 \\ 0 \end{bmatrix}.$$

This vector equation is equivalent to three scalar equations

$$\begin{cases} 1r + 1s = 1 \\ 2r + 0s = -1 \\ 3r + 1s = 0. \end{cases}$$

The solution of the first two equations easily gives $r = -1/2$ and $s = 3/2$. Note, they also satisfy the third equation.

The following examples illustrate how one can find the general solution for a single equation

$$a_{11}x_1 + a_{12}x_2 + a_{13}x_3 \quad = \quad d_1$$

$$\begin{bmatrix} a_{11} & a_{12} & a_{13} \end{bmatrix} \begin{bmatrix} x_1 \\ x_2 \\ x_3 \end{bmatrix} \quad = \quad [d_1].$$

Example 5.1.6. Find a particular solution of the equation $x_1 + 2x_2 + 3x_3 = 10$ by letting x_2 and x_3 equal to zero. Then x_1 must be equal to 10 and one particular solution vector is $\mathbf{x}_p = [10\ \ 0\ \ 0]^T$. In order to find the homogeneous solutions, let x_2 and x_3 be "free variables" with $x_2 = r$ and $x_3 = s$. Then the homogeneous equation $x_1 + 2x_2 + 3x_3 = 0$ gives a formula for $x_1 = -2r - 3s$. The vector form of these three equations is

$$\mathbf{x}_h = \begin{bmatrix} -2r - 3s \\ r \\ s \end{bmatrix} = r \begin{bmatrix} -2 \\ 1 \\ 0 \end{bmatrix} + s \begin{bmatrix} -3 \\ 0 \\ 1 \end{bmatrix}.$$

This suggests $\mathbf{y}_1 = \mathbf{a} = [-2 \ 1 \ 0]^T$ and $\mathbf{y}_2 = \mathbf{b} = [-3 \ 0 \ 1]^T$ form a basis for the nullspace of $A = [1 \ 2 \ 3]$. Clearly, both vectors satisfy the homogeneous equation and are not parallel and, so, in $\mathbb{R}^3$ they are linearly independent. By Theorem 5.1.2 the general solution is

$$\mathbf{x} = \mathbf{x}_h + \mathbf{x}_p = r \begin{bmatrix} -2 \\ 1 \\ 0 \end{bmatrix} + s \begin{bmatrix} -3 \\ 0 \\ 1 \end{bmatrix} + \begin{bmatrix} 10 \\ 0 \\ 0 \end{bmatrix}.$$

Example 5.1.7. Find a particular solution of the equation $2x_2 + x_3 = 10$ by letting $x_1 = 0$ and $x_3 = 0$ so that $x_2 = 10/2 = 5$ and $\mathbf{x}_p = [0 \ 5 \ 0]^T$. The homogeneous equation is $2x_2 + x_3 = 0$, and by letting $x_1 = r$ and $x_3 = s$ be the "free variables" we can solve for $x_2 = -s/2$. The vector form of this is

$$\mathbf{x}_h = \begin{bmatrix} r \\ -s/2 \\ s \end{bmatrix} = r \begin{bmatrix} 1 \\ 0 \\ 0 \end{bmatrix} + s \begin{bmatrix} 0 \\ -1/2 \\ 1 \end{bmatrix}.$$

Both column vectors on the right side are homogeneous solutions and are not parallel and, hence, they form a basis for the nullspace of $A = [0 \ 2 \ 1]$. By Theorem 5.1.2 the general solution is

$$\mathbf{x} = \mathbf{x}_h + \mathbf{x}_p = r \begin{bmatrix} 1 \\ 0 \\ 0 \end{bmatrix} + s \begin{bmatrix} 0 \\ -1/2 \\ 1 \end{bmatrix} + \begin{bmatrix} 0 \\ 5 \\ 0 \end{bmatrix}.$$

5.1.3 Solving Two Equations in $\mathbb{R}^3$

The set of points in $\mathbb{R}^3$ that satisfy two equations is equal to the possible intersection of two planes in $\mathbb{R}^3$. Either the planes never intersect, intersect in a line or the two planes are the same. Given a collection of algebraic equations it is not always evident which one of the three cases one has. If the planes are parallel and are not the same, then there is no solution and the algebraic equations are called *inconsistent*. For example, consider the following two equations

$$\begin{cases} x_1 + 2x_2 + 3x_3 = 1 \\ 2x_1 + 4x_2 + 6x_3 = 3. \end{cases}$$

By subtracting two times equation one from equation two we have a contradiction $0 = 3 - 2$! In terms of elementary matrices and the augmented matrix this is

$$E_{21}(-2) \, [A \ \mathbf{d}] = \begin{bmatrix} 1 & 0 \\ -2 & 1 \end{bmatrix} \begin{bmatrix} 1 & 2 & 3 & 1 \\ 2 & 4 & 6 & 3 \end{bmatrix}$$

$$= \begin{bmatrix} 1 & 2 & 3 & 1 \\ 0 & 0 & 0 & 1 \end{bmatrix}. \tag{5.1.1}$$

The last row is equivalent to $0x_1 + 0x_2 + 0x_3 = 1$! This means $\mathbf{d} = \begin{bmatrix} 1 & 3 \end{bmatrix}^T$ is not in the range space of the 2×3 matrix A.

In the next example the two planes are not parallel and, so, the solution set will be a line in $\mathbb{R}^3$. Example 2.5.3 illustrates that the direction vector of the line given by the intersection of the two planes could be found by taking the cross product of the two normal vectors. The following example gives an alternative way, which will be useful for higher dimensional systems.

Example 5.1.8. Find all the solutions of

$$\begin{cases} x_1 + 2x_2 + 3x_3 = 4 \\ 4x_1 + 5x_2 + 6x_3 = 7. \end{cases}$$

The augmented matrix is

$$[A \ \mathbf{d}] = \begin{bmatrix} 1 & 2 & 3 & 4 \\ 4 & 5 & 6 & 7 \end{bmatrix}.$$

Use an elementary row operation to generate an equivalent system

$$E_{21}(-4) \, [A \ \mathbf{d}] = \begin{bmatrix} 1 & 0 \\ -4 & 1 \end{bmatrix} \begin{bmatrix} 1 & 2 & 3 & 4 \\ 4 & 5 & 6 & 7 \end{bmatrix}$$

$$= \begin{bmatrix} 1 & 2 & 3 & 4 \\ 0 & -3 & -6 & -9 \end{bmatrix}. \tag{5.1.2}$$

Let $x_3 = t$ be the "free variable" so that $E_{21}(-4) \, [A \ \mathbf{d}]$ represents an equivalent system

$$\begin{bmatrix} 1 & 2 \\ 0 & -3 \end{bmatrix} \begin{bmatrix} x_1 \\ x_2 \end{bmatrix} = \begin{bmatrix} 4 \\ -9 \end{bmatrix} - t \begin{bmatrix} 3 \\ -6 \end{bmatrix}.$$

Find a particular solution by setting $x_3 = t = 0$ and solving for $x_2 = -9/(-3) = 3$ and $x_1 = 4 - 2(3) = -2$ and, hence, $\mathbf{x}_p = \begin{bmatrix} -2 & 3 & 0 \end{bmatrix}^T$. In order to find the homogeneous solution, replace the right side of the algebraic system by the zero vector $\mathbf{d} = \mathbf{0}$ and repeat the row operation to obtain

$$\begin{bmatrix} 1 & 2 \\ 0 & -3 \end{bmatrix} \begin{bmatrix} x_1 \\ x_2 \end{bmatrix} = \begin{bmatrix} 0 \\ 0 \end{bmatrix} - t \begin{bmatrix} 3 \\ -6 \end{bmatrix}.$$

Solve for $x_2 = 6t/(-3) = -2t$ and $x_1 = -3t - 2(-2t) = t$ so that $\mathbf{x}_h = [t \quad -2t \quad t]^T$. Since $E = E_{21}(-4)$ has an inverse matrix, one can use Theorem 5.1.2 to find the general solution

$$\mathbf{x} = \mathbf{x}_h + \mathbf{x}_p = t \begin{bmatrix} 1 \\ -2 \\ 1 \end{bmatrix} + \begin{bmatrix} -2 \\ 3 \\ 0 \end{bmatrix}.$$

Note, the direction vector $[1 \quad -2 \quad 1]^T$ is a multiple of the cross product of the two normal vectors $\mathbf{n} = [1 \quad 2 \quad 3]^T$ and $\hat{\mathbf{n}} = [4 \quad 5 \quad 6]^T$, which is $\mathbf{n} \times \hat{\mathbf{n}} = [-3 \quad 6 \quad -3]^T$.

Example 5.1.9. Find all the solutions of

$$\begin{cases} x_1 + 2x_2 + 3x_3 = 4 \\ \quad x_1 + 3x_2 = 5. \end{cases}$$

Use an elementary row operation to generate an equivalent system

$$\begin{aligned} E_{21}(-1) \, [A \ d] &= \begin{bmatrix} 1 & 0 \\ -1 & 1 \end{bmatrix} \begin{bmatrix} 1 & 2 & 3 & 4 \\ 1 & 3 & 0 & 5 \end{bmatrix} \\ &= \begin{bmatrix} 1 & 2 & 3 & 4 \\ 0 & 1 & -3 & 1 \end{bmatrix}. \end{aligned} \tag{5.1.3}$$

In this example let $x_3 = t$ be the "free variable" so that $E_{21}(-1) \, [A \ d]$ represents an equivalent system

$$\begin{bmatrix} 1 & 2 \\ 0 & 1 \end{bmatrix} \begin{bmatrix} x_1 \\ x_2 \end{bmatrix} = \begin{bmatrix} 4 \\ 1 \end{bmatrix} - t \begin{bmatrix} 3 \\ -3 \end{bmatrix}.$$

A particular solution can be found by setting $x_3 = t = 0$ and solving for $x_2 = 1$ and $x_1 = 4 - 2(1) = 2$ so that $\mathbf{x}_p = [2 \quad 1 \quad 0]^T$. In order to find the homogeneous solutions, replace the right side of the algebraic system by the zero vector $\mathbf{d} = \mathbf{0}$ and repeat the row operation to obtain

$$\begin{bmatrix} 1 & 2 \\ 0 & 1 \end{bmatrix} \begin{bmatrix} x_1 \\ x_2 \end{bmatrix} = \begin{bmatrix} 0 \\ 0 \end{bmatrix} - t \begin{bmatrix} 3 \\ -3 \end{bmatrix}.$$

Solve for $x_2 = 3t$ and $x_1 = -3t - 2(3t) = -9t$ so that $\mathbf{x}_h = [-9t \quad 3t \quad t]^T$. Since $E = E_{21}(-1)$ has an inverse matrix and by Theorem 5.1.2, the general solution is

$$\mathbf{x} = \mathbf{x}_h + \mathbf{x}_p = t \begin{bmatrix} -9 \\ 3 \\ 1 \end{bmatrix} + \begin{bmatrix} 2 \\ 1 \\ 0 \end{bmatrix}.$$

5.1.4   Application to Two-loop Circuit

The two-loop circuit in Subsection 3.1.5, see Figure 3.1.4, will be used to illustrate how the representation of the solution for the current equation is used. Let x_1 be the potential at the top node relative to the bottom node. The Kirchhoff voltage law gives three equations for the voltage drops in the three vertical portions of the loops

$$\begin{cases} x_1 + E_1 - R_1 i_1 = 0 \\ x_1 - R_3 i_3 = 0 \\ x_1 - (-i_2)R_2 + E_2 = 0. \end{cases}$$

The Kirchhoff current law requires the three currents to satisfy

$$-i_1 + i_2 - i_3 = 0.$$

The vector form of the four equations is

$$\begin{bmatrix} R_1 & 0 & 0 & -1 \\ 0 & R_2 & 0 & 1 \\ 0 & 0 & R_3 & -1 \\ -1 & 1 & -1 & 0 \end{bmatrix} \begin{bmatrix} i_1 \\ i_2 \\ i_3 \\ x_1 \end{bmatrix} = \begin{bmatrix} E_1 \\ -E_2 \\ 0 \\ 0 \end{bmatrix}.$$

Systems of this form, called the equilibrium equations, also appear in structure, fluid flow and optimizations models, and they will be discussed in more detail in Section 5.3.

The representation of the current solution to the fourth equation leads to a reduction in size of the larger algebraic problem. In order to find the homogeneous solution, let $i_2 = r$ and $i_3 = s$ be the "free variables" so that $i_1 = r - s$. The vector form of this is

$$\mathbf{i}_h = \begin{bmatrix} i_1 \\ i_2 \\ i_3 \end{bmatrix} = \begin{bmatrix} r - s \\ r \\ s \end{bmatrix} = r \begin{bmatrix} 1 \\ 1 \\ 0 \end{bmatrix} + s \begin{bmatrix} -1 \\ 0 \\ 1 \end{bmatrix}.$$

The two column vectors on the right side are homogeneous solutions.

The two column vectors are *nullspace* basis vectors, which can be recorded in a 3×2 matrix

$$N = \begin{bmatrix} 1 & -1 \\ 1 & 0 \\ 0 & 1 \end{bmatrix}.$$

This means $[-1\ \ 1\ \ -1]N = [0\ \ 0]$ or equivalently

$$([\,-1\ \ 1\ \ -1\,]\,N)^T = [\,0\ \ 0\,]^T$$

$$N^T \begin{bmatrix} -1 \\ 1 \\ -1 \end{bmatrix} = \begin{bmatrix} 0 \\ 0 \end{bmatrix}.$$

In order to reduce the size of the algebraic problem from four to two unknowns, put $\mathbf{i}_h$ into the top three equations to obtain

$$\begin{bmatrix} R_1 & 0 & 0 \\ 0 & R_2 & 0 \\ 0 & 0 & R_3 \end{bmatrix} (r\begin{bmatrix} 1 \\ 1 \\ 0 \end{bmatrix} + s\begin{bmatrix} -1 \\ 0 \\ 1 \end{bmatrix}) + \begin{bmatrix} -1 \\ 1 \\ -1 \end{bmatrix} x_1 = \begin{bmatrix} E_1 \\ -E_2 \\ 0 \end{bmatrix}$$

$$\begin{bmatrix} R_1 & 0 & 0 \\ 0 & R_2 & 0 \\ 0 & 0 & R_3 \end{bmatrix} \begin{bmatrix} 1 & -1 \\ 1 & 0 \\ 0 & 1 \end{bmatrix} \begin{bmatrix} r \\ s \end{bmatrix} + \begin{bmatrix} -1 \\ 1 \\ -1 \end{bmatrix} x_1 = \begin{bmatrix} E_1 \\ -E_2 \\ 0 \end{bmatrix}.$$

Multiply both sides by the transpose of the basis vectors N and use the fact that they are homogeneous solutions

$$N^T (\begin{bmatrix} R_1 & 0 & 0 \\ 0 & R_2 & 0 \\ 0 & 0 & R_3 \end{bmatrix} N \begin{bmatrix} r \\ s \end{bmatrix} + \begin{bmatrix} -1 \\ 1 \\ -1 \end{bmatrix} x_1) = N^T \begin{bmatrix} E_1 \\ -E_2 \\ 0 \end{bmatrix}$$

$$(N^T \begin{bmatrix} R_1 & 0 & 0 \\ 0 & R_2 & 0 \\ 0 & 0 & R_3 \end{bmatrix} N) \begin{bmatrix} r \\ s \end{bmatrix} + N^T \begin{bmatrix} -1 \\ 1 \\ -1 \end{bmatrix} x_1 = N^T \begin{bmatrix} E_1 \\ -E_2 \\ 0 \end{bmatrix}$$

$$\begin{bmatrix} R_1 + R_2 & -R_1 \\ -R_1 & R_1 + R_3 \end{bmatrix} \begin{bmatrix} r \\ s \end{bmatrix} + \begin{bmatrix} 0 \\ 0 \end{bmatrix} x_1 = \begin{bmatrix} E_1 - E_2 \\ -E_1 \end{bmatrix}$$

$$\begin{bmatrix} R_1 + R_2 & -R_1 \\ -R_1 & R_1 + R_3 \end{bmatrix} \begin{bmatrix} r \\ s \end{bmatrix} = \begin{bmatrix} E_1 - E_2 \\ -E_1 \end{bmatrix}.$$

The solution of this 2×2 system is easily found by Cramer's rule to be

$$r = \frac{E_1 R_3 - E_2(R_1 + R_3)}{R_1 R_2 + R_1 R_3 + R_2 R_3} \text{ and}$$

$$s = \frac{-E_2 R_1 - E_1 R_2}{R_1 R_2 + R_1 R_3 + R_2 R_3}.$$

The current vector is

$$\begin{bmatrix} i_1 \\ i_2 \\ i_3 \end{bmatrix} = r\begin{bmatrix} 1 \\ 1 \\ 0 \end{bmatrix} + s\begin{bmatrix} -1 \\ 0 \\ 1 \end{bmatrix}$$

$$= \begin{bmatrix} E_1(R_3 + R_2) - E_2 R_3 \\ E_1 R_3 - E_2(R_1 + R_3) \\ -R_2 E_1 - R_1 E_2 \end{bmatrix} \frac{1}{R_1 R_2 + R_1 R_3 + R_2 R_3}.$$

5.1.5 Exercises

1. Consider Example 5.1.1.

(a). Use the definition of cross product to compute the two normal vectors $\mathbf{n}$ and $\widehat{\mathbf{n}}$.

(b). Find r and s such that $r\mathbf{a} + s\mathbf{b} = \widehat{\mathbf{b}}$.

2.   Let a plane containing the origin be a linear combination of $\mathbf{a} = [1 \ \ 1 \ \ 1]^T$ and $\mathbf{b} = [2 \ \ 0 \ \ 1]^T$.
    (a).   Find the vector and algebraic equations that describe this plane.
    (b).   Show this plane is also a linear combination of $\widehat{\mathbf{a}} = [0 \ \ 2 \ \ 1]^T$ and $\widehat{\mathbf{b}} = [1 \ \ -1 \ \ 0]^T$.

3.   Find the general solution in $\mathbb{R}^3$ of $3x_1 + 2x_2 + x_3 = 6$.
4.   Find the general solution in $\mathbb{R}^3$ of $4x_1 + x_3 = 4$.
5.   Find the general solution in $\mathbb{R}^3$ of $2x_1 + x_2 = 10$.
6.   Find the general solution in $\mathbb{R}^3$ of $2x_1 + 3x_2 + 5x_3 = 30$.
7.   Use an elementary matrix and an augmented matrix to show the following equations are inconsistent (have no solution)

$$\begin{cases} 3x_1 + 2x_2 + x_3 = 2 \\ 9x_1 + 6x_2 + 3x_3 = 5. \end{cases}$$

8.   Find the general solution in $\mathbb{R}^3$ of

$$\begin{cases} 3x_1 + 2x_2 + x_3 = 2 \\ 9x_1 + 6x_2 + 3x_3 = 6. \end{cases}$$

9.   Find the general solution in $\mathbb{R}^3$ of

$$\begin{cases} x_1 + 2x_2 + x_3 = 2 \\ x_1 + 3x_3 = 3. \end{cases}$$

10.   Find the general solution in $\mathbb{R}^3$ of

$$\begin{cases} x_1 + 2x_2 + x_3 = 2 \\ x_1 + 3x_2 = 1. \end{cases}$$

11.   Find the general solution in $\mathbb{R}^3$ of

$$\begin{cases} x_1 + 2x_2 + x_3 = 2 \\ 2x_1 + 5x_2 + 3x_3 = 7. \end{cases}$$

12.   Consider the application to the two-loop circuit.
    (a).   Justify the 2×2 matrix and the vector equation.
    (b).   Use Cramer's rule to find the indicated solutions for r and s.

5.2   Row Echelon Form

This section presents a systematic way of representing solutions of $Ax = d$ where A is an $m \times n$ matrix with $m < n$. The main step is to use elementary row operations to transform the system to an equivalent system in row echelon form. The examples in the previous section illustrated this for $m = 1$ or 2 equations and $n = 3$ variables. This will allow us to identify the homogeneous and particular solutions. A number of interesting examples will be done by-hand calculations as well as by using the MATLAB commands rref() and null().

5.2.1 Solutions in $\mathbb{R}^4$

The solution in $\mathbb{R}^4$ of $x_1 + 2x_2 + 3x_3 = 10$ is different from the solution of the same equation in $\mathbb{R}^3$. In Example 5.1.5 we derived the solution in $\mathbb{R}^3$ to be

$$\mathbf{x} = r \begin{bmatrix} -2 \\ 1 \\ 0 \end{bmatrix} + s \begin{bmatrix} -3 \\ 0 \\ 1 \end{bmatrix} + \begin{bmatrix} 10 \\ 0 \\ 0 \end{bmatrix}.$$

In $\mathbb{R}^4$ the equation is $x_1 + 2x_2 + 3x_3 + 0x_4 = 10$ and there is a third free variable $x_4 = t$. Hence, the homogeneous solution in $\mathbb{R}^4$ must satisfy $x_1 + 2r + 3s = 0$, $x_2 = r$, $x_3 = s$ and $x_4 = t$. The vector form of the homogeneous solution is

$$\mathbf{x}_h = r \begin{bmatrix} -2 \\ 1 \\ 0 \\ 0 \end{bmatrix} + s \begin{bmatrix} -3 \\ 0 \\ 1 \\ 0 \end{bmatrix} + t \begin{bmatrix} 0 \\ 0 \\ 0 \\ 1 \end{bmatrix}.$$

The three vectors in $\mathbb{R}^4$ form a basis for the nullspace of $A = \begin{bmatrix} 1 & 2 & 3 & 0 \end{bmatrix}$. A particular solution is now a vector in $\mathbb{R}^4$ and is $\mathbf{x}_p = \begin{bmatrix} 10 & 0 & 0 & 0 \end{bmatrix}^T$ and, so, the general solution in $\mathbb{R}^4$ is

$$\mathbf{x} = r \begin{bmatrix} -2 \\ 1 \\ 0 \\ 0 \end{bmatrix} + s \begin{bmatrix} -3 \\ 0 \\ 1 \\ 0 \end{bmatrix} + t \begin{bmatrix} 0 \\ 0 \\ 0 \\ 1 \end{bmatrix} + \begin{bmatrix} 10 \\ 0 \\ 0 \\ 0 \end{bmatrix}.$$

The next example illustrates that increasing the number of variables can convert an inconsistent smaller variable problem into a problem with multiple solutions! The following system with two equations in $\mathbb{R}^3$ is inconsistent (has no solution)

$$\begin{cases} x_1 + 2x_2 + 3x_3 = 1 \\ 2x_1 + 4x_2 + 6x_3 = 3. \end{cases}$$

Modify these two equations by introducing a fourth variable

$$\begin{cases} x_1 + 2x_2 + 3x_3 + 4x_4 = 1 \\ 2x_1 + 4x_2 + 6x_3 + 9x_4 = 3. \end{cases} \tag{5.2.1}$$

Use an elementary matrix transformation on the augmented matrix to obtain the representation of an equivalent system

$$E_{21}(-2) \begin{bmatrix} A & \mathbf{d} \end{bmatrix} = \begin{bmatrix} 1 & 0 \\ -2 & 1 \end{bmatrix} \begin{bmatrix} 1 & 2 & 3 & 4 & 1 \\ 2 & 4 & 6 & 9 & 3 \end{bmatrix}$$

$$= \begin{bmatrix} 1 & 2 & 3 & 4 & 1 \\ 0 & 0 & 0 & 1 & 1 \end{bmatrix}. \tag{5.2.2}$$

Choose two "free variables" to be $x_2 = r$ and $x_3 = s$. Then in $\mathbb{R}^4$ an equivalent system is

$$\begin{bmatrix} 1 & 4 \\ 0 & 1 \end{bmatrix}\begin{bmatrix} x_1 \\ x_4 \end{bmatrix} = \begin{bmatrix} 1 \\ 1 \end{bmatrix} - r\begin{bmatrix} 2 \\ 0 \end{bmatrix} - s\begin{bmatrix} 3 \\ 0 \end{bmatrix}.$$

Set r and s equal to zero to find a particular solution $\mathbf{x}_p = [-3 \ \ 0 \ \ 0 \ \ 1]^T$. In order to find the homogeneous solution, consider

$$\begin{bmatrix} 1 & 4 \\ 0 & 1 \end{bmatrix}\begin{bmatrix} x_1 \\ x_4 \end{bmatrix} = \begin{bmatrix} 0 \\ 0 \end{bmatrix} - r\begin{bmatrix} 2 \\ 0 \end{bmatrix} - s\begin{bmatrix} 3 \\ 0 \end{bmatrix}.$$

Then $x_4 = 0$, $x_2 = r$, $x_3 = s$ and $x_1 = -2r - 3s$ or in vector notation the homogeneous solution is

$$\mathbf{x}_h = r\begin{bmatrix} -2 \\ 1 \\ 0 \\ 0 \end{bmatrix} + s\begin{bmatrix} -3 \\ 0 \\ 1 \\ 0 \end{bmatrix}.$$

Use Theorem 5.1.2 with $E = E_{21}(-2)$ to get the general solution in $\mathbb{R}^4$

$$\mathbf{x} = r\begin{bmatrix} -2 \\ 1 \\ 0 \\ 0 \end{bmatrix} + s\begin{bmatrix} -3 \\ 0 \\ 1 \\ 0 \end{bmatrix} + \begin{bmatrix} -3 \\ 0 \\ 0 \\ 1 \end{bmatrix}.$$

5.2.2   Row Echelon Form

In the previous examples we used row operations to transform the augmented matrix into another matrix with all zeros in the "lower left" portion of the matrix. The "free variables" were defined to be the variables not located at a "corner" (called a pivot) of the matrix. These were illustrated by the transformed matrices in (5.1.1), (5.1.2), (5.1.3) and (5.2.2). A more precise definition of the transformed matrix and one that generalizes beyond two equations is row echelon form, which is found by a sequence of row operations.

Definitions 5.2.1. The $m \times (n + 1)$ augmented matrix $[A \ \ \mathbf{d}]$ has been transformed into a *row echelon form matrix* by a product of elementary matrices E if and only if

$$E \ [A \ \ \mathbf{d}] = B \text{ where}$$

(i). any non-zero row in B has more leading zero components than the previous row and

(ii). any row below a zero row must also be a zero row.

In terms of the components b_{ij} of B this means $b_{ij} = 0$ for $j < n_i$, $b_{in_i} \neq 0$ and $n_{i-1} < n_i$. The b_{in_i} are called the *pivots*, the variables x_{n_i} are called *fixed* and the remaining variables are called *free*. The number of pivots in B is called the *rank of* $[A \ \ \mathbf{d}]$, and the number of pivots in the first n columns of B is called the *rank of* A.

Example 5.2.1. In the above system in (5.2.1) with two equations and four variables the augmented matrix is

$$[A \ \mathbf{d}] = \begin{bmatrix} 1 & 2 & 3 & 4 & 1 \\ 2 & 4 & 6 & 9 & 3 \end{bmatrix}.$$

The row echelon form matrix is

$$E_{21}(-2) \, [A \ \mathbf{d}] = \begin{bmatrix} 1 & 2 & 3 & 4 & 1 \\ 0 & 0 & 0 & 1 & 1 \end{bmatrix} \text{ where}$$

the two pivots are $b_{1n_1} = 1$ with $n_1 = 1$ and $b_{2n_2} = 1$ with $n_2 = 4$. The fixed variables are x_1 and x_4, and the free variables are x_2 and x_3. The rank of the 2×4 matrix A is two.

Example 5.2.2. Consider the following system with three equations and four unknowns

$$\begin{cases} x_1 + 2x_2 + x_3 + 3x_4 = 1 \\ 2x_1 + 4x_2 + x_3 + 7x_4 = 3 \\ x_3 + x_4 = 7. \end{cases}$$

The augmented matrix is

$$[A \ \mathbf{d}] = \begin{bmatrix} 1 & 2 & 1 & 3 & 1 \\ 2 & 4 & 1 & 7 & 3 \\ 0 & 0 & 1 & 1 & 7 \end{bmatrix}.$$

Subtract 2 times row 1 from row 2 and then add row 2 to row 3 to get

$$E_{32}(1) \, E_{21}(-2) \, [A \ \mathbf{d}] = \begin{bmatrix} 1 & 2 & 1 & 3 & 1 \\ 0 & 0 & -1 & 1 & 1 \\ 0 & 0 & 0 & 2 & 8 \end{bmatrix}.$$

The three pivots are $b_{1n_1} = 1$ with $n_1 = 1$, $b_{2n_2} = -1$ with $n_2 = 3$ and $b_{3n_3} = 2$ with $n_3 = 4$. The fixed variables are x_1, x_3 and x_4, and the free variable is $x_2 = r$. The rank of the 3×4 matrix A is three.

The row echelon form matrix is a short way of writing the equivalent system

$$\begin{bmatrix} 1 & 2 & 3 \\ 0 & -1 & 1 \\ 0 & 0 & 2 \end{bmatrix} \begin{bmatrix} x_1 \\ x_3 \\ x_4 \end{bmatrix} = \begin{bmatrix} 1 \\ 1 \\ 8 \end{bmatrix} - r \begin{bmatrix} 2 \\ 0 \\ 0 \end{bmatrix}.$$

A particular solution is found by setting the free variable $r = 0$ and solving the upper triangular system for the fixed variables. Since the diagonal components are the pivots, the upper triangular solve can be done to yield a particular solution $\mathbf{x}_p = [-14 \ 0 \ 3 \ 4]^T$. In order to find the homogeneous solution, replace the right side $\mathbf{d}$ by the zero vector and solve

$$\begin{bmatrix} 1 & 2 & 3 \\ 0 & -1 & 1 \\ 0 & 0 & 2 \end{bmatrix} \begin{bmatrix} x_1 \\ x_3 \\ x_4 \end{bmatrix} = -r \begin{bmatrix} 2 \\ 0 \\ 0 \end{bmatrix}.$$

This gives $x_4 = 0$, $x_3 = 0$, $x_2 = r$ and $x_1 = -2r$ or in vector notation the homogeneous solution is

$$
\mathbf{x}_h = r \begin{bmatrix} -2 \\ 1 \\ 0 \\ 0 \end{bmatrix}.
$$

By Theorem 5.1.2 with $E_{32}(1)$ $E_{21}(-2)$ the general solution in $\mathbb{R}^4$ is

$$
\mathbf{x} = r \begin{bmatrix} -2 \\ 1 \\ 0 \\ 0 \end{bmatrix} + \begin{bmatrix} -14 \\ 0 \\ 3 \\ 4 \end{bmatrix}.
$$

Example 5.2.3. Consider the following system with three equations and four unknowns

$$
\begin{cases}
2x_1 + x_2 + x_3 + 3x_4 = 4 \\
2x_1 + 4x_2 + 2x_3 + 11x_4 = 10 \\
3x_2 + x_3 + 8x_4 = 6.
\end{cases}
$$

The augmented matrix is

$$
[A \ \mathbf{d}] = \begin{bmatrix} 2 & 1 & 1 & 3 & 4 \\ 2 & 4 & 2 & 11 & 10 \\ 0 & 3 & 1 & 8 & 6 \end{bmatrix}.
$$

Subtract row 1 from row 2 and then subtract row 2 from row 3 to get

$$
E_{32}(-1) \ E_{21}(-1) \ [A \ \mathbf{d}] = \begin{bmatrix} 2 & 1 & 1 & 3 & 4 \\ 0 & 3 & 1 & 8 & 6 \\ 0 & 0 & 0 & 0 & 0 \end{bmatrix}.
$$

The two pivots are $b_{1n_1} = 2$ with $n_1 = 1$ and $b_{2n_2} = 3$ with $n_2 = 2$. The fixed variables are x_1 and x_2, and the free variables are $x_3 = r$ and $x_4 = s$. The rank of the 3×4 matrix A is two.

The row echelon form matrix is a short way of writing the equivalent system

$$
\begin{bmatrix} 2 & 1 \\ 0 & 3 \end{bmatrix} \begin{bmatrix} x_1 \\ x_2 \end{bmatrix} = \begin{bmatrix} 4 \\ 6 \end{bmatrix} - r \begin{bmatrix} 1 \\ 1 \end{bmatrix} - s \begin{bmatrix} 3 \\ 8 \end{bmatrix}.
$$

A particular solution is found by setting the free variables $r = 0$ and $s = 0$ and solving the upper triangular system for the fixed variables. A particular solution is $\mathbf{x}_p = [1 \ 2 \ 0 \ 0]^T$. In order to find the homogeneous solution, replace the right side $\mathbf{d}$ by the zero vector and solve

$$
\begin{bmatrix} 2 & 1 \\ 0 & 3 \end{bmatrix} \begin{bmatrix} x_1 \\ x_2 \end{bmatrix} = -r \begin{bmatrix} 1 \\ 1 \end{bmatrix} - s \begin{bmatrix} 3 \\ 8 \end{bmatrix}.
$$

This gives $x_4 = s$, $x_3 = r$, $x_2 = (-1/3)r + (-3/8)s$ and $x_1 = (-2/6)r + (-1/6)s$ so that the homogeneous solution is

$$\mathbf{x}_h = r \begin{bmatrix} -2/6 \\ -1/3 \\ 1 \\ 0 \end{bmatrix} + s \begin{bmatrix} -1/6 \\ -8/3 \\ 0 \\ 1 \end{bmatrix}.$$

Use Theorem 5.1.2 with $E = E_{32}(-1)\, E_{21}(-1)$ to obtain the general solution in $\mathbb{R}^4$

$$\mathbf{x} = r \begin{bmatrix} -2/6 \\ -1/3 \\ 1 \\ 0 \end{bmatrix} + s \begin{bmatrix} -1/6 \\ -8/3 \\ 0 \\ 1 \end{bmatrix} + \begin{bmatrix} 1 \\ 2 \\ 0 \\ 0 \end{bmatrix}.$$

5.2.3   General Solution of $A\mathbf{x} = \mathbf{d}$

Consider the algebraic system $A\mathbf{x} = \mathbf{d}$ where A is an $m \times n$ matrix with $m < n$. Let E be a product of elementary matrices such that

$$E\,[A \quad \mathbf{d}] = [EA \quad E\mathbf{d}] = B$$

is in row echelon form. Since E has an inverse, the solution sets of the algebraic problems $A\mathbf{x} = \mathbf{d}$ and $EA\mathbf{x} = E\mathbf{d}$ are equal. However, there may not be a solution. Reconsider Example 5.2.3 where the third equation has been changed to $3x_2 + x_3 + 8x_4 = 7$. Then $E\,[A \quad \mathbf{d}] = [EA \quad E\mathbf{d}] = B$ becomes

$$E_{32}(-1)\, E_{21}(-1)\, [A \quad \mathbf{d}] = B = \begin{bmatrix} 2 & 1 & 1 & 3 & 4 \\ 0 & 3 & 1 & 8 & 6 \\ 0 & 0 & 0 & 0 & 1 \end{bmatrix}.$$

The last row implies a contradiction

$$0x_1 + 0x_2 + 0x_3 + 0x_4 = 1!$$

The rank of the 3×4 matrix A is two, and the rank of the 3×5 augmented matrix $[A \quad \mathbf{d}]$ is three.

Theorem 5.2.1 *The algebraic system $A\mathbf{x} = \mathbf{d}$ is inconsistent (has no solution) if and only if any row, say row i, in the row echelon form matrix B has all zeros to the left of component $b_{i,n+1}$ and $b_{i,n+1}$ is not zero.*

The existence of *particular solution* requires that there be some vector $\mathbf{x}_p$ such that $A\mathbf{x}_p = \mathbf{d}$. This means $\mathbf{d}$ must be a linear combination of the column vectors of A. The *homogeneous solutions of* $A\mathbf{x}_h = \mathbf{0}$ are linear combinations of column vectors obtained from the free variables given by the row echelon matrix. The *multiple solutions* are all homogeneous solutions plus a particular solution.

Outline for Solving $A\mathbf{x} = \mathbf{d}$.

1. Find the augmented matrix $[A\ \mathbf{d}]$.
2. Use elementary matrices to find the row echelon matrix $E[A\ \mathbf{d}]$.
3. There is no solution if the row echelon form matrix is inconsistent.
4. Use the pivots to find the fixed and free variables.
5. Find a particular solution of $EA\mathbf{x}_p = E\mathbf{d}$.
    Set the free variables equal to zero and
    solve for the fixed variables.
6. Find the homogeneous solutions of $EA\mathbf{x}_h = E\mathbf{0} = \mathbf{0}$.
    Replace $E\mathbf{d}$ by the zero vector and
    solve for the fixed variables in terms of the free variables.
7. The general solution is the linear combination of the homogeneous
    solution plus a particular solution.

Example 5.2.4. This example is more complicated since it has four equations and six unknowns

$$\begin{cases} x_1 + x_2 + x_3 + x_4 + x_5 + x_6 = 1 \\ x_1 + 2x_2 + x_4 + x_6 = 1 \\ x_2 + x_3 + 2x_4 + x_5 + x_6 = 2 \\ 2x_5 + x_6 = 4. \end{cases}$$

The following elementary row operations and the augmented matrix generate the row echelon form matrix

$$[A\ \mathbf{d}] \;=\; \begin{bmatrix} 1 & 1 & 1 & 1 & 1 & 1 & 1 \\ 1 & 2 & 0 & 1 & 0 & 1 & 1 \\ 0 & 1 & 1 & 2 & 1 & 1 & 2 \\ 0 & 0 & 0 & 0 & 2 & 1 & 4 \end{bmatrix}$$

$$E_{32}(-1)\, E_{21}(-1)\, [A\ \mathbf{d}] \;=\; \begin{bmatrix} 1 & 1 & 1 & 1 & 1 & 1 & 1 \\ 0 & 1 & -1 & 0 & -1 & 0 & 0 \\ 0 & 0 & 2 & 2 & 2 & 1 & 2 \\ 0 & 0 & 0 & 0 & 2 & 1 & 4 \end{bmatrix}.$$

The free variables are $x_4 = r$ and $x_6 = s$, and the row echelon matrix gives the following system

$$\begin{bmatrix} 1 & 1 & 1 & 1 \\ 0 & 1 & -1 & -1 \\ 0 & 0 & 2 & 2 \\ 0 & 0 & 0 & 2 \end{bmatrix} \begin{bmatrix} x_1 \\ x_2 \\ x_3 \\ x_5 \end{bmatrix} = \begin{bmatrix} 1 \\ 0 \\ 2 \\ 4 \end{bmatrix} - r \begin{bmatrix} 1 \\ 0 \\ 2 \\ 0 \end{bmatrix} - s \begin{bmatrix} 1 \\ 0 \\ 1 \\ 1 \end{bmatrix}.$$

By setting the two free variables to zero and solving the upper triangular system, we find a particular solution $\mathbf{x}_p = [-1\ 1\ -1\ 0\ 2\ 0]^T$. Set the vector $\mathbf{d}$ equal to the zero vector and solve the following for the homogeneous solutions

$$
\begin{bmatrix} 1 & 1 & 1 & 1 \\ 0 & 1 & -1 & -1 \\ 0 & 0 & 2 & 2 \\ 0 & 0 & 0 & 2 \end{bmatrix} \begin{bmatrix} x_1 \\ x_2 \\ x_3 \\ x_5 \end{bmatrix} = -r \begin{bmatrix} 1 \\ 0 \\ 2 \\ 0 \end{bmatrix} - s \begin{bmatrix} 1 \\ 0 \\ 1 \\ 1 \end{bmatrix}.
$$

Use Theorem 5.1.2 with $E = E_{32}(-1)\, E_{21}(-1)$ to find the general solution in $\mathbb{R}^6$

$$
\mathbf{x} = \mathbf{x}_h + \mathbf{x}_p = r \begin{bmatrix} 1 \\ -1 \\ -1 \\ 1 \\ 0 \\ 0 \end{bmatrix} + s \begin{bmatrix} 0 \\ -1/2 \\ 0 \\ 0 \\ -1/2 \\ 1 \end{bmatrix} + \begin{bmatrix} -1 \\ 1 \\ -1 \\ 0 \\ 2 \\ 0 \end{bmatrix}.
$$

The nullspace of A has dimension equal to two, and the dimension of the range space is four.

5.2.4 Reduced Row Echelon Form and MATLAB

Given an augmented matrix $[A \ \mathbf{d}]$ the row echelon matrix is not unique! As an illustration consider the algebraic system

$$
\begin{cases} x_1 + 2x_2 + 3x_3 = 1 \\ 4x_1 + 5x_2 + 6x_3 = 3. \end{cases}
$$

The augmented matrix is

$$
[A \ \mathbf{d}] = \begin{bmatrix} 1 & 2 & 3 & 1 \\ 4 & 5 & 6 & 3 \end{bmatrix}.
$$

One row echelon form matrix is

$$
E_{21}(-4)\, [A \ \mathbf{d}] = \begin{bmatrix} 1 & 2 & 3 & 1 \\ 0 & -3 & -6 & -1 \end{bmatrix}.
$$

Another row echelon form matrix is obtained by dividing row 2 by -3

$$
E_2(1/(-3))\, E_{21}(-4)\, [A \ \mathbf{d}] = \begin{bmatrix} 1 & 2 & 3 & 1 \\ 0 & 1 & 2 & 1/3 \end{bmatrix}.
$$

A third one is obtained by using row 2 to get a zero in the 12-component

$$
E_{12}(-2)\, E_2(1/(-3))\, E_{21}(-4)\, [A \ \mathbf{d}] = \begin{bmatrix} 1 & 0 & -1 & 1/3 \\ 0 & 1 & 2 & 1/3 \end{bmatrix}.
$$

This is an example of a reduced row echelon matrix, which can be shown to be unique. Moreover, the general solution to the algebraic solution of the original system is easy to construct from

$$
\begin{bmatrix} 1 & 0 \\ 0 & 1 \end{bmatrix} \begin{bmatrix} x_1 \\ x_2 \end{bmatrix} = \begin{bmatrix} 1/3 \\ 1/3 \end{bmatrix} - r \begin{bmatrix} -1 \\ 2 \end{bmatrix} \quad \text{where } x_3 = r.
$$

In this case the general solution is

$$\begin{bmatrix} x_1 \\ x_2 \\ x_3 \end{bmatrix} = \begin{bmatrix} 1/3 \\ 1/3 \\ 0 \end{bmatrix} + r \begin{bmatrix} 1 \\ -2 \\ 1 \end{bmatrix}.$$

Definition 5.2.2. The $m \times (n+1)$ augmented matrix $[A \ \mathbf{d}]$ has been transformed into *reduced row echelon form matrix* by a product of elementary matrices E if and only if

$$E\,[A \ \mathbf{d}] = B \text{ where}$$

(i). $E\,[A \ \mathbf{d}]$ has row echelon form,
(ii).   the pivots are all equal to one ($b_{in_i} = 1$) and
(iii).   zeros are above the pivots ($b_{kn_i} = 0$ for all $k < i$).

The MATLAB commands rref() and null() will easily compute the reduced row echelon matrix and the nullspace basis, respectively.

```
>> [A d]
        1 2 3 1
        4 5 6 3
>> rref([A d])
        1.0000  0.0000 -1.0000  0.3333
        0.0000  1.0000  2.0000  0.3333
>> N = null(A,'r')
    N =
        1
       -2
        1
>> A*N
        0
        0
```

The by-hand computations for Example 5.2.4 are easy to verify.

```
>> [A d]
        1 1 1 1 1 1 1
        1 2 0 1 0 1 1
        0 1 1 2 1 1 2
        0 0 0 0 2 1 4
>> rref([A d])
        1 0 0 -1 0  0 -1
        0 1 0  1 0 .5  1
        0 0 1  1 0  0 -1
        0 0 0  0 1 .5  2
>> N = null(A, 'r')
    N =
```

$$\begin{array}{rr} 1 & 0 \\ -1 & -.5 \\ -1 & 0 \\ 1 & 0 \\ 0 & -.5 \\ 0 & 1 \end{array}$$

>> A*N

$$\begin{array}{cc} 0 & 0 \\ 0 & 0 \\ 0 & 0 \\ 0 & 0 \end{array}$$

5.2.5 Exercises

1. Consider the solution of $3x_1 + 2x_2 + x_3 = 12$ in $\mathbb{R}^4$.
   (a). Find the particular solution in $\mathbb{R}^4$.
   (b). Find the homogeneous solutions in $\mathbb{R}^4$.
   (c). Find the general solution in $\mathbb{R}^4$.
2. Consider the row echelon form for some $[A \; \mathbf{d}]$

$$\begin{bmatrix} 1 & 2 & 3 & 6 & 1 & 1 \\ 0 & 0 & 1 & 2 & 1 & 2 \\ 0 & 0 & 0 & 0 & 2 & 3 \\ 0 & 0 & 0 & 0 & 0 & 1 \end{bmatrix}.$$

   (a). Find the pivots, fixed and free variables, and the ranks of A and $[A \; \mathbf{d}]$.
   (b). Does the algebraic system $A\mathbf{x} = \mathbf{d}$ have a solution?
3. Consider the solution in $\mathbb{R}^4$ of

$$\begin{cases} 3x_1 + 2x_2 + x_3 + x_4 = 6 \\ 9x_1 + 6x_2 + 3x_3 + 4x_4 = 19. \end{cases}$$

   (a). Find a row echelon matrix.
   (b). Find the particular solution in $\mathbb{R}^4$.
   (c). Find the homogeneous solutions in $\mathbb{R}^4$.
   (d). Find the general solution in $\mathbb{R}^4$.
4. Consider the solution in $\mathbb{R}^4$ of

$$\begin{cases} 2x_1 + x_2 + x_3 + 3x_4 = 1 \\ 2x_1 + 3x_2 + 2x_3 + 3x_4 = 4 \\ 4x_2 + x_4 = 7. \end{cases}$$

   (a). Find a row echelon matrix.
   (b). Find the particular solution in $\mathbb{R}^4$.
   (c). Find the homogeneous solutions in $\mathbb{R}^4$.
   (d). Find the general solution in $\mathbb{R}^4$.

5.   Consider the solution in $\mathbb{R}^4$ of

$$\begin{cases} 2x_1 + x_2 + x_3 + 3x_4 = 4 \\ 2x_1 + 4x_2 + 2x_3 + 11x_4 = 10 \\ 6x_2 + 2x_3 + 16x_4 = 12. \end{cases}$$

(a).   Find a row echelon matrix.
(b).   Find the reduced row echelon matrix.
(c).   Find the general solution in $\mathbb{R}^4$.

6.   Consider the solution in $\mathbb{R}^4$ of

$$\begin{cases} x_1 + x_2 + x_3 + x_4 = 1 \\ 2x_1 + 3x_3 + 4x_4 = 2 \\ 2x_2 - 3x_3 + 2x_4 = 6. \end{cases}$$

(a).   Find a row echelon matrix.
(b).   Find the reduced row echelon matrix.
(c).   Find the general solution in $\mathbb{R}^4$.

7.   Consider the solution Example 5.2.4. By-hand calculations find the reduced row echelon matrix.

8.   Use the MATLAB commands rref() and null() to confirm any calculations done in exercises 2-7.

5.3   Nullspaces and Equilibrium Equations

In Subsection 5.1.4 an application was given to a two-loop circuit, and the solution was obtained by using the nullspace associated with the current equation at the top node in Figure 3.1.4. The resulting equation has the form

$$\begin{bmatrix} R_1 & 0 & 0 & -1 \\ 0 & R_2 & 0 & 1 \\ 0 & 0 & R_3 & -1 \\ -1 & 1 & -1 & 0 \end{bmatrix} \begin{bmatrix} x_1 \\ x_2 \\ x_3 \\ y_1 \end{bmatrix} = \begin{bmatrix} f_1 \\ f_2 \\ f_3 \\ g_1 \end{bmatrix}$$

where the first three unknowns are the currents and the fourth unknown is the potential at the top node relative to the ground, see Figure 3.1.4. This system is a special case of the equilibrium equations. In this section the general equilibrium equations will be defined and solved by both block Gauss elimination and nullspace methods. Applications to more complicated circuits, structures and fluid flows models will be outlined. Additional material on equilibrium equations can be found in chapter two of [8].

Definition 5.3.1. Let A be an $n \times n$ matrix and E be an $m \times n$ matrix with $m < n$. The *equilibrium equations* can be written as

$$\begin{aligned} A\mathbf{x} + E^T\mathbf{y} &= \mathbf{f} \text{ and} \\ E\mathbf{x} &= \mathbf{g} \end{aligned}$$

or using 0 as an $m \times m$ matrix of zeros

$$\left[\begin{array}{cc} A & E^T \\ E & 0 \end{array} \right] \left[\begin{array}{c} \mathbf{x} \\ \mathbf{y} \end{array} \right] = \left[\begin{array}{c} \mathbf{f} \\ \mathbf{g} \end{array} \right].$$

In the above two-loop circuit algebraic system A is the 3×3 diagonal matrix with the three resistor values, and E is the 1×3 row matrix with coefficients given by the Kirchhoff current law. Typically, $E\mathbf{x} = \mathbf{g}$ represents some sort of balance or conservation of quantities in $\mathbf{x}$ such as currents, mass or forces. The top equation $A\mathbf{x} + E^T\mathbf{y} = \mathbf{f}$ may be an implementation of an empirical law such as voltage drops, momentum or deformation from forces.

5.3.1 Block Gauss Elimination Method

In this section assume A is *symmetric* $(A = A^T)$ *positive definite* $(\mathbf{x} \neq 0$ implies $\mathbf{x}^T A \mathbf{x} > 0)$, and E has full row rank (the row echelon form has exactly m non-zero pivots). In other words, the columns of E^T are linearly independent and, hence, $E^T\mathbf{y} = \mathbf{0}$ implies $\mathbf{y} = \mathbf{0}$. By combining these two assumptions we can conclude by Theorem 3.7.3 that both A and $EA^{-1}E^T$ have inverse matrices.

Solve the first equation for $\mathbf{x} = A^{-1}(\mathbf{f} - E^T\mathbf{y})$ and then put this into the second equation

$$\begin{aligned} E\mathbf{x} &= \mathbf{g} \\ EA^{-1}(\mathbf{f} - E^T\mathbf{y}) &= \mathbf{g}. \end{aligned}$$

This gives an equation for $\mathbf{y}$ where $EA^{-1}E^T$ is an $m \times m$ matrix

$$EA^{-1}E^T\mathbf{y} = EA^{-1}\mathbf{f} - \mathbf{g}.$$

Once $\mathbf{y}$ is known, $\mathbf{x}$ is computed from the first equation

$$\mathbf{x} = A^{-1}(\mathbf{f} - E^T\mathbf{y}).$$

The block Gauss elimination method may also be described by a block row operation, see Section 3.6,

$$\left[\begin{array}{cc} I_n & 0 \\ -EA^{-1} & I_m \end{array} \right] \left[\begin{array}{cc} A & E^T \\ E & 0 \end{array} \right] \left[\begin{array}{c} \mathbf{x} \\ \mathbf{y} \end{array} \right] = \left[\begin{array}{cc} I_n & 0 \\ -EA^{-1} & I_m \end{array} \right] \left[\begin{array}{c} \mathbf{f} \\ \mathbf{g} \end{array} \right]$$

$$\left[\begin{array}{cc} A & E^T \\ 0 & -EA^{-1}E^T \end{array} \right] \left[\begin{array}{c} \mathbf{x} \\ \mathbf{y} \end{array} \right] = \left[\begin{array}{c} \mathbf{f} \\ -EA^{-1}\mathbf{f} + \mathbf{g} \end{array} \right].$$

Block Gauss Elimination Method for Equilibrium Equations.
1. Solve $A\mathbf{z} = \mathbf{f}$ and $AZ = E^T$ so that $\mathbf{z} = A^{-1}\mathbf{f}$ and $Z = A^{-1}E^T$.
2. Compute $E\mathbf{z} - \mathbf{g}$ and EZ.
3. Solve $EZ\mathbf{y} = E\mathbf{z} - \mathbf{g}$ so that $\mathbf{y} = (EA^{-1}E^T)^{-1}(EA^{-1}\mathbf{f} - \mathbf{g})$.
4. Let $\mathbf{x} = \mathbf{z} - Z\mathbf{y} = A^{-1}\mathbf{f} - A^{-1}E^T\mathbf{y}$.

Example 5.3.1. Consider a particular case of the two-loop circuit

$$
\begin{bmatrix}
2 & 0 & 0 & -1 \\
0 & 3 & 0 & 1 \\
0 & 0 & 4 & -1 \\
-1 & 1 & -1 & 0
\end{bmatrix}
\begin{bmatrix}
x_1 \\
x_2 \\
x_3 \\
y_1
\end{bmatrix}
=
\begin{bmatrix}
9 \\
-9 \\
0 \\
0
\end{bmatrix}.
$$

Since A is a diagonal matrix, the solves for $\mathbf{z}$ and Z are easy

$$
\mathbf{z} =
\begin{bmatrix}
9/2 \\
-9/3 \\
0/4
\end{bmatrix}
\text{ and } Z =
\begin{bmatrix}
-1/2 \\
1/3 \\
-1/4
\end{bmatrix}.
$$

$E\mathbf{z} - \mathbf{g} = -1(9/2) + 1(-9/3) - 1(0/4) - 0 = -15/2$ and $EZ = -1(-1/2) + 1(1/3) - 1(-1/4) = 13/12$. Then the solution of $EZ\mathbf{y} = E\mathbf{z} - \mathbf{g}$ is also easy to compute $\mathbf{y} = (-15/2)/(13/12) = -90/13$. The last step is

$$
\begin{aligned}
\mathbf{x} &= \mathbf{z} - Z\mathbf{y} \\
&=
\begin{bmatrix}
9/2 \\
-9/3 \\
0/4
\end{bmatrix}
-
\begin{bmatrix}
-1/2 \\
1/3 \\
-1/4
\end{bmatrix}
(-90/13) =
\begin{bmatrix}
27/26 \\
-18/26 \\
-45/26
\end{bmatrix}.
\end{aligned}
$$

Example 5.3.2. If there are more rows in E, then the solves become more complicated as in

$$
\begin{bmatrix}
2 & -1 & 0 & 1 & 0 \\
-1 & 2 & -1 & -1 & 1 \\
0 & -1 & 2 & 0 & 1 \\
1 & -1 & 0 & 0 & 0 \\
0 & 1 & 1 & 0 & 0
\end{bmatrix}
\begin{bmatrix}
x_1 \\
x_2 \\
x_3 \\
y_1 \\
y_2
\end{bmatrix}
=
\begin{bmatrix}
1 \\
2 \\
3 \\
4 \\
5
\end{bmatrix}.
$$

Here A is tridiagonal and E has two rows

$$
A =
\begin{bmatrix}
2 & -1 & 0 \\
-1 & 2 & -1 \\
0 & -1 & 2
\end{bmatrix}
\text{ and } E =
\begin{bmatrix}
1 & -1 & 0 \\
0 & 1 & 1
\end{bmatrix}.
$$

The inverse of A is

$$
A^{-1} = (1/4)
\begin{bmatrix}
3 & 2 & 1 \\
2 & 4 & 2 \\
1 & 2 & 3
\end{bmatrix}
$$

and, hence, $z = A^{-1}\mathbf{f}$ and $Z = A^{-1}E^T$ are

$$
z = (1/4)
\begin{bmatrix}
10 \\
16 \\
14
\end{bmatrix}
\text{ and } Z = (1/4)
\begin{bmatrix}
1 & 3 \\
-2 & 6 \\
-1 & 5
\end{bmatrix}.
$$

Now $EZ\mathbf{y} = E\mathbf{z} - \mathbf{g}$ becomes

$$\begin{bmatrix} .75 & -.75 \\ -.75 & 2.75 \end{bmatrix} \begin{bmatrix} y_1 \\ y_2 \end{bmatrix} = \begin{bmatrix} -5.5 \\ 2.5 \end{bmatrix}$$

whose solution is $y_1 = -8.8333$ and $y_2 = -1.500$. The final step requires the computation of $\mathbf{x} = \mathbf{z} - Z\mathbf{y}$

$$\begin{bmatrix} x_1 \\ x_2 \\ x_3 \end{bmatrix} = \begin{bmatrix} 2.5 \\ 4.0 \\ 3.5 \end{bmatrix} + \begin{bmatrix} .25 & .75 \\ -.5 & 1.5 \\ -.25 & 1.25 \end{bmatrix} \begin{bmatrix} -8.3333 \\ -1.5000 \end{bmatrix} = \begin{bmatrix} 5.8333 \\ 1.8333 \\ 3.1667 \end{bmatrix}.$$

5.3.2 Nullspace Method for Equilibrium Equations

The nullspace method is patterned on how we solved the two-loop circuit problem in Section 5.1. Again, assume A is symmetric positive definite and E has full row rank. The full row rank assumption gives the existence of a particular solution $\mathbf{x}_p$ so that $E\mathbf{x}_p = \mathbf{g}$. Moreover, there is a basis for the nullspace with $n - m$ vectors. Both these can be constructed as illustrated in the previous sections from the row echelon form of the augmented matrix $[E \ \mathbf{g}]$.

Let N be the $n \times (n-m)$ matrix formed by the basis vectors for the nullspace of E. This implies E times each column in N is a zero vector and, consequently, $EN = 0$. The homogeneous solutions are linear combinations of the columns in N, which may be written as a matrix-vector product $N\mathbf{x}_0$ where the coefficients in the linear combination are the components of $\mathbf{x}_0$. The general solution of $E\mathbf{x}_p = \mathbf{g}$ is

$$\mathbf{x} = \mathbf{x}_h + \mathbf{x}_p = N\mathbf{x}_0 + \mathbf{x}_p.$$

The next step is to put this representation for $\mathbf{x}$ into the first equation

$$A(N\mathbf{x}_0 + \mathbf{x}_p) + E^T\mathbf{y} = \mathbf{f}.$$

Now multiply both sides by N^T and use $EN = 0$ giving $(EN)^T = 0^T$, $N^T E^T = 0$ and

$$\begin{aligned} N^T(A(N\mathbf{x}_0 + \mathbf{x}_p) + E^T\mathbf{y}) &= N^T\mathbf{f} \\ N^T A N\mathbf{x}_0 + N^T A\mathbf{x}_p + N^T E^T\mathbf{y} &= N^T\mathbf{f} \\ N^T A N\mathbf{x}_0 + N^T A\mathbf{x}_p + 0 &= N^T\mathbf{f}. \end{aligned}$$

Thus, one must solve

$$N^T A N\mathbf{x}_0 = N^T\mathbf{f} - N^T A\mathbf{x}_p.$$

The above two assumptions imply the $(n - m) \times (n - m)$ matrix $N^T A N$ has an inverse.

Finally one can solve for $\mathbf{y}$ by multiplying the first equation by E

$$E(A(N\mathbf{x}_0 + \mathbf{x}_p) + E^T\mathbf{y}) = E\mathbf{f}.$$

This requires the solution of

$$EE^T \mathbf{y} = E\mathbf{f} - E(A(N\mathbf{x}_0 + \mathbf{x}_p)).$$

The $m \times m$ matrix EE^T has an inverse because the columns in E^T are linearly independent.

Nullspace Method for Solving the Equilibrium Equations.

1. Find a row echelon form of the augmented matrix $[E \ \mathbf{g}]$.
2. Find a particular solution $\mathbf{x}_p$.
3. Find a basis for the nullspace and N.
4. Compute $\mathbf{w} = N^T(\mathbf{f} - A\mathbf{x}_p)$ and $W = N^T A$.
5. Solve $WN\mathbf{x}_0 = w$ so that $\mathbf{x}_0 = (N^T A N)^{-1}(N^T \mathbf{f} - N^T A\mathbf{x}_p)$.
6. Solve $EE^T \mathbf{y} = E(\mathbf{f} - A(N\mathbf{x}_0 + \mathbf{x}_p))$.

Example 5.3.3. Reconsider the two-loop circuit problem in Example 5.3.1. The 1×3 matrix $E = [-1 \ 1 \ -1]$ already is in row echelon form. Since $\mathbf{g}$ is zero, the particular solution is zero. The nullspace basis vectors are the columns in

$$N = \begin{bmatrix} 1 & -1 \\ 1 & 0 \\ 0 & 1 \end{bmatrix}.$$

Next compute $\mathbf{w}$ and W

$$\mathbf{w} = N^T(\mathbf{f} - A\mathbf{x}_p) = \begin{bmatrix} 1 & -1 \\ 1 & 0 \\ 0 & 1 \end{bmatrix}^T \left(\begin{bmatrix} 9 \\ -9 \\ 0 \end{bmatrix} - A\mathbf{0} \right) = \begin{bmatrix} 0 \\ -9 \end{bmatrix}$$

$$W = N^T A = \begin{bmatrix} 1 & -1 \\ 1 & 0 \\ 0 & 1 \end{bmatrix}^T \begin{bmatrix} 2 & 0 & 0 \\ 0 & 3 & 0 \\ 0 & 0 & 4 \end{bmatrix} = \begin{bmatrix} 2 & 3 & 0 \\ -2 & 0 & 4 \end{bmatrix}.$$

The solution of

$$WN\mathbf{x}_0 = w$$

$$\begin{bmatrix} 2 & 3 & 0 \\ -2 & 0 & 4 \end{bmatrix} \begin{bmatrix} 1 & -1 \\ 1 & 0 \\ 0 & 1 \end{bmatrix} \begin{bmatrix} x_1^0 \\ x_2^0 \end{bmatrix} = \begin{bmatrix} 0 \\ -9 \end{bmatrix}$$

$$\begin{bmatrix} 5 & -2 \\ -2 & 6 \end{bmatrix} \begin{bmatrix} x_1^0 \\ x_2^0 \end{bmatrix} = \begin{bmatrix} 0 \\ -9 \end{bmatrix}.$$

is $x_1^0 = -18/26$ and $x_2^0 = -45/26$. The $\mathbf{x}$ vector is

$$\mathbf{x} = N\mathbf{x}_0 + \mathbf{x}_p$$

$$= \begin{bmatrix} 1 & -1 \\ 1 & 0 \\ 0 & 1 \end{bmatrix} \begin{bmatrix} -18/26 \\ -45/26 \end{bmatrix} + \begin{bmatrix} 0 \\ 0 \\ 0 \end{bmatrix} = \begin{bmatrix} 27/26 \\ -18/26 \\ -45/26 \end{bmatrix}.$$

Example 5.3.4. Reconsider Example 5.3.2 where the augmented matrix is in row echelon form

$$[E \ \mathbf{g}] = \begin{bmatrix} 1 & -1 & 0 & 4 \\ 0 & 1 & 1 & 5 \end{bmatrix}.$$

The free variable is $x_3 = r$ and this is equivalent to

$$\begin{bmatrix} 1 & -1 \\ 0 & 1 \end{bmatrix} \begin{bmatrix} x_1 \\ x_2 \end{bmatrix} = \begin{bmatrix} 4 \\ 5 \end{bmatrix} - r \begin{bmatrix} 0 \\ 1 \end{bmatrix}.$$

Set $r = 0$ and solve for the particular solution $\mathbf{x}_p = [9 \ \ 5 \ \ 0]^T$. Find the homogeneous solution by solving

$$\begin{bmatrix} 1 & -1 \\ 0 & 1 \end{bmatrix} \begin{bmatrix} x_1 \\ x_2 \end{bmatrix} = \begin{bmatrix} 0 \\ 0 \end{bmatrix} - r \begin{bmatrix} 0 \\ 1 \end{bmatrix}.$$

This gives $x_2 = -r$, $x_1 = -r$ and $x_3 = r$ or in vector form

$$\mathbf{x}_h = \begin{bmatrix} -1 \\ -1 \\ 1 \end{bmatrix} r \text{ and } N = \begin{bmatrix} -1 \\ -1 \\ 1 \end{bmatrix}.$$

Next compute $\mathbf{w} = N^T(\mathbf{f} - A\mathbf{x}_p)$ and $W = N^T A$.

$$\begin{aligned}
\mathbf{w} &= N^T(\mathbf{f} - A\mathbf{x}_p) \\
&= \begin{bmatrix} -1 \\ -1 \\ 1 \end{bmatrix}^T \left(\begin{bmatrix} 1 \\ 2 \\ 3 \end{bmatrix} - \begin{bmatrix} 2 & -1 & 0 \\ -1 & 2 & -1 \\ 0 & -1 & 2 \end{bmatrix} \begin{bmatrix} 9 \\ 5 \\ 0 \end{bmatrix} \right) \\
&= -1(-12) - 1(1) + 1(8) = 19 \text{ and}
\end{aligned}$$

$$\begin{aligned}
W &= N^T A \\
&= \begin{bmatrix} -1 \\ -1 \\ 1 \end{bmatrix}^T \begin{bmatrix} 2 & -1 & 0 \\ -1 & 2 & -1 \\ 0 & -1 & 2 \end{bmatrix} = \begin{bmatrix} -1 & -2 & 3 \end{bmatrix}.
\end{aligned}$$

Solve $WN\mathbf{x}_0 = w$

$$\begin{bmatrix} -1 & -2 & 3 \end{bmatrix} \begin{bmatrix} -1 \\ -1 \\ 1 \end{bmatrix} \mathbf{x}_0 = 19 \text{ so that } \mathbf{x}_0 = 19/6.$$

Then $\mathbf{x} = N\mathbf{x}_0 + \mathbf{x}_p$ is as follows

$$\mathbf{x} = \begin{bmatrix} -1 \\ -1 \\ 1 \end{bmatrix} (19/6) + \begin{bmatrix} 9 \\ 5 \\ 0 \end{bmatrix} = \begin{bmatrix} 35/6 \\ 11/6 \\ 19/6 \end{bmatrix}.$$

Find $\mathbf{y}$ by solving $EE^T\mathbf{y} = E^T(\mathbf{f} - A(N\mathbf{x}_0 + \mathbf{x}_p))$

$$\begin{bmatrix} 1 & -1 & 0 \\ 0 & 1 & 1 \end{bmatrix} \begin{bmatrix} 1 & 0 \\ -1 & 1 \\ 0 & 1 \end{bmatrix} \begin{bmatrix} y_1 \\ y_2 \end{bmatrix} = \begin{bmatrix} 1 & -1 & 0 \\ 0 & 1 & 1 \end{bmatrix} \begin{bmatrix} -53/6 \\ 44/6 \\ -9/6 \end{bmatrix}$$

$$\begin{bmatrix} 2 & -1 \\ -1 & 2 \end{bmatrix} \begin{bmatrix} y_1 \\ y_2 \end{bmatrix} = \begin{bmatrix} -97/6 \\ 35/6 \end{bmatrix}$$

whose solution is $y_1 = -159/18$ and $y_2 = -27/18$.

5.3.3   Application to Three-loop Circuit

Consider the three-loop circuit in Subsection 3.6.3, see Figure 3.6.1. The matrix equation for the five currents $\mathbf{x}$ and two potentials at the top nodes $\mathbf{y}$ is

$$\begin{bmatrix} R_1 & 0 & 0 & 0 & 0 & -1 & 0 \\ 0 & R_2 & 0 & 0 & 0 & 1 & -1 \\ 0 & 0 & R_3 & 0 & 0 & -1 & 0 \\ 0 & 0 & 0 & R_4 & 0 & 0 & -1 \\ 0 & 0 & 0 & 0 & R_5 & 0 & 1 \\ -1 & 1 & -1 & 0 & 0 & 0 & 0 \\ 0 & -1 & 0 & -1 & 1 & 0 & 0 \end{bmatrix} \begin{bmatrix} x_1 \\ x_2 \\ x_3 \\ x_4 \\ x_5 \\ y_1 \\ y_2 \end{bmatrix} = \begin{bmatrix} E_1 \\ -E_2 \\ 0 \\ 0 \\ -E_3 \\ 0 \\ 0 \end{bmatrix} .$$

This is an equilibrium equation with A being a 5×5 diagonal matrix and the augmented matrix being

$$[E \ \mathbf{g}] = \begin{bmatrix} -1 & 1 & -1 & 0 & 0 & 0 \\ 0 & -1 & 0 & -1 & 1 & 0 \end{bmatrix} .$$

Here we outline the nullspace method, which can be viewed as a dimension reduction scheme where the size of the matrix solves gets smaller. The following row operations transform the augmented matrix to reduced row echelon form

$$E_{12}(1)\, E_2(-1)\, E_1(-1)\, [E \ \mathbf{g}] = \begin{bmatrix} 1 & 0 & 1 & 1 & -1 & 0 \\ 0 & 1 & 0 & 1 & -1 & 0 \end{bmatrix} .$$

There are three free variables $x_3 = r$, $x_4 = s$ and $x_5 = t$, and this implies

$$\begin{bmatrix} 1 & 0 \\ 0 & 1 \end{bmatrix} \begin{bmatrix} x_1 \\ x_2 \end{bmatrix} = \begin{bmatrix} 0 \\ 0 \end{bmatrix} - r \begin{bmatrix} 1 \\ 0 \end{bmatrix} - s \begin{bmatrix} 1 \\ 1 \end{bmatrix} - t \begin{bmatrix} -1 \\ -1 \end{bmatrix} .$$

Then the homogeneous solutions are

$$\mathbf{x}_h = r \begin{bmatrix} -1 \\ 0 \\ 1 \\ 0 \\ 0 \end{bmatrix} + s \begin{bmatrix} -1 \\ -1 \\ 0 \\ 1 \\ 0 \end{bmatrix} + t \begin{bmatrix} 1 \\ 1 \\ 0 \\ 0 \\ 1 \end{bmatrix} .$$

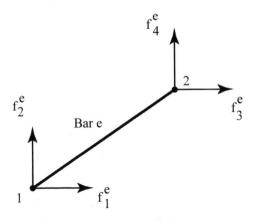

Figure 5.3.1: Bar e with Four Forces

The nullspace matrix is

$$N = \begin{bmatrix} -1 & -1 & 1 \\ 0 & -1 & 1 \\ 1 & 0 & 0 \\ 0 & 1 & 0 \\ 0 & 0 & 1 \end{bmatrix}.$$

The matrix for the solve in step five is only 3×3, which is smaller than the 7×7 matrix in the equilibrium equation,

$$N^T A N = \begin{bmatrix} R_1 + R_3 & R_1 & -R_1 \\ R_1 & R_1 + R_2 + R_4 & -R_1 - R_2 \\ -R_1 & -R_1 - R_2 & R_1 + R_2 + R_5 \end{bmatrix}.$$

5.3.4   Application to Six-bar Truss

The six-bar truss in Subsection 3.4.4, see Figure 3.4.1, assumed there was no deformation of the bars when an external force is applied to one or more nodes. If the force is large, then there may be some significant movement of the nodes resulting in extension or compression of the connecting bars. The model of this has the form of the equilibrium equations. The first equation is an enhanced version of Hooke's law, which we will not go into. The second equation manages the assembly of the local forces and requires the sum of the forces at each node to be a zero force vector.

Assume the structure deforms only in one plane so that at each end of the bar (often called an element) there is a two-dimensional force vector, see Figure 5.3.1. Let the superscript e denote the bar or element number, list the force components in the horizontal direction by odd number subscripts and the vertical forces by the even subscripts so that for each bar there is a four-dimensional force vector

$$\mathbf{f}^e = \begin{bmatrix} f_1^e & f_2^e & f_3^e & f_4^e \end{bmatrix}^T.$$

Using the same labels as in Figure 3.4.1, the sum of the forces at each node gives

Node 1: $[f_3^1 \ f_4^1] + [f_3^2 \ f_4^2] + [0 \ -w] = [0 \ 0],$

Node 2: $[f_1^2 \ f_2^2] + [f_1^3 \ f_2^3] + [f_1^4 \ f_2^4] = [0 \ 0]$ and

Node 3: $[f_1^1 \ f_2^1] + [f_3^3 \ f_4^3] + [f_3^5 \ f_4^5] + [f_3^6 \ f_4^6] = [0 \ 0].$

Now there are still three two-dimensional vector equations, but there are 18 local force components. This can be written as $E\mathbf{x} = \mathbf{g}$ where E is an 6×18 matrix and $\mathbf{g}$ is a external force vector with six components. In $E\mathbf{x} = \mathbf{g}$ the two equations from the two-dimensional vector equation at node one are in rows one and two, and the remaining four rows are obtained from the vector equations at nodes two and three where

$$E = \begin{bmatrix} 0 & 0 & 1 & 0 & 0 & 0 & 1 & 0 & 0 & 0 & 0 & 0 & 0 & 0 & 0 & 0 & 0 & 0 \\ 0 & 0 & 0 & 1 & 0 & 0 & 0 & 1 & 0 & 0 & 0 & 0 & 0 & 0 & 0 & 0 & 0 & 0 \\ 0 & 0 & 0 & 0 & 1 & 0 & 0 & 0 & 1 & 0 & 0 & 0 & 1 & 0 & 0 & 0 & 0 & 0 \\ 0 & 0 & 0 & 0 & 0 & 1 & 0 & 0 & 0 & 1 & 0 & 0 & 0 & 1 & 0 & 0 & 0 & 0 \\ 1 & 0 & 0 & 0 & 0 & 0 & 0 & 0 & 0 & 1 & 0 & 0 & 0 & 1 & 0 & 1 & 0 \\ 0 & 1 & 0 & 0 & 0 & 0 & 0 & 0 & 0 & 0 & 1 & 0 & 0 & 0 & 1 & 0 & 1 \end{bmatrix},$$

$$\mathbf{g} = \begin{bmatrix} 0 & w & 0 & 0 & 0 & 0 \end{bmatrix}^T \text{ and}$$

$$\mathbf{x} = \begin{bmatrix} f_1^1 & f_2^1 & f_3^1 & f_4^1 & f_1^2 & f_2^2 & f_3^2 & f_4^2 & f_1^3 & f_2^3 & f_3^3 & f_4^3 & f_3^4 & f_4^4 & f_3^5 & f_4^5 & f_3^6 & f_4^6 \end{bmatrix}^T.$$

The corresponding augmented matrix may be transformed to a row echelon form matrix by simply moving rows five and six to rows one and two. The development of the first equation in the equilibrium equations is beyond the scope of this course, but this illustration should give the reader an appreciation for the role of row echelon forms and nullspaces.

5.3.5   Application to Fluid Flow

Models for fluid flow have two basic parts. The first part describes the momentum in the directions of the flow, and the second part tracks the change in mass. When a volume is broken into union of small cubes, then these parts form the first and second equations in the equilibrium equations. In the case of an incompressible liquid such as water, any mass moving into a fixed volume must also leave the volume. The incompressibility condition is similar to Kirchhoff's current law and to the balance of forces at each node of a truss.

In order to keep the discussion as simple as possible, consider fluid flow in a plane, and the rate of change of mass for four adjacent squares or cells as indicated in Figure 5.3.2. Each cell has a horizontal and vertical component

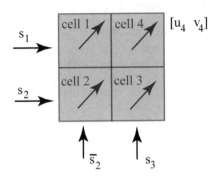

Figure 5.3.2: Fluid Flow in Four Cells

that are proportional to the velocity components in the fluid. Let cell i have horizontal and vertical components given by u_i and v_i where $i = 1, 2, 3$ and 4. For each cell the mass in minus the mass out yields four equations for the eight unknown components. For the four cells depicted in Figure 5.3.2 with incoming mass for cells 1-3, we have the following four equations

$$
\begin{array}{ll}
\text{Cell 1:} & s_1 - u_1 + v_2 - v_1 = 0, \\
\text{Cell 2:} & s_2 - u_2 + \bar{s}_2 - v_2 = 0, \\
\text{Cell 3:} & u_2 - u_3 + s_3 - v_3 = 0 \text{ and} \\
\text{Cell 4:} & u_1 - u_4 + v_3 - v_4 = 0.
\end{array}
$$

This can be written as $E\mathbf{x} = \mathbf{g}$ where E is an 4×8 matrix and $\mathbf{x}$ has components that are proportional to the velocity components

$$
\begin{bmatrix}
1 & 0 & 0 & 0 & 1 & -1 & 0 & 0 \\
0 & 1 & 0 & 0 & 0 & 1 & 0 & 0 \\
0 & -1 & 1 & 0 & 0 & 0 & 1 & 0 \\
-1 & 0 & 0 & 1 & 0 & 0 & -1 & 1
\end{bmatrix}
\begin{bmatrix}
u_1 \\ u_2 \\ u_3 \\ u_4 \\ v_1 \\ v_2 \\ v_3 \\ v_4
\end{bmatrix}
=
\begin{bmatrix}
s_1 \\ s_2 + \bar{s}_2 \\ s_3 \\ 0
\end{bmatrix}.
$$

The augmented matrix $[E \ \ \mathbf{g}]$ can be transformed into a reduced row echelon form by adding row 1 to row 4 and then row 2 to row 3. The vertical components will be the free variables, and the resulting representation of the unknowns can be used in the momentum equations for the fluid flow. This does indicate an important role for nullspaces in computational fluid flow modeling.

5.3.6   Solving Equilibrium Equations Using MATLAB

The MATLAB codes E_eq_bge.m and E_eq_null.m implement the block Gauss elimination and nullspace methods for the two-loop circuit problem. They can easily be modified to do more complicated problems. The following solves the three-loop circuit problem with the same parameters as in Subsection 3.6.3, Example 3.6.4.

```
>> A =
         1. 0  0  0  0
         0 .1  0  0  0
         0  0  3. 0  0
         0  0  0  2. 0
         0  0  0  0  4.
>> E =
        -1  1 -1  0  0
         0 -1  0 -1  1
>> f =
         10
        -20
         0
         0
        -30
>> g =
         0
         0
>> N = null(E,'r')
    N =
        -1 -1  1
         0 -1  1
         1  0  0
         0  1  0
         0  0  1
>> xp = E\g;
>> z = (N'*A*N)\(N'*(f - A*xp));
>> x = xp + N*z;
>> y = (E*E')\(E*(f - A*x));
>> sol_null = [x;y]
    sol_null =
         -5.2290
        -10.3053
         -5.0763
          1.8702
         -8.4351
        -15.2290
          3.7405
```

5.3.7 Exercises

1. Find the solution of the following equilibrium equation using the block Gauss elimination method

$$
\begin{bmatrix}
1 & 0 & 0 & 2 \\
0 & 2 & 0 & 1 \\
0 & 0 & 3 & -1 \\
2 & 1 & -1 & 0
\end{bmatrix}
\begin{bmatrix}
x_1 \\
x_2 \\
x_3 \\
y_1
\end{bmatrix}
=
\begin{bmatrix}
1 \\
-6 \\
12 \\
0
\end{bmatrix}.
$$

2. Find the solution of the following equilibrium equation using the block Gauss elimination method

$$
\begin{bmatrix}
3 & -1 & 0 & 1 & 0 \\
-1 & 3 & -1 & 1 & 2 \\
0 & -1 & 3 & 0 & 1 \\
1 & 1 & 0 & 0 & 0 \\
0 & 2 & 1 & 0 & 0
\end{bmatrix}
\begin{bmatrix}
x_1 \\
x_2 \\
x_3 \\
y_1 \\
y_2
\end{bmatrix}
=
\begin{bmatrix}
1 \\
2 \\
3 \\
4 \\
5
\end{bmatrix}.
$$

3. Find the solution of the equilibrium equation in exercise 1 using the nullspace method.

4. Find the solution of the equilibrium equation in exercise 2 using the nullspace method.

5. Use MATLAB to confirm any calculations in exercises 1-4.

6. Modify the six-bar truss problem to have external forces at both nodes one and three. Find the new equation $[E \ \ g]$ and its reduced row echelon form.

7. In the fluid flow problem consider six cells by attaching two cells to the right of cells three and four. Find the new equation $[E \ \ g]$ and its reduced row echelon form.

Chapter 6

Linear Initial Value Problems

First and second order linear initial value problems are studied. Initial value problems require finding a function of time that satisfies an equation with derivatives and an initial condition. Examples include the motion of projectiles in Subsection 2.3.3, mixing tanks in Subsection 3.7.4 and population models in Subsection 4.4.3. Additional applications include time dependent models of tuned circuits and resonant mass-spring systems. Particular solutions are found by both undetermined coefficients and variation of parameters. The system formulations are developed for the second order problems, and numerical solutions are given by the MATLAB differential equation solvers. Chapters six and seven are an introduction to differential equations, and this topic is covered in more detail in [1] and [7]

6.1 First Order Linear

A *first order linear initial value problem* for an unknown function $x = x(t)$ with given $f(t)$ is

$$\frac{dx}{dt} = x' = ax + f(t) \text{ and } x(0) = x_0.$$

This can be written in differential operator form where $L(x) \equiv x' - ax$ and $L(x) = f(t)$. The operator L is called *linear* because

$$
\begin{aligned}
L(x + y) &= (x + y)' - a(x + y) \\
&= x' - ax + y' - ay \\
&= L(x) + L(y)
\end{aligned}
$$

243

and for all real numbers c

$$
\begin{aligned}
L(cx) &= (cx)' - a(cx) \\
&= c(x' - ax) \\
&= cL(x).
\end{aligned}
$$

The objective is to find $x(t)$ given a, an initial condition $x(0)$ and the function $f(t)$.

When $f(t)$ is a constant, the solution is easy to find by defining $z \equiv ax + f$ and noting $z' = ax' + 0$ and so that $x' = ax + f$ becomes $(z/a)' = z$. In this case $z(t) = z(0)e^{at}$ and, hence,

$$
\begin{aligned}
ax(t) + f &= (ax(0) + f)e^{at} \\
x(t) &= (x(0) + f/a)e^{at} - f/a.
\end{aligned}
$$

If $f(t)$ is not a constant, then the solution process is more complicated.

6.1.1   Motivating Applications

A simple model for the vertical velocity v of a *projectile* is derived from Newton's law of motion

$$ mv' = -cv - mg $$

where m is the mass, g is the gravitational constant and $-cv$ is the air resistance. In this case $x = v$, $a = -c/m$, $f(t) = -g$ and $x(0) = v(0)$ is the initial velocity. The velocity as a function of these parameters is

$$ v(t) = (v(0) + mg/c)e^{-(c/m)t} - mg/c. $$

Another application is to *circuits*. Suppose the circuit has two components in series and an imposed voltage $V(t) = 110\sin(120\pi t)$. Let one component be a resistor R with voltage drop IR and the other component be a capacitor C with voltage drop Q/C where Q is the charge and $Q' = I$. Kirchhoff's voltage law gives

$$ Q'R + (1/C)Q = V(t) $$

where $x = Q$, $a = -1/(RC)$, $f(t) = V(t)/R$ and $x(0) = Q(0)$ is the initial charge. Since $f(t)$ is now a function of time, the solution is a little more difficult to find, see Example 6.1.4.

A third application is modeling the amount $A(t)$ of a chemical in a single *mixing tank* with fixed volume V with incoming concentration $C(t)$ with flow rate R and outgoing concentration $A(t)/V$ and flow rate R. The rate of change of $A(t)$ is the rate incoming minus the rate outgoing

$$ A' = RC(t) - R(A/V). $$

Here $x = A$, $a = -A/V$, $f(t) = RC(t)$ and $x(0) = A(0)$ is the initial amount. If the incoming concentration $C(t)$ varies with time, then the solution step will be more challenging.

Example 6.1.1. Consider the single mixing tank model with $R = 6$, $C = .1$, $V = 1000$ and zero initial amount. The initial value problem is

$$A' = .6 - .006A \text{ and } A(0) = 0.$$

We wish to verify $A(t) = -100e^{-.006t} + 100$ is the solution. The left and right sides of the differential equation must be equal and the solution evaluated at time $t = 0$ must also be zero:

left side $= A' = (-100e^{-.006t} + 100)' = -100(e^{-.006t})' = -100e^{-.006t}(-.006)$,

right side $= .6 - .006(-100e^{-.006t} + 100) = -.006(-100e^{-.006t})$ and

$A(0) = -100e^{-.006(0)} + 100 = -100(1) + 100 = 0.$

It is always a good idea to check or verify a proposed solution is truly a solution.

6.1.2 Homogeneous Solutions $L(x_h) = 0$

The solution of $x' = ax + 0$ is called the *homogeneous solution* of $x' = ax + f(t)$. From the properties of the exponential function, the homogeneous solution must have the form

$$x_h(t) = Ce^{at}.$$

Any *particular solution* of $x' = ax + f(t)$ is denoted by $x_p = x_p(t)$ so that $L(x_p) = f$ and $L(x_h) = 0$. Since L is a linear operator,

$$
\begin{aligned}
L(x_h + x_p) &= L(Ce^{at} + x_p) \\
&= CL(e^{at}) + L(x_p) \\
&= 0 + f.
\end{aligned}
$$

Thus, the *general solution* of $x' = ax + f(t)$ has the form

$$x(t) = Ce^{at} + x_p(t).$$

The constant C in the homogeneous solution is determined from the initial condition and the particular solution. In Example 6.1.1 the homogeneous solution is $-100e^{-.006t}$ and the particular solution is 100. If $f(t)$ is not a constant, then the particular solution will not be a constant and will be harder to find.

6.1.3 Solutions of $L(x_p) = f$ by Undetermined Coefficients

The method of undetermined coefficients works for some classes of $f(t)$, and it is an "educated" guess for the form of the solution. The simplest class is if $f(t)$ is a polynomial of t, and in this case one speculates that the particular solution may also be a polynomial. The coefficients in the polynomial particular solution must be chosen so that the differential equation holds for all values of t. Other classes of $f(t)$ are sine, cosine and exponential functions, and the particular solution for these are illustrated in the following examples.

Example 6.1.2. Consider the initial value problem

$$x' = 3x + 2t + 1 \text{ and } x(0) = 10.$$

Here $f(t) = 2t + 1$ and $a = 3$. Then the homogeneous and particular solutions
have the form

$$x_h = Ce^{3t} \text{ and } x_p = c_1 t + c_0.$$

First, find the coefficients in the particular solution by requiring x_p to satisfy
the differential equation

$$
\begin{aligned}
(c_1 t + c_0)' &= 3(c_1 t + c_0) + 2t + 1 \\
(0)t + (c_1 1 + 0)1 &= (3c_1 + 2)t + (3c_0 + 1)1.
\end{aligned}
$$

Since this must hold for all time t, the coefficients of t and 1 must be equal

$$
\begin{aligned}
0 &= 3c_1 + 2 \text{ and} \\
c_1 &= 3c_0 + 1.
\end{aligned}
$$

The solution is $c_1 = -2/3$ and $c_0 = -5/9$ giving $x_p = -(2/3)t - 5/9$. The
general solution is

$$x = x_h + x_p = Ce^{3t} - (2/3)t - 5/9.$$

Second, find the constant C in the homogeneous solution by requiring

$$
\begin{aligned}
x(0) &= 10 \\
Ce^{3(0)} - (2/3)0 - 5/9 &= 10 \\
C - 5/9 &= 10.
\end{aligned}
$$

Then $C = 95/9$ and $x = (95/9)e^{3t} - (2/3)t - 5/9$. If $f(t)$ were a quadratic
polynomial, then the particular solution must have the form

$$x_p = c_2 t^2 + c_1 t + c_0.$$

Example 6.1.3. Consider the initial value problem

$$x' = ax + 4e^{2t} \text{ and } x(0) = 10.$$

Since $f(t)$ is an exponential function, it is reasonable to speculate that the
particular solution should have the form

$$x_p = ce^{2t}.$$

Choose the coefficient so that the differential equation holds

$$
\begin{aligned}
(ce^{2t})' &= a(ce^{2t}) + 4e^{2t} \\
ce^{2t} 2 &= a(ce^{2t}) + 4e^{2t}.
\end{aligned}
$$

Since this must hold for all time t, $c2 = ac + 4$. Provided $a \neq 2$, one can solve
for c to get $c = 4/(2 - a)$. The general solution is

$$x = x_h + x_p = Ce^{at} + (4/(2 - a))e^{2t}.$$

The initial condition gives

$$
\begin{aligned}
x(0) &= 10 \\
Ce^{a(0)} + (4/(2-a))e^{2(0)} &= 10 \\
C + 4/(2-a) &= 10.
\end{aligned}
$$

If $a = 3$, then the solution of the initial value problem is $x = 14e^{3t} + (-4)e^{2t}$. If $a = 2$, then one must make a more elaborate guess for the particular solution such as $x_p = cte^{2t}$.

Example 6.1.4. If $f(t)$ is a sine or cosine function, then the particular solution might be a linear combination of the sine and cosine functions. Consider the initial value problem

$$
x' = ax + 10\sin(t) \text{ and } x(0) = 20.
$$

Assume the particular solution has the form

$$
x_p = c\sin(t) + d\cos(t).
$$

Choose the coefficients so that the differential equation holds

$$
\begin{aligned}
(c\sin(t) + d\cos(t))' &= a\,(c\sin(t) + d\cos(t)) + 10\sin(t) \\
c\cos(t) - d\sin(t) &= (ac + 10)\sin(t) + ad\cos(t).
\end{aligned}
$$

Match the coefficients of $\cos(t)$ and $\sin(t)$ to obtain

$$
\begin{aligned}
c &= ad \text{ and} \\
-d &= ac + 10.
\end{aligned}
$$

Solve these to get $d = -10/(a^2 + 1)$ and $c = -10a/(a^2 + 1)$. If $a = 2$, then the general solution is

$$
x = Ce^{2t} - 4\sin(t) - 2\cos(t)
$$

and the initial condition gives

$$
\begin{aligned}
x(0) &= 20 \\
Ce^{2(0)} - 4\sin(0) - 2\cos(0) &= 20 \\
C - 0 - 2 &= 20.
\end{aligned}
$$

The solution of the initial value problem is $x = 22e^{2t} - 4\sin(t) - 2\cos(t)$.

6.1.4 Solutions of $L(x_p) = f$ by Variation of Parameters

The particular solution will be assumed to have of the form

$$
x_p(t) = e^{at}v(t)
$$

where the function $v(t)$ (the variation of parameter) will be chosen so that the differential equation holds. Use the product rule to compute the derivative

$$
\begin{aligned}
\left(e^{at}v(t)\right)' &= a\left(e^{at}v(t)\right) + f(t) \\
\left(e^{at}\right)'v(t) + e^{at}(v(t))' &= a\left(e^{at}v(t)\right) + f(t) \\
e^{at}av(t) + e^{at}v'(t) &= a\left(e^{at}v(t)\right) + f(t).
\end{aligned}
$$

Solve for the derivative of $v(t)$

$$
v'(t) = e^{-at}f(t).
$$

Integrate this to solve for $v(t)$ and the particular solution

$$
x_p(t) = e^{at}v(t) = e^{at}\int^t e^{-a\tau}f(\tau)d\tau.
$$

This gives a compact formula for the particular solution. The integral exists if the $f(t)$ is continuous. In many cases the integral may exist, but it may not be easy to compute.

Example 6.1.5. Reconsider Example 6.1.2 where $a = 3$ and $f(t) = 2t + 1$ and use integration by parts to compute the integral

$$
\begin{aligned}
x_p(t) &= e^{at}\int^t e^{-a\tau}f(\tau)d\tau \\
&= e^{at}\int^t e^{-a\tau}(2\tau + 1)d\tau \\
&= e^{at}\int^t e^{-a\tau}2\tau d\tau + e^{at}\int^t e^{-a\tau}d\tau \\
&= e^{at}(2/a^2(-at - 1))e^{-at} + e^{at}(-1/a)e^{-at} \\
&= (-2/a)t - (2 + a)/a^2.
\end{aligned}
$$

This agrees with the method of undetermined coefficients that was used in Example 6.1.2. Note, the lower limit in the integral can be any constant, because the difference in the integral is part of the homogeneous solution.

The results of the above discussion are summarized in the following theorem.

Theorem 6.1.1 *(First Order Linear Initial Value Problem) Consider the initial value problem $L(x) = x' - ax = f\ (t)$ and $x(0) = x_0$. If $f(t)$ is continuous on the interval $[0\ \ T]$, then for t in this interval the solution of the differential equation is*

$$
\begin{aligned}
x(t) &= Ce^{at} + e^{at}\int^t e^{-a\tau}f(\tau)d\tau \\
&= e^{at}(C + \int^t e^{-a\tau}f(\tau)d\tau).
\end{aligned}
$$

The solution of the initial value problem can be found by requiring C to be chosen so that $x(0) = x_0$. Moreover, this is the only solution to the initial value problem.

6.1.5 Solutions Using int() and dsolve()

MATLAB can do symbolic operations, which is in contrast to numerical computation. The MATLAB command int() does symbolic integration such as in Example 6.1.5 where integration by parts was required.

```
>> syms t a
>> sol = int('exp(-a*t)*(2*t+1)','t')
>>      sol = -1/a*(-2/a*(-exp(-a*t)*a*t-exp(-a*t))+exp(-a*t))
>> simple(sol)
>>      -exp(-a*t)*(2*a*t + 2 + a)/a^2
```

The MATLAB command dsolve() finds the symbolic solution of many initial value problems such as in Example 6.1.5.

```
>> syms t a
>> solde = dsolve('Dx = a*x + 2*t + 1', 'x(0) = 10')
>>      solde = -2/a*t - 2/a^2 - 1/a + (2 + a + 10*a^2)/a^2*exp(a*t)
```

Both the above symbolic calculations agree with the by-hand calculations done in Example 6.1.5.

6.1.6 Exercises

1. As in Example 6.1.1 verify $x = (95/9)e^{3t} - (2/3)t - 5/9$ is a solution to Example 6.1.2.
2. As in Example 6.1.1 verify $x = 14e^{3t} + (-4)e^{2t}$ is a solution to Example 6.1.3.
3. Consider the initial value problem $x' = 5x + 3t + 2$ and $x(0) = 5$.
    (a).   Find the homogeneous solution.
    (b).   Use undetermined coefficients to find the particular solution.
    (c).   Find the solution to the initial value problem.
4. Consider the initial value problem $x' = 3x + 3t^2 + 2t + 1$ and $x(0) = 10$.
    (a).   Find the homogeneous solution.
    (b).   Use undetermined coefficients to find the particular solution (see the remark at the end of Example 6.1.2).
    (c).   Find the solution to the initial value problem.
5. Consider the initial value problem $x' = 2x + 10e^{3t}$ and $x(0) = 5$.
    (a).   Find the homogeneous solution.
    (b).   Use undetermined coefficients to find the particular solution.
    (c).   Find the solution to the initial value problem.
6. Consider the initial value problem $x' = 2x + 4e^{2t}$ and $x(0) = 12$.
    (a).   Find the homogeneous solution.

(b).   Use undetermined coefficients to find the particular solution
(see the remark at the end of Example 6.1.3).
(c).   Find the solution to the initial value problem.

7.   Consider the initial value problem $x' = 5x + 10\cos(t)$ and $x(0) = 1$.
(a).   Find the homogeneous solution.
(b).   Use undetermined coefficients to find the particular solution.
(c).   Find the solution to the initial value problem.

8.   Consider the initial value problem $x' = 5x + 10\cos(2t) + 5\sin(2t)$ and
$x(0) = 1$.
(a).   Find the homogeneous solution.
(b).   Use undetermined coefficients to find the particular solution.
(c).   Find the solution to the initial value problem.

9.   Use variation of parameters as in Theorem 6.1.1 to find the solution to
the initial value problem in exercise 3.

10.   Use variation of parameters as in Theorem 6.1.1 to find the solution to
the initial value problem in exercise 5.

11.   Use the MATLAB commands int() and dsolve() to confirm any by-hand
calculations in exercises 3-10.

12.   Let $x(t)$ and $y(t)$ satisfy the same initial value problem

$$\begin{aligned} x' &= ax + f(t) \text{ and } x(0) = x_0 \\ y' &= ay + f(t) \text{ and } y(0) = x_0. \end{aligned}$$

(a). Show $z \equiv x - y$ must satisfy $z' = az$ and $z(0) = 0$.
(b). Show there is only one solution to the initial value problem, that is,
$x = y$.

6.2   Second Order Linear

Second order initial value problems require one to find a function of time $x(t)$
that satisfies a differential equation with two time derivatives and two initial
conditions. We will focus on *second order linear initial value problems* with
constant coefficients

$$ax'' + bx' + cx = f(t) \text{ with } x(0) = x_0 \text{ and } x'(0) = v_0.$$

The motion of a projectile as illustrated in Subsection 2.3.3 is an example where
$x(t)$ is the position, $x'(t)$ is the velocity in one direction, $x''(t)$ is the acceleration
and the differential equation is derived from Newton's law of motion. Two
important applications are the motion of mass attached to a spring and the
electrical current in a series LRC circuit. These are precursors to models of
cars and trucks with several springs and shock absorbers and to models of
tuning circuits for a radio or television.

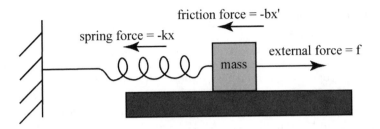

Figure 6.2.1: Mass-Spring System

6.2.1 Motivating Application to Mass-Spring

Consider a mass, which is attached to a spring as indicated in Figure 6.2.1. Let $x = x(t)$ be the displacement from equilibrium, which is the position with no motion and no external force. The mass moves in the horizontal direction and is subject to three forces: $-kx$ from the spring for "small values" of $x(t)$, $-bx'$ due the friction force given by gravity and the solid surface, and an external force f. By Newton's law of motion the mass times acceleration is the sum of the three forces

$$mx'' = f - kx - bx'.$$

If the initial position $x(0)$ and initial velocity $x'(0)$ are given, then the objective is to find $x(t)$ for all time $t > 0$. A similar system is a model for a car or truck, but in this case the mass is the vehicle, the system is vertical and the friction force is replaced by a force from a shock absorber. If either the shock absorber is too small or too woren, then the coefficient b will be relatively small.

Example 6.2.1. Assume there are no friction and external forces so that the model is

$$mx'' = 0 - kx - 0x' \text{ and } x(0) = x_0 \text{ and } x'(0) = v_0.$$

If the initial position or the initial velocity is not zero, then the mass will have a periodic motion. In fact, one can easily verify from the first and second derivatives and the chain rule that two possible solutions are

$$x_1(t) \equiv \sin(\sqrt{k/m}t) \text{ and } x_2(t) \equiv \cos(\sqrt{k/m}t).$$

The first and second derivatives of $x_1(t)$ are

$$x_1'(t) = \frac{d}{dt}\sin(\sqrt{k/m}t) = \cos(\sqrt{k/m}t)\frac{d}{dt}\sqrt{k/m}t = \cos(\sqrt{k/m}t)\sqrt{k/m}$$

$$x_1''(t) = \frac{d}{dt}\cos(\sqrt{k/m}t)\sqrt{k/m} = -\sin(\sqrt{k/m}t)\sqrt{k/m}\sqrt{k/m}.$$

This means $mx_1'' = -kx_1$. The verification for $x_2(t)$ is similar. Because the derivative operation is linear, the most general solution of $mx'' + kx = 0$ is a linear combination of $x_1(t)$ and $x_2(t)$

$$x(t) = c_1 \sin(\sqrt{k/mt}) + c_2 \cos(\sqrt{k/mt}).$$

The initial conditions can be used to determine the values of the coefficients. As an illustration suppose $m = 1$, $k = 4$ so that $\sqrt{k/m} = 2$, and with $x(0) = 2$ and $x'(0) = 1$. The initial conditions mean at time equal to zero the mass is two units the right of the equilibrium position and is set in motion with a velocity of unit one moving to the right. In order to solve for the coefficients, we must use the two initial conditions

$$
\begin{aligned}
x(0) &= c_1 \sin(2(0)) + c_2 \cos(2(0)) = 2 \\
&= c_1 0 + c_2 1 = 2 \text{ and} \\
x'(0) &= c_1 \cos(2(0)) 2 - c_2 \sin(2()) 2 = 1 \\
&= c_1 2 + c_2 0 = 1.
\end{aligned}
$$

Note, $x'(0)$ means first compute the derivative and then evaluate the derivative at time equal to zero. This algebraic system in matrix form is

$$\begin{bmatrix} 0 & 1 \\ 2 & 0 \end{bmatrix} \begin{bmatrix} c_1 \\ c_2 \end{bmatrix} = \begin{bmatrix} 2 \\ 1 \end{bmatrix}$$

and is easy to solve. So, the solution of the second order initial value problem is

$$x(t) = (1/2)\sin(2t) + 2\cos(2t).$$

6.2.2   Solution of the Homogeneous System: $f(t) = 0$

In Example 6.1.1 the solution of $mx'' + kx = 0$ was simply stated, and we verified that it was a linear combination of sine and cosine functions. In the remainder of this and the next sections we will present a method for discovering the form of the solutions to the *homogeneous second order differential equation*

$$ax'' + bx' + cx = f(t) = 0.$$

When the coefficients are constants, we assume the solution has the form of an exponential function

$$x = e^{rt} \text{ where } r \text{ is possibly a complex number.}$$

In order to find r, we put this exponential function into the differential equation

$$
\begin{aligned}
a\left(e^{rt}\right)'' + b\left(e^{rt}\right)' + ce^{rt} &= 0 \\
ae^{rt}r^2 + be^{rt}r + ce^{rt} &= 0 \\
(ar^2 + br + c)e^{rt} &= 0.
\end{aligned}
$$

Whether or not r is real or complex, the exponential function is

$$e^{rt} = e^{\text{Re}(r)t + i\,\text{Im}(r)t} = e^{\text{Re}(r)t}e^{i\,\text{Im}(r)t}$$

is defined and not zero, see Section 1.2. Thus, the quadratic factor must be zero, which is often called the *auxiliary or characteristic* equation,

$$ar^2 + br + c = 0.$$

By the quadratic formula there are three cases to consider: two distinct real roots, repeated real roots and complex roots. As we shall see, in all three cases this leads to two homogeneous solutions $x_1(t)$ and $x_2(t)$ that are not multiples of each other.

Definition 6.2.1. The differential operator $L(x)$ is *linear* if and only if

$$L(x + y) = L(x) + L(y) \text{ and } L(\widehat{c}x) = \widehat{c}L(x).$$

The second order differential operator $L(x) \equiv ax'' + bx' + cx$ is linear because

$$
\begin{aligned}
L(x + y) &= a(x + y)'' + b(x + y)' + c(x + y) \\
&= a(x'' + y'') + b(x' + y') + c(x + y) \\
&= L(x) + L(y)
\end{aligned}
$$

and for all real numbers $\widehat{c}$

$$
\begin{aligned}
L(\widehat{c}x) &= a(\widehat{c}x)'' + b(\widehat{c}x)' + c(\widehat{c}x) \\
&= a\widehat{c}x'' + b\widehat{c}x' + c\widehat{c}x \\
&= \widehat{c}L(x).
\end{aligned}
$$

Because of the linear property, any linear combination of the two homogeneous solutions will also be a homogeneous solution

$$
\begin{aligned}
L(c_1x_1 + c_2x_2) &= L(c_1x_1) + L(c_2x_2) \\
&= c_1L(x_1) + c_2L(x_2) \\
&= c_10 + c_20 = 0.
\end{aligned}
$$

Example 6.2.1 illustrates this where the characteristic equation is $1r^2 + 0r + 4 = 0$ and has conjugate imaginary roots $r = \pm i2$, which leads to the sine and cosine functions as outlined in the Section 6.3. The next two subsections discuss the distinct real and repeated real root cases.

6.2.3 Homogeneous Solution: Distinct Real Roots

Consider the distinct real solutions of $ar^2 + br + c = 0$ where $b^2 - 4ac > 0$ and

$$
\begin{aligned}
r_1 &= (-b + \sqrt{b^2 - 4ac})/2a \text{ and} \\
r_2 &= (-b - \sqrt{b^2 - 4ac})/2a.
\end{aligned}
$$

The two homogeneous solutions are

$$x_1(t) = e^{r_1 t} \text{ and } x_2(t) = e^{r_2 t}.$$

Because $r_1 \neq r_2$, they are not multiples of each other, and $x(t) = c_1 e^{r_1 t} + c_1 e^{r_2 t}$ is the general homogeneous solution.

Example 6.2.2. The second order differential equation $x'' + 5x' - 6x = 0$ has the characteristic equation

$$r^2 + 5r - 6 = 0$$

whose solutions are

$$r = (-5 \pm \sqrt{5^2 - 4(1)(-6)})/2 = (-5 \pm \sqrt{49})/2 = 1 \text{ or } -6.$$

Therefore, the solution of the differential equation is

$$x(t) = c_1 e^t + c_2 e^{-6t}.$$

Suppose the initial conditions are $x(0) = 1$ and $x'(0) = 2$ so that the coefficient must satisfy

$$
\begin{aligned}
x(0) &= c_1 e^0 + c_2 e^{-6(0)} = c_1 + c_2 = 1 \text{ and} \\
x'(0) &= c_1 e^0 + c_2 e^{-6(0)}(-6) = c_1 - 6c_1 = 2.
\end{aligned}
$$

Note, this algebraic system may be written in matrix form

$$
\begin{bmatrix} 1 & 1 \\ 1 & -6 \end{bmatrix}
\begin{bmatrix} c_1 \\ c_2 \end{bmatrix}
= \begin{bmatrix} 1 \\ 2 \end{bmatrix},
$$

and the solution is $c_1 = -8/(-7)$ and $c_2 = 1/(-7)$. The solution of the initial value problem is

$$x(t) = (8/7)e^t - (1/7)e^{-6t}.$$

6.2.4   Homogeneous Solution: Repeated Real Roots

Consider the repeated real solutions of $ar^2 + br + c = 0$ where $b^2 - 4ac = 0$ so that $r_1 = r_2 = -b/2a$. Using the first order differential equation as a guide, choose two possible homogeneous solutions to be

$$x_1(t) = e^{(-b/2a)t} \text{ and } x_2(t) = te^{(-b/2a)t}.$$

We need to verify $x_2(t) = te^{(-b/2a)t}$ is really a homogeneous solution of

$$ax'' + bx' + cx = 0$$

when $b^2 - 4ac = 0$. Insert $x_2(t) = te^{(-b/2a)t}$ into this differential equation

$$
\begin{aligned}
L(te^{(-b/2a)t}) &= a\left(te^{(-b/2a)t}\right)'' + b(te^{(-b/2a)t})' + c(te^{(-b/2a)t}) \\
&= a(te^{(-b/2a)t}(-b/2a)^2 + 2e^{(-b/2a)t}(-b/2a)) \\
&\quad b(1e^{(-b/2a)t} + te^{(-b/2a)t}(-b/2a)) + c(te^{(-b/2a)t}) \\
&= e^{(-b/2a)t}[(a2(-b/2a) + b) + t(a(-b/2a)^2 + b(-b/2a) + c)] \\
&= e^{(-b/2a)t}[0 + t(0)] = 0.
\end{aligned}
$$

Example 6.2.3. The second order differential equation $x'' + 4x' + 4x = 0$ has the characteristic equation

$$r^2 + 4r + 4 = 0$$

whose solution is

$$r = (-4 \pm \sqrt{4^2 - 4(1)(4)})/2 = (4 \pm \sqrt{0})/2 = -2 \text{ and } -2.$$

The solution of the differential equation is

$$x(t) = c_1 e^{-2t} + c_2 t e^{-2t}.$$

If the initial conditions are $x(0) = 1$ and $x'(0) = 3$, then

$$
\begin{aligned}
x(0) &= c_1 e^{-2(0)} + c_2(0)e^{-2(0)} = c_1 + c_2 0 = 1 \text{ and} \\
x'(0) &= c_1 e^{-2(0)}(-2) + c_2(1e^{-2(0)} + (0)e^{-2(0)}(-2)) \\
&= c_1(-2) + c_2 = 3.
\end{aligned}
$$

The matrix version of the algebraic system is

$$
\begin{bmatrix} 1 & 0 \\ -2 & 1 \end{bmatrix} \begin{bmatrix} c_1 \\ c_2 \end{bmatrix} = \begin{bmatrix} 1 \\ 3 \end{bmatrix}
$$

whose solution is $c_1 = 1$ and $c_2 = 5$. The solution of the initial value problem is

$$x(t) = e^{-2t} + 5te^{-2t}.$$

6.2.5 Homogeneous Initial Value Problem

Let $x_1(t)$ and $x_2(t)$ be homogeneous solutions so that $x(t) = c_1 x_1(t) + c_2 x_2(t)$ is also a homogeneous solution. We want to be able to choose the coefficients such that the initial conditions $x(0) = x_0$ and $x'(0) = v_0$ hold. This results in the following algebraic problem

$$
\begin{aligned}
x(0) &= c_1 x_1(0) + c_2 x_2(0) \text{ and} \\
x'(0) &= c_1 x_1'(0) + c_2 x_2'(0).
\end{aligned}
$$

The matrix form is

$$
\begin{bmatrix} x_1(0) & x_2(0) \\ x_1'(0) & x_2'(0) \end{bmatrix} \begin{bmatrix} c_1 \\ c_2 \end{bmatrix} = \begin{bmatrix} x_0 \\ v_0 \end{bmatrix}.
$$

The 2×2 matrix

$$
W(0) \equiv \begin{bmatrix} x_1(0) & x_2(0) \\ x_1'(0) & x_2'(0) \end{bmatrix}
$$

is called the *Wronskian* matrix associated with the differential equation. If $\det(W(0)) \neq 0$, then the solution of algebraic equation exists and is unique, see Theorem 3.7.3. For the distinct real case

$$
\det(W(0)) = \det\left(\begin{bmatrix} 1 & 1 \\ r_1 & r_2 \end{bmatrix} \right) = r_2 - r_1 \neq 0.
$$

For the repeated real case

$$\det(W(0)) = \det(\begin{bmatrix} 1 & 0 \\ -b/2a & 1 \end{bmatrix}) = 1 \neq 0.$$

Thus, one can solve the initial value problem for the both the distinct and repeated real solutions of the characteristic equation. These results are summarized in the following theorem

Theorem 6.2.1 *(Homogeneous Linear Initial Value Problem) Let $x_1(t)$ and $x_2(t)$ be homogeneous solutions of the differential equation $L(x) = ax'' + bx' + cx = 0$. Consider the initial value problem given by the differential equation and the initial conditions $x(0) = x_0$ and $x'(0) = v_0$. L is a linear operator and, so, $c_1 x_1(t) + c_2 x_2(t)$ is also a homogeneous solution. If $\det(W(0)) = x_1(0)x_2'(0) - x_2(0)x_1'(0) \neq 0$, then one can find the coefficients such that the initial conditions hold, and the solution of the initial value problem is unique.*

6.2.6   Solutions Using MATLAB Command dsolve()

The MATLAB command dsolve() can also be used to solve linear second order initial value problems. The following illustrate how Examples 6.2.1-6.2.3 can be done using dsolve(), and note the results agree with the above by-hand calculations.

```
>> syms m k t
% example 1
>> dsolve('m*D2x + k*x = 0','t')
      C1*sin((k/m)^1/2*t) + C2*cos((k/m)^1/2*t)
% example 1 with initial conditions
>> dsolve('D2x + 4*x = 0','x(0) = 2','Dx(0) = 1','t')
      2*cos(2*t) + 1/2*sin(2*t)
% example 2 with initial conditions
>> dsolve('D2x + 5*Dx - 6*x = 0','x(0) = 1','Dx(0) = 2','t')
      8/7*exp(t) - 1/7*exp(-6*t)
% example 3 with initial conditions
>> dsolve('D2x + 4*Dx +4*x = 0','x(0) = 1','Dx(0) = 3','t')
      exp(-2*t) + 5*exp(-2*t)*t
```

6.2.7   Exercises

1.   Consider Example 6.2.1.
   (a).   Show $x_2(t)$ is a homogeneous solution.
   (b).   Verify $x(t) = (1/2)\sin(2t) + 2\cos(2t)$ is the solution of the initial value problem.
2.   Verify $x(t) = (8/7)e^t - (1/7)e^{-6t}$ is the solution of the initial value problem in Example 6.2.2.

3.   Verify $x(t) = e^{-2t} + 5te^{-2t}$ is the solution of the initial value problem in Example 6.2.3.

4.   Consider the initial value problem $x'' - 3x' + 2x = 0$ with $x(0) = 1$ and $x'(0) = 2$.
     (a).   Find the homogeneous solutions.
     (b).   Find the solution to the initial value problem.

5.   Consider the initial value problem $2x'' + 2x' - 12x = 0$ with $x(0) = -1$ and $x'(0) = 1$.
     (a).   Find the homogeneous solutions.
     (b).   Find the solution to the initial value problem.

6.   Consider the initial value problem $x'' + 2x' + x = 0$ with $x(0) = 2$ and $x'(0) = 1$.
     (a).   Find the homogeneous solutions.
     (b).   Find the solution to the initial value problem.

7.   Consider the initial value problem $3x'' - 24x' + 48x = 0$ with $x(0) = -2$ and $x'(0) = 3$.
     (a).   Find the homogeneous solutions.
     (b).   Find the solution to the initial value problem.

8.   Use the MATLAB command dsolve() to confirm any by-hand calculations in exercises 4-7.

9.   Let $x(t)$ and $y(t)$ be solutions of the initial value problem

$$\begin{aligned} ax'' + bx' + cx &= 0 \text{ with } x(0) = x_0 \text{ and } x'(0) = v_0 \\ ay'' + by' + cy &= 0 \text{ with } y(0) = x_0 \text{ and } y'(0) = v_0. \end{aligned}$$

     (a).   Define $z(t) \equiv x(t) - y(t)$ and show

$$az'' + bz' + cz = 0 \text{ with } z(0) = 0 \text{ and } z'(0) = 0.$$

     (b).   Let $W(0)$ be the Wronskian associated with the differential equation. Assume $\det(W(0)) \neq 0$ and show $z(t)$ must be zero and, hence, the solution of the initial value problem is unique.

6.3   Homogeneous and Complex Solution

As in the preceding section the homogeneous solution of $ax'' + bx' + cx = 0$ has the form e^{rt}. The r may be a complex number and must be chosen so that the differential equation holds. This leads to the characteristic equation $ar^2 + br + c = 0$, which can be solved by the quadratic formula. The distinct and repeated real root cases were considered in the previous section. Here we assume $b^2 - 4ac < 0$ where the roots are complex numbers, and complex roots will lead to homogeneous solutions that are products of exponential and sine or cosine functions. An application to damped (non-zero friction or resistive forces) mass-springs will illustrate the importance of this topic.

6.3.1   Homogeneous Solutions: Complex Roots

In this section we use Euler's formula for the complex exponential function, see Section 1.2,

$$\begin{aligned}
e^{iy} &= \cos(y) + i\sin(y) \text{ and} \\
e^{z} &= e^{x+iy} = e^x e^{iy} = e^x(\cos(y) + i\sin(y)).
\end{aligned}$$

The sine and cosine functions can be expressed as complex linear combinations of e^{iy} and e^{-iy}

$$\begin{aligned}
\cos(y) &= \frac{1}{2}e^{iy} + \frac{1}{2}e^{-iy} \text{ and} \\
\sin(y) &= \frac{1}{2i}e^{iy} - \frac{1}{2i}e^{-iy}.
\end{aligned}$$

We will use these when x is replaced by the real part rt, and y is replaced by the imaginary part of rt.

When $b^2 - 4ac < 0$, the solutions of the characteristic equation can be expressed as

$$\begin{aligned}
r &= \alpha \pm i\beta \text{ where} \\
\alpha &\equiv -b/2a \text{ and } \beta \equiv \sqrt{-b^2 + 4ac}/2a.
\end{aligned}$$

Euler's formula then gives two possible homogeneous solutions

$$\begin{aligned}
\widehat{x}_1(t) &= e^{rt} = e^{(\alpha+i\beta)t} = e^{\alpha t}(\cos(\beta t) + i\sin(\beta t)) \text{ and} \\
\widehat{x}_2(t) &= e^{rt} = e^{(\alpha-i\beta)t} = e^{\alpha t}(\cos(\beta t) - i\sin(\beta t)).
\end{aligned}$$

These are complex valued functions, but appropriate linear combinations generate two real valued homogeneous solutions

$$\begin{aligned}
x_1(t) &= (1/2)\widehat{x}_1(t) + (1/2)\widehat{x}_2(t) = e^{\alpha t}\cos(\beta t) \text{ and} \\
x_2(t) &= (1/(2i))\widehat{x}_1(t) - (1/(2i))\widehat{x}_2(t) = e^{\alpha t}\sin(\beta t).
\end{aligned}$$

One can easily show the Wronskian at time equal to zero has non-zero determinant

$$\det(W(0)) = \det\left(\begin{bmatrix} x_1(0) & x_2(0) \\ x_1'(0) & x_2'(0) \end{bmatrix}\right) = \det\left(\begin{bmatrix} 1 & 0 \\ \alpha & \beta \end{bmatrix}\right) = \beta \neq 0,$$

and, therefore, one can find c_1 and c_2 such that $c_1 x_1(t) + c_2 x_2(t)$ will satisfy given initial values $x(0)$ and $x'(0)$.

Example 6.3.1. Again consider the undamped mass-spring system $mx'' + kx = 0$ where $a = m$, $b = 0$ and $c = k$ so that $b^2 - 4ac = 0 - 4mk < 0$. Then $\alpha = 0$ and $\beta = \sqrt{4mk}/2m = \sqrt{k/m}$ and

$$\begin{aligned}
x_1(t) &= e^{0t}\cos(\sqrt{k/mt}) = \cos(\sqrt{k/mt}) \text{ and} \\
x_2(t) &= e^{0t}\sin(\sqrt{k/mt}) = \sin(\sqrt{k/mt}).
\end{aligned}$$

Suppose the mass is initially at rest and is one unit to the left of the equilibrium position, that is, $x(0) = -1$ and $x'(0) = 0$. Find c_1 and c_2 so that $c_1 x_1(t) + c_2 x_2(t)$ satisfies these initial conditions

$$c_1 \cos(\sqrt{k/m}0) + c_2 \sin(\sqrt{k/m}0) = c_1 1 + c_2 0 = -1 \text{ and}$$
$$c_1 \sin(\sqrt{k/m}0)(-\sqrt{k/m}) + c_2 \cos(\sqrt{m/k}0)\sqrt{k/m} = c_1 0 + c_2 \sqrt{k/m} = 0.$$

The matrix version of this algebraic system is

$$\begin{bmatrix} 1 & 0 \\ 0 & \sqrt{k/m} \end{bmatrix} \begin{bmatrix} c_1 \\ c_2 \end{bmatrix} = \begin{bmatrix} -1 \\ 0 \end{bmatrix}$$

whose solution is $c_1 = -1$ and $c_2 = 0$. The solution of the corresponding initial value problem is

$$x(t) = (-1)\cos(\sqrt{k/mt}) + (0)\sin(\sqrt{k/mt}) = -\cos(\sqrt{k/mt}).$$

Example 6.3.2. Consider the initial value problem $x'' + 6x' + 25x = 0$ with $x(0) = 2$ and $x'(0) = 3$. First, the characteristic equation is $r^2 + 6r + 25 = 0$ and the solution is

$$r = \frac{-6 \pm \sqrt{6^2 - 4(1)25}}{2(1)} = -3 \pm 4i.$$

Second, note this is the complex root case where $\alpha = -3$ and $\beta = 4$ and, thus, the homogeneous solutions are

$$x_1(t) = e^{-3t}\cos(4t) \text{ and}$$
$$x_2(t) = e^{-3t}\sin(4t).$$

Third, find c_1 and c_2 so that $x(t) = c_1 x_1(t) + c_2 x_2(t)$ satisfies the given initial conditions. Here we must compute the derivative of $x(t)$ by using the product rule

$$\begin{aligned} x' &= (c_1 e^{-3t}\cos(4t) + c_2 e^{-3t}\sin(4t))' \\ &= c_1(e^{-3t}(-3)\cos(4t) + e^{-3t}\sin(4t)(-4)) + \\ &\quad c_2(e^{-3t}(-3)\sin(4t) + e^{-3t}\cos(4t)(4)). \end{aligned}$$

The following algebraic system must be solved

$$\begin{aligned} c_1 e^{-3(0)}\cos(4(0)) + c_2 e^{-3(0)}\sin(4(0)) &= \\ c_1 1 + c_2 0 &= 2 \text{ and} \\ c_1(-3 + 0) + c_2(0 + 4) &= 3. \end{aligned}$$

The matrix form of this is

$$\begin{bmatrix} 1 & 0 \\ -3 & 4 \end{bmatrix} \begin{bmatrix} c_1 \\ c_2 \end{bmatrix} = \begin{bmatrix} 2 \\ 3 \end{bmatrix}.$$

The algebraic solution is $c_1 = 2$ and $c_2 = 9/4$, and the solution to the initial value problem is

$$x(t) = 2e^{-3t}\cos(4t) + (9/4)e^{-3t}\sin(4t).$$

Note, the mass-spring oscillates, but amplitudes of the oscillations rapidly decrease to zero. This is a result of the friction forces, which remove energy from the mass-spring system.

6.3.2   Application to the Damped Mass-Spring System

Consider the model given by $mx'' + bx' + kx = 0$ where the mass m and spring constant k are fixed and the friction (damping) constant b is positive with small to large values. As b goes from small to large, the nature of the homogeneous solution will change due to changing characteristic equation roots from complex, repeated to distinct real roots. The three types of homogeneous solutions are

$$b^2 - 4mk \;\; < \;\; 0 \; (underdamped):$$
$$e^{(-b/2m)t}\cos(\sqrt{4mk - b^2}t) \text{ and } e^{(-b/2m)t}\sin(\sqrt{4mk - b^2}t)$$
$$b^2 - 4mk \;\; = \;\; 0 \; (critically\ damped):$$
$$e^{(-b/2m)t} \text{ and } te^{(-b/2m)t}$$
$$b^2 - 4mk \;\; > \;\; 0 \; (overdamped):$$
$$e^{(-b/2m-\sqrt{b^2-4mk}/2m)t} \text{ and } e^{(-b/2m+\sqrt{b^2-4mk}/2m)t}.$$

Because m, b and k are all positive, the coefficients of time t in the above exponential function are all negative. This implies the homogeneous solution will approach zero as time increases, which is not the case if b is zero.

Example 6.3.3. Let $m = 1$, $k = 1$ and vary $b = 1, 2$ and 4. Consider the initial value problem

$$x'' + bx' + x = 0 \text{ with } x(0) = 2 \text{ and } x'(0) = 0.$$

The characteristic equation is $r^2 + br + 1 = 0$ whose solutions are

$$r = \frac{-b \pm \sqrt{b^2 - 4}}{2}.$$

Let $b = 1$ so that the characteristic equation has complex roots $r = -1/2 \pm i\sqrt{3}/2$ and the homogeneous solutions have the form

$$x(t) = c_1 e^{(-1/2)t}\cos((\sqrt{3}/2)t) + c_2 e^{(-1/2)t}\sin((\sqrt{3}/2)t).$$

As in Example 6.3.2 the initial conditions are used to find the coefficients

$$c_1 e^{(-1/2)(0)}\cos(0) + c_2 e^{(-1/2)(0)}\sin(0) \;\; = \;\; c_1 1 + c_2 0 = 2 \text{ and}$$
$$c_1(-1/2 + 0) + c_2(0 + \sqrt{3}/2) \;\; = \;\; 0.$$

The matrix form of this is

$$\begin{bmatrix} 1 & 0 \\ -1/2 & \sqrt{3}/2 \end{bmatrix} \begin{bmatrix} c_1 \\ c_2 \end{bmatrix} = \begin{bmatrix} 2 \\ 0 \end{bmatrix}.$$

The algebraic solution is $c_1 = 2$ and $c_2 = 2/\sqrt{3}$, and the solution to the initial value problem is

$$x(t) = 2e^{(-1/2)t}\cos((\sqrt{3}/2)t) + (2/\sqrt{3})e^{(-1/2)t}\sin((\sqrt{3}/2)t).$$

Let $b = 2$ and in this case the characteristic equation has repeated real roots $r = -1$ and -1 and the homogeneous solution is

$$x(t) = c_1 e^{-t} + c_2 t e^{-t}.$$

Imposing the initial conditions gives

$$\begin{aligned} c_1 1 + c_2 0 &= 2 \text{ and} \\ c_1(-1) + c_2(1+0) &= 0 \end{aligned}$$

whose matrix form is

$$\begin{bmatrix} 1 & 0 \\ -1 & 1 \end{bmatrix} \begin{bmatrix} c_1 \\ c_2 \end{bmatrix} = \begin{bmatrix} 2 \\ 0 \end{bmatrix}.$$

The algebraic solution is $c_1 = 2$ and $c_2 = 2$, and the solution to the initial value problem is

$$x(t) = 2e^{-t} + 2te^{-t}.$$

Let $b = 4$, which illustrates the distinct real roots case with $r = -2 \pm \sqrt{3}$, and the general homogeneous solution is

$$x(t) = c_1 e^{(-2+\sqrt{3})t} + c_2 e^{(-2-\sqrt{3})t}.$$

The initial conditions are used to find the coefficients as follows

$$\begin{aligned} c_1 1 + c_2 1 &= 2 \text{ and} \\ c_1(-2+\sqrt{3}) + c_2(-2-\sqrt{3}) &= 0. \end{aligned}$$

Or, in terms of matrices

$$\begin{bmatrix} 1 & 1 \\ -2+\sqrt{3} & -2-\sqrt{3} \end{bmatrix} \begin{bmatrix} c_1 \\ c_2 \end{bmatrix} = \begin{bmatrix} 2 \\ 0 \end{bmatrix}.$$

By Cramer's rule the solution is $c_1 = 2(-2-\sqrt{3})/(-2\sqrt{3}) = 2/\sqrt{3}+1$ and $c_2 = -2(-2+\sqrt{3})/(-2\sqrt{3}) = -2/\sqrt{3}+1$. The solution to the initial value problem is

$$x(t) = (2/\sqrt{3}+1)e^{(-2+\sqrt{3})t} + (-2/\sqrt{3}+1)e^{(-2-\sqrt{3})t}.$$

6.3.3   MATLAB Commands dsolve() and ezplot()

The MATLAB command dsolve() is used to find the symbolic solutions to Example 6.3.3 for $b = 0, 1, 2$ and 4. Then the MATLAB command ezplot() is used to generate Figure 6.3.1.

```
>> % b = 0
>> solb0 = dsolve('D2x + 0*Dx + x = 0','x(0) = 2','Dx(0) = 0', 't')
   solb0 = 2*cos(t)
>> % b =1
>> solb1 = dsolve('D2x + 1*Dx + x = 0','x(0) = 2','Dx(0) = 0', 't')
   solb1 = 2*exp(-1/2*t)*cos(1/2*3^(1/2)*t)
           + 2/3*3^(1/2)*exp(-1/2*t)*sin(1/2*3^(1/2)*t)
>> % b = 2
>> solb2 = dsolve('D2x + 2*Dx + x = 0','x(0) = 2','Dx(0) = 0', 't')
   solb2 = 2*exp(-*t) + 2*exp(-t)*t
>> % b = 4
>> solb4 = dsolve('D2x + 4*Dx + x = 0','x(0) = 2','Dx(0) = 0', 't')
   solb4 = (2/3*3^(1/2) + 1)*exp((-2 + 3^(1/2))*t)
           + 1/3*(-2 + 3^(1/2))*3^(1/2)*exp(-(2 + 3^(1/2))*t)
>> % use ezplot to create the curves on the same graph
>> ezplot(solb4, [0 8])
>> hold on
>> ezplot(solb2, [0 8])
>> ezplot(solb1, [0 8])
>> ezplot(solb0, [0 8])
```

6.3.4   Exercises

1.   Consider the homogeneous solutions $x_1(t) = e^{\alpha t}\cos(\beta t)$ and $x_2(t) = e^{\alpha t}\sin(\beta t)$.

   (a).   Compute their derivatives.

   (b).   Find the Wronskian matrix $W(0)$.

   (c).   Show $\det(W(0)) = \beta$.

2.   Verify $x(t) = -\cos(\sqrt{k/m}\,t)$ is the solution for Example 6.3.1.

3.   Consider the initial value problem $3x'' + 27x = 0$ with $x(0) = 2$ and $x'(0) = -1$.

   (a).   Find the homogeneous solutions of the differential equation.

   (b).   Find the solution to the initial value problem.

4.   Consider the initial value problem $x'' + 2x' + 2x = 0$ with $x(0) = 1$ and $x'(0) = -1$.

   (a).   Find the homogeneous solutions of the differential equation.

   (b).   Find the solution to the initial value problem.

5.   Consider the initial value problem $2x'' + 4x' + 6x = 0$ with $x(0) = 1$ and $x'(0) = 2$.

   (a).   Find the homogeneous solutions of the differential equation.

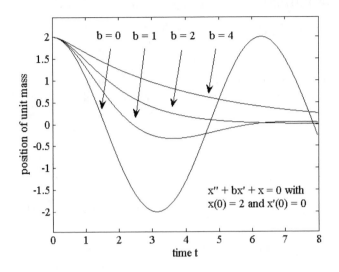

Figure 6.3.1: Variable Damped Mass-Spring

(b). Find the solution to the initial value problem.

6. Consider Example 6.3.3. Verify the proposed functions are really solutions for the three choices of b.

7. Consider the initial value problem $x'' + 4x' + kx = 0$ with $x(0) = 3$ and $x'(0) = 0$. Find the solutions for variable $k = 1, 4$ and 10.

8. Consider the initial value problem $mx'' + 4x' + 1x = 0$ with $x(0) = 3$ and $x'(0) = 0$. Find the solutions for variable $m = 1, 4$ and 10.

9. Use MATLAB's command dsolve() to confirm any calculations done in exercises $3 - 5, 7$ and 8.

6.4 Nonhomogeneous Differential Equations

In this section we focus on finding particular solutions of second order linear differential equations $L(x) = ax'' + bx' + cx = f(t) \neq 0$. As in the first order linear problems discussed in Section 6.1, this can be done by either the undetermined coefficients or variation of parameters methods. The undetermined coefficients method works for $f(t)$ of a certain type and will be illustrated by the first five examples. The variation of parameters method works for any continuous function $f(t)$ and will be presented in the Section 6.5. An important application to resonance of a forced mass-spring systems will be given.

6.4.1   Particular Solutions by Undetermined Coefficients

Any function $x(t) = x_p(t)$ that satisfies the second order differential equation

$$L(x) = ax'' + bx + cx = f(t)$$

is called a *particular solution*. Because L is a linear operator, see Definition 6.2.1, any homogeneous solution plus a particular solution will also satisfy $L(x) = f(t)$

$$
\begin{aligned}
L(c_1 x_1 + c_2 x_2 + x_p) &= L(c_1 x_1) + L(c_2 x_2) + L(x_p) \\
&= c_1 L(x_1) + c_2 L(x_2) + L(x_p) \\
&= c_1 0 + c_2 0 + f(t) = f(t).
\end{aligned}
$$

The *general solution* is denoted by $c_1 x_1 + c_2 x_2 + x_p$ where x_1 is not a multiple of x_2.

Theorem 6.4.1 *(Properties of $L(x)$) Let $L(x) \equiv ax'' + bx + cx$ where a, b and c are constants. If $L(x_1) = 0$, $L(x_2) = 0$ and $L(x_p) = f(t)$, then*

1. $L(x)$ *is linear operator,*
2. $L(c_1 x_1 + c_2 x_2) = 0,$
3. $L(c_1 x_1 + c_2 x_2 + x_p) = f(t)$ *and*
4. $L(x + y) = f + g$ *when* $L(x) = f$ *and* $L(y) = g$.

In this section we will focus on using the undetermined coefficient method for finding particular solutions. The undetermined coefficient method involves making an educated guess about the form of a particular solution. This discussion will be limited to cases where the right side of the differential equation is a polynomial, an exponential, sine or cosine function. If $f(t)$ is a linear or quadratic polynomial, then one might expect a particular solution to be polynomial where the coefficients must be chosen so that the differential equation holds for all time t. See Examples 6.4.1 and 6.4.2 for the polynomial case and Examples 6.4.3 and 6.4.4 for exponential functions. In Example 6.4.5 and the application to the forced mass-spring problem we will find particular solutions when the sine and cosine functions are in the right side of the differential equation.

Example 6.4.1. Consider the differential equation $x'' + 3x' + 2x = 3t + 1$ where the homogeneous solutions are e^{-t} and e^{-2t}. Since the right side is a first order polynomial, assume the particular solution is also linear polynomial

$$x_p = a_1 t + a_0,$$

and choose a_1 and a_0 so that the differential equation holds. This means

$$
\begin{aligned}
(a_1 t + a_1)'' + 3(a_1 t + a_0)' + 2(a_1 t + a_0) &= 3t + 1 \\
0 + 3a_1 + 2(a_1 t + a_0) &= 3t + 1 \\
(2a_1)t + (3a_1 + 2a_0)1 &= 3t + 1.
\end{aligned}
$$

So, match the coefficients of t and 1 to get

$$2a_1 = 3 \text{ and } 3a_1 + 2a_0 = 1.$$

The matrix form is

$$\begin{bmatrix} 0 & 2 \\ 2 & 3 \end{bmatrix} \begin{bmatrix} a_0 \\ a_1 \end{bmatrix} = \begin{bmatrix} 3 \\ 1 \end{bmatrix}$$

whose solution is $a_1 = 3/2$ and $a_0 = -7/4$. The particular solution of the differential equation is $x_p = (3/2)t + (-7/4)$, and the general solution is the homogeneous plus a particular solution

$$x = x_h + x_p = c_1 e^{-t} + c_2 e^{-2t} + (3/2)t + (-7/4).$$

Example 6.4.2. Consider the differential equation $x'' + 2x' + 2x = 2t^2 + 3t + 2$ where the homogeneous solutions are $e^{-t}\cos(t)$ and $e^{-t}\sin(t)$. Assume the particular solution is also a quadratic polynomial

$$x_p = a_2 t^2 + a_1 t + a_0.$$

Choose a_2, a_1 and a_0 so that the differential equation holds

$$
\begin{aligned}
\left(a_2 t^2 + a_1 t + a_0\right)'' + 2\left(a_2 t^2 + a_1 t + a_0\right)' + 2\left(a_2 t^2 + a_1 t + a_0\right) &= 2t^2 + 3t + 2 \\
2a_2 + 4a_2 t + 2a_1 + 2\left(a_2 t^2 + a_1 t + a_0\right) &= 2t^2 + 3t + 2 \\
(2a_2)t^2 + (4a_2 + 2a_1)t + (2a_2 + 2a_1 + 2a_0)1 &= 2t^2 + 3t + 2.
\end{aligned}
$$

Match the coefficients of t^2, t and 1 to get three algebraic equations, which may be written in matrix form as

$$\begin{bmatrix} 2 & 2 & 2 \\ 0 & 2 & 4 \\ 0 & 0 & 2 \end{bmatrix} \begin{bmatrix} a_0 \\ a_1 \\ a_2 \end{bmatrix} = \begin{bmatrix} 2 \\ 3 \\ 2 \end{bmatrix}.$$

The algebraic solution is easily found and gives a particular solution $x_p = t^2 + (-1/2)t + (1/2)$. The general solution of the differential equations is

$$x = c_1 e^{-t}\cos(t) + c_2 e^{-t}\sin(t) + t^2 + (-1/2)t + (1/2).$$

Example 6.4.3. Consider the same differential operator as in Example 6.4.1, but suppose it has a different right side $x'' + 3x' + 2x = 10e^{3t}$. Assume a particular solution has the form

$$x_p = ae^{3t},$$

and choose the coefficient so that the differential equation holds

$$
\begin{aligned}
(ae^{3t})'' + 3(ae^{3t})' + 2(ae^{3t}) &= 10e^{3t} \\
ae^{3t}9 + 3ae^{3t}3 + 2(ae^{3t}) &= 10e^{3t} \\
(20a)e^{3t} &= 10e^{3t}.
\end{aligned}
$$

Match the coefficient of the exponential function to find $a = 10/20 = 1/2$. The particular solution of the differential equation is $x_p = (1/2)e^{3t}$, and the general solution is the homogeneous plus a particular solution

$$x = x_h + x_p = c_1 e^{-t} + c_2 e^{-2t} + (1/2)e^{3t}.$$

At this point observe Examples 6.4.1 and 6.4.3 can be combined to generate more particular solutions by making use of the linear property of the differential operator $L(x) = ax'' + bx' + cx$. If $L(x) = f$ and $L(y) = g$, then $L(x + y) = L(x) + L(y) = f + g$ so that a particular solution of $L(z) = f + g$ is $z = x + y$. In Example 6.4.1 $f = 3t + 1$ with $x = (3/2)t + (-7/4)$ and in Example 6.4.3 $g = 10e^{3t}$ with $y = (1/2)e^{3t}$. Therefore, a particular solution of

$$x'' + 3x' + 2x = 3t + 1 + 10e^{3t}$$

is $x + y = (3/2)t + (-7/4) + (1/2)e^{3t}$.

Example 6.4.4. Consider the same differential operator as in Example 6.4.1, but suppose the right side is one of the homogeneous solutions $x'' + 3x' + 2x = 5e^{-t}$. In this case the particular solution can not be a multiple of e^{-t}, and another guess is to assume a particular has the form

$$x_p = ate^{-t}.$$

Find the coefficient by requiring this to satisfy the differential equation

$$
\begin{aligned}
(ate^{-t})'' + 3(ate^{-t})' + 2(ate^{-t}) &= 5e^{-t} \\
(a(-2e^{-t} + te^{-t})) + 3(a(1e^{-t} + te^{-t}(-1))) + 2(ate^{-t}) &= 5e^{-t} \\
ae^{-t} + (0)te^{-t} &= 5e^{-t}.
\end{aligned}
$$

Thus, $a = 5$, a particular solution is $x_p = 5te^{-t}$ and the general solution is

$$x = x_h + x_p = c_1 e^{-t} + c_2 e^{-2t} + 5te^{-t}.$$

Example 6.4.5. Consider sine and cosine functions in the right side $x'' + 3x' + 2x = 2\cos(t) + 3\sin(t) = f(t)$. Assume the particular solution is a linear combination of $\cos(t)$ and $\sin(t)$

$$x_p = a\cos(t) + b\sin(t).$$

Put this into the differential equation and match the coefficients of $\cos(t)$ and $\sin(t)$

$$
\begin{aligned}
(a\cos(t) + b\sin(t))'' + 3(a\cos(t) + b\sin(t))' + 2(a\cos(t) + b\sin(t)) &= f(t) \\
-a\cos(t) - b\sin(t) - 3a\sin(t) + 3b\cos(t) + 2(a\cos(t) + b\sin(t)) &= f(t) \\
(-a + 3b + 2a)\cos(t) + (-b - 3a + 2b)\sin(t) &= f(t).
\end{aligned}
$$

This gives two algebraic equations whose matrix form is

$$\begin{bmatrix} 1 & 3 \\ -3 & 1 \end{bmatrix} \begin{bmatrix} a \\ b \end{bmatrix} = \begin{bmatrix} 2 \\ 3 \end{bmatrix}.$$

Cramer's rule can be used to solve for $a = -7/10$ and $b = 9/10$. A particular solution is $x_p = (-7/10)\cos(t) + (9/10)\sin(t)$, and the general solution is

$$x = x_h + x_p = c_1 e^{-t} + c_2 e^{-2t} + (-7/10)\cos(t) + (9/10)\sin(t).$$

6.4.2 Nonhomogeneous Initial Value Problems

Let $L(x) = ax'' + bx' + cx$ where a, b and c are constants. Homogeneous solutions of $L(x_h) = 0$, $x_1(t)$ and $x_2(t)$, were derived in the previous two sections, and some particular solutions of $L(x_p) = f(t)$ were found via undetermined coefficients for some classes of $f(t)$. Because L is a linear operator, $L(x_h + x_p) = L(x_h) + L(x_p) = 0 + f(t)$. The general solution is the linear combination of the two homogeneous solutions plus a particular solution

$$x(t) = c_1 x_1(t) + c_2 x_2(t) + x_p(t). \tag{6.4.1}$$

The solution of the *initial value problem* (IVP) $L(x) = f(t)$ with $x(0) = x_0$ and $x'(0) = v_0$ is found by choosing c_1 and c_2 in the general solution so that the initial conditions are satisfied. This reduces to two algebraic equations

$$\begin{aligned} x(0) &= c_1 x_1(0) + c_2 x_2(0) + x_p(0) \text{ and} \\ x'(0) &= c_1 x_1'(0) + c_2 x_2'(0) + x_p'(0). \end{aligned}$$

The matrix form of this system is

$$\begin{bmatrix} x_1(0) & x_2(0) \\ x_1'(0) & x_2'(0) \end{bmatrix} \begin{bmatrix} c_1 \\ c_2 \end{bmatrix} = \begin{bmatrix} x_0 - x_p(0) \\ v_0 - x_p'(0) \end{bmatrix}. \tag{6.4.2}$$

The 2×2 matrix is the Wronskian

$$W(0) \equiv \begin{bmatrix} x_1(0) & x_2(0) \\ x_1'(0) & x_2'(0) \end{bmatrix},$$

and in order to solve for the coefficients, it must have non-zero determinant. Using Theorem 3.7.3 and the discussion in Section 5.1, this is equivalent to the column vectors in $W(0)$ being linearly independent. These results are summarized in the following theorem.

Theorem 6.4.2 (*Second Order Linear Initial Value Problem*) *Consider the nonhomogeneous initial value problem* $L(x) = ax'' + bx' + cx = f(t)$ *with* $x(0) = x_0$ *and* $x'(0) = v_0$. *Let* $x_1(t)$ *and* $x_2(t)$ *be homogeneous solutions and* $x_p(t)$ *be a particular solution. If* $\det(W(0)) \neq 0$, *then the nonhomogeneous initial value problem has a solution given by (6.4.1) and the algebraic solution of (6.4.2).*

268   CHAPTER 6.   LINEAR INITIAL VALUE PROBLEMS

Outline for Solving Nonhomogeneous Second Order Linear IVP.
1. Find two homogeneous solutions.
2. Find a particular solution.
3. Form the general solution $x(t) = c_1 x_1(t) + c_2 x_2(t) + x_p(t)$.
4. Find the coefficients by solving equation (6.4.2).

Example 6.4.2 with initial conditions $x(0) = 2$ **and** $x'(0) = 1$. Steps 1-3 have been done and yield

$$x = c_1 e^{-t} \cos(t) + c_2 e^{-t} \sin(t) + t^2 + (-1/2)t + (1/2).$$

The particular solution is $x_p(t) = t^2 + (-1/2)t + (1/2)$ so that $x_p(0) = 1/2$ and $x_p'(0) = -1/2$. The algebraic system in (6.4.2) is

$$\begin{bmatrix} 1 & 0 \\ -1 & 1 \end{bmatrix} \begin{bmatrix} c_1 \\ c_2 \end{bmatrix} = \begin{bmatrix} 2 - 1/2 \\ 1 - (-1/2) \end{bmatrix}$$

whose solution is $c_1 = 3/2$ and $c_2 = 3$. Thus, the solution to the initial value problem is

$$x = (3/2)e^{-t} \cos(t) + 3e^{-t} \sin(t) + t^2 + (-1/2)t + (1/2).$$

Example 6.4.4 with initial conditions $x(0) = 1$ **and** $x'(0) = 2$. Steps 1-3 have been done and yield

$$x = c_1 e^{-t} + c_2 e^{-2t} + 5te^{-t}.$$

The particular solution is $x_p(t) = 5te^{-t}$ giving $x_p(0) = 0$ and $x_p'(0) = 5$. The algebraic system in (6.4.2) is

$$\begin{bmatrix} 1 & 1 \\ -1 & -2 \end{bmatrix} \begin{bmatrix} c_1 \\ c_2 \end{bmatrix} = \begin{bmatrix} 1 - 0 \\ 2 - 5 \end{bmatrix}.$$

The solution is found by using Cramer's rule to be $c_1 = 1/(-1) = -1$ and $c_2 = -2/(-1) = 2$. Thus, the solution to the initial value problem is

$$x = -e^{-t} + 2e^{-2t} + 5te^{-t}.$$

6.4.3   Application to Forced Mass-Spring

The forced mass-spring model has an external force $f(t) \neq 0$ as well as forces from the spring and from friction. The corresponding initial value problem is $mx'' + bx' + kx = f(t)$ with $x(0) = x_0$ and $x'(0) = v_0$. In order to keep the discussion as clear as possible, let $m = 1$, $b = 0$, $k = 1$ and $f(t) = \sin(wt)$ with $x(0) = 0$ and $x'(0) = 0$. The natural frequency, which is given by the homogeneous solutions $\cos(\sqrt{k/m}t) = \cos(t)$ and $\sin(\sqrt{k/m}t) = \sin(t)$, is $\sqrt{k/m}/(2\pi) = 1/(2\pi)$, and the frequency of the external force is $w/(2\pi)$. The objective of this subsection is to understand what happens to the motion of the mass as w approaches 1.

Find the particular solution when $w \neq 1$. In this case $f(t) = \sin(wt)$ is not a homogeneous solution, and we speculate the particular solution has the form

$$x_p(t) = a\cos(wt) + b\sin(wt).$$

Put this into the differential equation $x'' + x = \sin(wt)$ and match coefficients of the sine and cosine function

$$
\begin{aligned}
(a\cos(wt) + b\sin(wt))'' + a\cos(wt) + b\sin(wt) &= 0\cos(wt) + 1\sin(wt) \\
-w^2 a\cos(wt) - w^2 b\sin(wt) + a\cos(wt) + b\sin(wt) &= 0\cos(wt) + 1\sin(wt) \\
(1 - w^2)a\cos(wt) + (1 - w^2)b\sin(wt) &= 0\cos(wt) + 1\sin(wt).
\end{aligned}
$$

This implies $(1 - w^2)a = 0$ and $(1 - w^2)b = 1$ so that the particular solution is

$$x_p(t) = \frac{1}{1 - w^2}\sin(wt).$$

Since the denominator goes to zero as w approaches one, this indicates some special event!

Next impose the initial conditions on the general solution

$$x(t) = c_1\cos(t) + c_2\sin(t) + \frac{1}{1 - w^2}\sin(wt).$$

Use the zero initial position and velocity

$$
\begin{aligned}
x(0) &= c_1\cos(0) + c_2\sin(0) + \frac{1}{1 - w^2}\sin(w0) = c_1 = 0 \text{ and} \\
x'(0) &= c_1(-\sin(0)) + c_2\cos(0) + \frac{1}{1 - w^2}\cos(w0)w = c_2 + \frac{w}{1 - w^2} = 0.
\end{aligned}
$$

Thus, the solution to the initial value problem is

$$x(t) = \frac{-w}{1 - w^2}\sin(t) + \frac{1}{1 - w^2}\sin(wt).$$

In order to investigate what happens as w approaches one, note the general solution has an indeterminate form with respect to w

$$x(t) = \frac{-w\sin(t) + \sin(wt)}{1 - w^2}.$$

Write this solution as a function of both t and w, and compute the limit with respect to w using L'Hôpital's rule

$$
\begin{aligned}
\lim_{w \to 1} x(t, w) &= \lim_{w \to 1} \frac{\frac{d}{dw}(-w\sin(t) + \sin(wt))}{\frac{d}{dw}(1 - w^2)} \\
&= \lim_{w \to 1} \frac{-\sin(t) + \cos(wt)\,t}{-2w} \\
&= \frac{1}{2}\sin(t) + \frac{t\cos(t)}{-2}.
\end{aligned}
$$

Since the amplitude of $\cos(t)$ is $t/(-2)$, the spring eventually will be deformed so much that the model is no longer valid. This is known as pure *resonance*, and it can be controlled by inserting a suitable amount of resistive or damping force, as is illustrated in the next subsection.

The solution of the above limiting case can also be derived directly by letting $w = 1$ and finding the solution to $x'' + x = \sin(t)$. Since $f(t) = \sin(t)$ is also a particular solution, we speculate the particular solution has the form

$$x_p(t) = t(a\cos(t) + b\sin(t)).$$

By placing this into the differential equation and collecting like terms we have

$$(0)t\sin(t) + (0)t\cos(t)(-b+b) - 2b\sin(t) - 2a\cos(t) = 1\sin(t) + 0\cos(t).$$

This gives the same particular solution as well as the same solution to the initial value problem.

6.4.4   Damped Forced Mass-Spring Using MATLAB

Consider the forced mass-spring when the resistive or damping force is not zero. Use the MATLAB command dsolve() to find the solutions for $b = 0, 1, 2$ and 4 of the nonhomogeneous initial value problem

$$x'' + bx' + x = \sin(t) \text{ with } x(0) = 0 \text{ and } x'(0) = 0.$$

The MATLAB command ezplot is then used to graph the four curves, see Figure 6.4.1. Note, for large values of time and positive b the homogeneous solutions have essentially decayed to zero so that the curves in this figure reflect the particular solutions $(-1/2)t\cos(t)$ for $b = 0$, $-cos(t)$ for $b = 1$, $(-1/2)cos(t)$ for $b = 2$ and $(-1/4)cos(t)$ for $b = 4$

```
>> % let b = 0
>> solb0 = dsolve('D2x+0*Dx+x = sin(t)', 'x(0) = 0', Dx(0) = 0','t')
>> simple(solb0)
      -1/2*cos(t)*t + 1/2*sin(t)
>> % let b = 1
>> solb1 = dsolve('D2x+1*Dx+x = sin(t)', 'x(0) = 0', Dx(0) = 0','t')
>> simple(solb1)
      exp(-1/2*t)*cos(1/2*3^(1/2)*t) - cos(t) +
          1/3*exp(-1/2*t)*cos(1/2*3^(1/2)*t)*3^(1/2
>> % let b = 2
>> solb2 = dsolve('D2x+2*Dx+x = sin(t)', 'x(0) = 0', Dx(0) = 0','t')
>> simple(solb2)
      -1/2*cos(t) + 1/2*exp(-t) + 1/2*exp(-t)*t
>> % let b = 4
>> solb4 = dsolve('D2x+4*Dx+x = sin(t)', 'x(0) = 0', Dx(0) = 0','t')
>> simple(solb4)
```

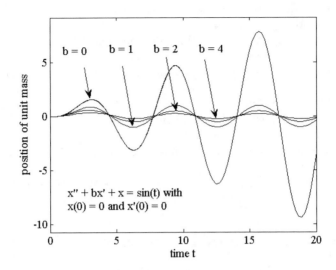

Figure 6.4.1: Forced Mass-Spring

$$-1/4*\cos(t) + (1/8 + 1/12*3^\wedge(1/2))*\exp((3^\wedge(1/2) - 2)*t) +$$
$$(1/24*(3^\wedge(1/2)-2)*3^\wedge(1/2))*\exp(-(2+3^\wedge(1/2))*t)$$

```
>> use ezplot to create the curves on the same graph
>> ezplot(solb4, [0 20])
>> hold on
>> ezplot(solb2, [0 20])
>> ezplot(solb1, [0 20])
>> ezplot(solb0, [0 20])
```

6.4.5 Exercises

1. Find the homogeneous and particular solutions for $x'' + 5x' + 4x = 3t + 4$.

2. Find the homogeneous and particular solutions for $x'' + 2x' + 2x = t^2 + 2t + 3$.

3. Find the homogeneous and particular solutions for $x'' + 5x' + 4x = 3e^{2t}$.

4. Find the homogeneous and particular solutions for $x'' + 5x' + 4x = 2e^{-4t}$.

5. Find the homogeneous and particular solutions for $x'' + 4x = \sin(t) + 3\cos(t)$.

6. Find the homogeneous and particular solutions for $x'' + 4x = \sin(2t)$.

7. Solve the initial value problem for the differential equation in Example 6.4.1 using the initial conditions $x(0) = 1$ and $x'(0) = 2$.

8. Solve the initial value problem for the differential equation in Example 6.4.3 using the initial conditions $x(0) = 0$ and $x'(0) = -1$.

9.   Find the solution to the initial value problem $x'' + 5x' + 4x = 6e^{2t} + 3t + 4$ with $x(0) = 0$ and $x'(0) = 0$.

10.   Find the solution to the initial value problem $x'' + 4x = 2e^{2t} + t + 4$ with $x(0) = 0$ and $x'(0) = 1$.

11.   Consider the force mass-spring problem $x'' + x = \cos(wt)$ with $x(0) = 0$ and $x'(0) = 0$.

(a).   Find the particular solution when $w \neq 1$.

(b).   Solve the initial value problem when $w \neq 1$.

(c).   Solve the initial value problem when $w = 1$.

12.   Confirm any by-hand calculations in exercises 1-11 using MATLAB's command dsolve().

13.   Use MATLAB's commands dsolve() and ezplot() to solve and graph the $x'' + bx' + x = \cos(t)$ with $x(0) = 0$ and $x'(0) = 0$ for $b = 0, 1, 2$ and 4.

6.5   System Form of Linear Second Order

This section begins with an application to LRC series circuits where a model for the current is also a second order linear differential equation. The right side can be a very complicated function, and the method of undetermined coefficients may not yield a particular solution. Possible alternatives are to use the variation of parameters method to find the particular solution or to use a suitable numerical method. Both these alternatives require the second order linear differential equation to be written as a first order system of differential equations. The numerical method that is implemented in the MATLAB command ode45() is a variation of the Runge-Kutta scheme and uses variable time steps.

6.5.1   Application to a Series LRC Circuit

In the model for the forced mass-spring the damping or friction coefficient b of the velocity x' was used to control the magnitude of the oscillations of the mass. In a car or truck this is similar to having effective shock absorbers. The model for the current in a series LRC circuit with an external source is also a second order differential equation with sine and cosine functions of time in the right side. In a tuning circuit one wants to maximize the current by adjusting the inductors or the capacitors. The series LRC circuit is a primitive tuning circuit, and for a simple external source one can derive an exact formula for the maximum current. In more complicated external sources, such as from an AM (amplitude modulation) radio station, the initial value problem must be solved numerically.

A series LRC circuit has three components, an inductor L, resistor R and capacitor C, and an external source $E(t)$. In an AM radio the external source is from the antenna and is a function with terms equal to products of sine and cosine functions. Figure 6.5.1 is a depiction of the series circuit and the corresponding differential equation model. The voltage drop across the resistor is $V_R = RI$, the voltage drop across the inductor is $V_L = LI'$ and the voltage

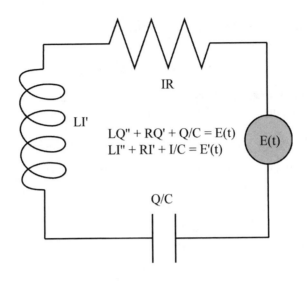

LI'

LQ" + RQ' + Q/C = E(t)
LI" + RI' + I/C = E'(t)

IR

E(t)

Q/C

Figure 6.5.1: Series LRC Circuit

drop across the capacitor is given by $Q = CV_C$ where Q is the charge and $Q' = I$ is the current. The model is derived from Kirchhoff voltage law, which requires the sum of the voltage drops be equal to the imposed voltage

$$V_L + V_R + V_C = E(t)$$
$$LI' + RI + Q/C = E(t).$$

Compute the derivative of both sides with respect to time to obtain a model for the current

$$LI'' + RI' + I/C = E'(t).$$

The objective is to fix the resistance R, and to vary the inductance L or the capacitance C so that the current I will have a maximum amplitude. In tuning a radio the amplitudes will vary with the frequency of the various stations. The desired station will have a large amplitude relative to other stations.

Example 6.5.1. Let $E(t) = E_0 \sin(\gamma t)$ where E_0 is a constant (unmodulated carrier wave) and $\gamma = f_s 2\pi$ with f_s equals the frequency of the carrier wave. The model for the current is

$$LI'' + RI' + I/C = E'(t) = E_0 \gamma \cos(\gamma t)$$

where the coefficients are positive constants. The homogeneous solutions $I_1(t)$ and $I_2(t)$ are derived as in the previous sections from the characteristic algebraic equation. Because the L, R and C are positive, regardless of which of the three cases (distinct, repeated or complex roots), the homogeneous solutions will decay to zero and are called the *transient* solutions. The signal that remains significant is the particular solution $I_p(t)$. The method of undetermined coefficients suggest that the particular solution should be a linear combination of the $\cos(\gamma t)$ and $\sin(\gamma t)$

$$I_p(t) = a\cos(\gamma t) + b\sin(\gamma t).$$

By placing this into the differential equation, computing the derivatives and collecting like terms we have for $A \equiv -La\gamma^2 + Rb\gamma + a/C$ and $B \equiv -Lb\gamma^2 - Ra\gamma + b/C$

$$A\cos(\gamma t) + B\sin(\gamma t) = E_0\gamma\cos(\gamma t) + 0\sin(\gamma t).$$

Match the coefficients of $\cos(\gamma t)$ and $\sin(\gamma t)$ to obtain two algebraic equations whose matrix form is

$$\begin{bmatrix} -L\gamma^2 + 1/C & R\gamma \\ -R\gamma & -L\gamma^2 + 1/C \end{bmatrix}\begin{bmatrix} a \\ b \end{bmatrix} = \begin{bmatrix} \gamma E_0 \\ 0 \end{bmatrix}.$$

Cramer's rule easily gives the solution

$$a = \gamma E_0 \frac{-L\gamma^2 + 1/C}{(-L\gamma^2 + 1/C)^2 + (R\gamma)^2} \text{ and}$$

$$b = \gamma E_0 \frac{R\gamma}{(-L\gamma^2 + 1/C)^2 + (R\gamma)^2}.$$

Let θ be such that $\tan(\theta) = (-L\gamma^2 + 1/C)/(R\gamma)$ and write the particular solution as

$$\begin{aligned} I_p(t) &= \gamma E_0 [\frac{-L\gamma^2 + 1/C}{(-L\gamma^2 + 1/C)^2 + (R\gamma)^2}\cos(\gamma t) + \\ &\qquad \frac{R\gamma}{(-L\gamma^2 + 1/C)^2 + (R\gamma)^2}\sin(\gamma t)] \\ &= \frac{\gamma E_0}{\sqrt{(-L\gamma^2 + 1/C)^2 + (R\gamma)^2}}[\sin(\theta)\cos(\gamma t) + \cos(\theta)\sin(\gamma t)] \\ &= \frac{\gamma E_0}{\sqrt{(-L\gamma^2 + 1/C)^2 + (R\gamma)^2}}\sin(\theta + \gamma t). \end{aligned}$$

The amplitude is the coefficient of $\sin(\theta + \gamma t)$ and will be a maximum if the denominator is a minimum. Since the resistance is fixed, the denominator will be a minimum when $-L\gamma^2 + 1/C = 0$. Thus, the current will have a maximum amplitude when

$$-L\gamma^2 + 1/C = 0 \text{ or}$$

$$\gamma = \frac{1}{\sqrt{LC}}.$$

If $\gamma = f_s2\pi$ and f_s is the desired frequency, then L and C should be adjusted so that $\gamma = f_s2\pi = 1/\sqrt{LC}$. As will be illustrated by the MATLAB numerical calculations at the end of this section, these results also approximate the maximum current for the modulated signal.

6.5.2 Systems and Variation of Parameters

If the right side of the second order differential equation is too complicated, the undetermined coefficient method may not be able to find a particular solution. The variation of parameters method, which was used in Section 6.1 for first order differential equations, can be generalized to find a particular solution for second order differential equations. However, they will need to be reformulated as a first order system where there are two first order differential equations and two unknowns for the function and its derivative.

Consider the linear second order differential equation with constant coefficients and $a \neq 0$

$$ax'' + bx' + cx = f(t). \qquad (6.5.1)$$

Let y be the derivative of x and, thus,

$$\begin{aligned} x' &= y \text{ and} \\ y' &= x'' = (f(t) - bx' - cx)/a \\ &= f(t)/a - by/a - cx/a. \end{aligned}$$

This can be written in vector form as a *first order system of differential equations*

$$\left[\begin{array}{c} x \\ y \end{array}\right]' = \left[\begin{array}{c} x' \\ y' \end{array}\right] = \left[\begin{array}{cc} 0 & 1 \\ -c/a & -b/a \end{array}\right]\left[\begin{array}{c} x \\ y \end{array}\right] + \left[\begin{array}{c} 0 \\ f(t)/a \end{array}\right]. \qquad (6.5.2)$$

Let $x_1(t)$ and $x_2(t)$ be homogeneous solutions $(f(t) = 0)$ of the second order problem in (6.5.1). In terms of vector equation in (6.5.2) this means

$$\left[\begin{array}{c} x_1 \\ x_1' \end{array}\right]' = \left[\begin{array}{cc} 0 & 1 \\ -c/a & -b/a \end{array}\right]\left[\begin{array}{c} x_1 \\ x_1' \end{array}\right] \text{ and}$$

$$\left[\begin{array}{c} x_2 \\ x_2' \end{array}\right]' = \left[\begin{array}{cc} 0 & 1 \\ -c/a & -b/a \end{array}\right]\left[\begin{array}{c} x_2 \\ x_2' \end{array}\right].$$

Use the vector notation

$$\mathbf{x}_1 \equiv \left[\begin{array}{c} x_1 \\ x_1' \end{array}\right], \ \mathbf{x}_2 \equiv \left[\begin{array}{c} x_2 \\ x_2' \end{array}\right], \ \mathbf{f} \equiv \left[\begin{array}{c} 0 \\ f(t)/a \end{array}\right] \text{ and } A \equiv \left[\begin{array}{cc} 0 & 1 \\ -c/a & -b/a \end{array}\right].$$

The two vectors $\mathbf{x}_1$ and $\mathbf{x}_2$ satisfy

$$\mathbf{x}_1' = A\mathbf{x}_1 \text{ and } \mathbf{x}_2' = A\mathbf{x}_2. \qquad (6.5.3)$$

The augmentation of the vectors $\mathbf{x}_1$ and $\mathbf{x}_2$ is the Wronskian at time t

$$W(t) \equiv [\mathbf{x}_1 \ \mathbf{x}_2] = \left[\begin{array}{cc} x_1 & x_2 \\ x_1' & x_2' \end{array}\right].$$

The two vector equations in (6.5.3) may be written as a 2×2 matrix equation

$$W(t)' \equiv [\mathbf{x}_1' \ \mathbf{x}_2'] = [A\mathbf{x}_1 \ A\mathbf{x}_2] = A[\mathbf{x}_1 \ \mathbf{x}_2] = AW(t). \qquad (6.5.4)$$

The general homogeneous vector solution of (6.5.2) can be written as

$$\mathbf{x}_h = \mathbf{x}_1 c_1 + \mathbf{x}_2 c_2 = \begin{bmatrix} x_1 & x_2 \\ x_1' & x_2' \end{bmatrix} \begin{bmatrix} c_1 \\ c_2 \end{bmatrix} = W(t)\mathbf{c}.$$

Following the first order differential equation variation of parameters method, we speculate that a vector particular solution is also a linear combination of the column vectors in $W(t)$, but now the coefficients depend on time and the vector particular solution has the form

$$\mathbf{x}_p = \mathbf{x}_1 v_1(t) + \mathbf{x}_2 v_2(t) = \begin{bmatrix} x_1 & x_2 \\ x_1' & x_2' \end{bmatrix} \begin{bmatrix} v_1(t) \\ v_2(t) \end{bmatrix} = W(t)\mathbf{v}(t).$$

Choose the components of the $\mathbf{v}(t)$ so that the nonhomogeneous system (6.5.2) is satisfied

$$\begin{aligned} \mathbf{x}_p' &= A\mathbf{x}_p + \mathbf{f} \\ (W(t)\mathbf{v}(t))' &= A(W(t)\mathbf{v}(t)) + \mathbf{f} \\ (\begin{bmatrix} x_1 \\ x_1' \end{bmatrix} v_1(t) + \begin{bmatrix} x_2 \\ x_2' \end{bmatrix} v_2(t))' &= (AW(t))\mathbf{v}(t) + \mathbf{f}. \end{aligned}$$

Now apply the product rule to each component of the two dimensional vectors

$$\begin{aligned} \mathbf{x}_1' v_1(t) + \mathbf{x}_1 v_1'(t) + \mathbf{x}_2' v_2(t) + \mathbf{x}_2 v_2'(t) &= (AW(t))\mathbf{v}(t) + \mathbf{f} \\ W(t)'\mathbf{v}(t) + W(t)\mathbf{v}'(t) &= (AW(t))\mathbf{v}(t) + \mathbf{f}. \end{aligned}$$

Apply the matrix equation in (6.5.4) to get

$$(AW(t))\mathbf{v}(t) + W(t)\mathbf{v}'(t) = (AW(t))\mathbf{v}(t) + \mathbf{f}.$$

This means the derivatives of the variable coefficients must be a solution of

$$\begin{aligned} W(t)\mathbf{v}'(t) &= \mathbf{f} \\ \begin{bmatrix} x_1 & x_2 \\ x_1' & x_2' \end{bmatrix} \begin{bmatrix} v_1'(t) \\ v_2'(t) \end{bmatrix} &= \begin{bmatrix} 0 \\ f(t)/a \end{bmatrix}. \end{aligned}$$

Cramer's rule easily gives

$$v_1'(t) = \frac{\det(\begin{bmatrix} 0 & x_2 \\ f(t)/a & x_2' \end{bmatrix})}{\det(W(t))} = \frac{(-f(t)/a)x_2}{\det(W(t))} \text{ and}$$

$$v_2'(t) = \frac{\det(\begin{bmatrix} x_1 & 0 \\ x_1' & f(t)/a \end{bmatrix})}{\det(W(t))} = \frac{(f(t)/a)x_1}{\det(W(t))}.$$

The vector particular solution is found by integrating both of these to give the vector version of the *variation of parameters particular solution* formula

$$\mathbf{x}_p = \mathbf{x}_1 v_1(t) + \mathbf{x}_2 v_2(t) = \mathbf{x}_1 \int \frac{(-f(t)/a)x_2}{\det(W(t))} + \mathbf{x}_2 \int \frac{(f(t)/a)x_1}{\det(W(t))}. \tag{6.5.5}$$

The first component of this vector equation is denoted by $x_p = x_1 v_1(t) + x_2 v_2(t)$. The integrations can be difficult, but they could be approximated by numerical integrations.

Example 6.5.2. Find the particular solution $x'' - 4x = t + 1$ using the variation of parameters formula in (6.5.5). The two homogeneous solutions have the form e^{rt} where r solves the characteristic equation $r^2 - 4 = 0$. This gives $x_1 = e^{2t}$ and $x_2 = e^{-2t}$ so that

$$\det(W(t)) = \det\left(\begin{bmatrix} e^{2t} & e^{-2t} \\ e^{2t}2 & e^{-2t}(-2) \end{bmatrix}\right) = -4.$$

Use integration by part to find

$$v_1(t) = \int \frac{(-f(t)/a)x_2}{\det(W(t))} = \int \frac{(-(t+1)/1)e^{-2t}}{-4} = e^{-2t}(-t/8 - 3/16) \text{ and}$$

$$v_2(t) = \int \frac{(f(t)/a)x_1}{\det(W(t))} = \int \frac{(t+1)/1e^{2t}}{-4} = e^{2t}(-t/8 - 1/16).$$

The particular solution is

$$\begin{aligned} x_p &= x_1 v_1(t) + x_2 v_2(t) \\ &= e^{2t}(e^{-2t}(-t/8 - 3/16)) + e^{-2t}(e^{2t}(-t/8 - 1/16)) \\ &= (t+1)/(-4). \end{aligned}$$

The integrations seem to make variation of parameter approach for this problem more difficult; however, the variation of parameters formula in (6.5.5) is more general as long as the integrals exist, as is the case if the integrands are continuous. Moreover, this approach generalizes to more complicated systems of differential equations and will be further studied in Subsection 7.3.3.

6.5.3 Numerical Solution Using MATLAB's ode45()

In this section we will numerically approximate the current in the series LRC circuit where there is a modulated imposed signal

$$\begin{aligned} E(t) &= 100\sin(f_a 2\pi t)\sin(f_s 2\pi t) \text{ where} \\ \text{audio frequency} &= f_a \ll f_s = \text{station frequency.} \end{aligned}$$

In the first part of this section we were able to find the exact particular solution for the unmodulated signal

$$E(t) = 100\sin(f_s 2\pi t),$$

and in this case the maximum amplitude of the current was determined to be

$$\frac{\gamma 100}{\sqrt{(-L\gamma^2 + 1/C)^2 + (R\gamma)^2}}$$

where $\gamma = f_s 2\pi$. If $\gamma = 710$, $R = 400$, and L and C are chosen so that $-L\gamma^2 + 1/C \approx 0$, then the maximum amplitude is approximately

$$\frac{\gamma 100}{\sqrt{(0)^2 + (400\gamma)^2}} = \frac{100}{400} = 0.25.$$

If $L = 2$ and $C = 10^{-6}$, then the tuned station is $\gamma = 1/\sqrt{2(10^{-6})} = 10^3/\sqrt{2} \approx 707.1$. A significant variation from $L = 2$ and $C = 10^{-6}$ causes the amplitude to decrease as long as γ is at this tuned signal and the signal is unmodulated. The numerical calculations will illustrate the tuning formula $-L\gamma^2 + 1/C = 0$ giving an approximation for tuning even if the signal is modulated.

The MATLAB command ode45() is a variation of the Runge-Kutta numerical method and uses variable time steps. It requires three input parameters: the character representation of the function file containing the right side of the system, the row vector with the initial and final times and the row vector with the initial charge and current. The call to ode45() is contained in the MATLAB code file lrc.m. The MATLAB function file yplrc.m contains the right side of the vector equation (6.5.2). In the file yplrc.m the charge is depicted by the four character symbol y(1), and the current is depicted by the four character symbol y(2). In yplrc.m the first and second components of the right side of the vector equation (6.5.2) are denoted by the eight character symbols yplrc(1) and yplrc(2), respectively.

MATLAB Code lrc.m

```
% This file has a call to ode45().
% It requires a second file called yplrc.m,
% which models a series LRC tuning circuit.
clear;
t0 = 0;
tf = .3;              % final scaled time
y0 = [0 0];           % initial charge and current
[t y] = ode45('yplrc', [t0 tf], y0);
plot(t, y(:,2))
title( 'Series LRC Circuit')
xlabel('time')
ylabel('current')
```

MATLAB Code yplrc.m

```
function yplrc = yplrc(t,y)
L = 20; R = 400; C = 10^(-6);
fa = 63/(2*pi);              % scaled audio frequency
fs = 710/(2*pi);             % scaled station frequency
E0 = 100*sin(2*pi*fa.*t);    % modulated signal
% E0 = 100;                  % unmodulated signal
yplrc(1) = y(2);
yplrc(2) = (E0*sin(2*pi*fs.*t) - R*y(2) - (1/C)*y(1))/L;
```

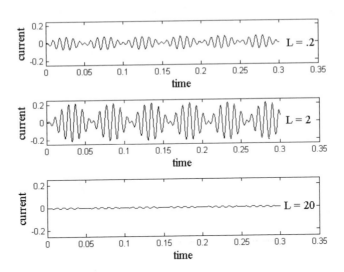

Figure 6.5.2: Tuned Circuit with Modulated Signal

$$\text{yplrc} = [\text{yplrc}(1) \ \text{yplrc}(2)]';$$

The code was executed three times with three values of inductance $L = 0.2$, 2 and 20. The currents are given in Figure 6.5.2, where the largest maximum amplitude corresponds to $L = 2$ and is consistent with the unmodulated analysis.

6.5.4 Exercises

1. In Example 6.5.1 derive the algebraic system and solution for the coefficients a and b.

2. Use Example 6.5.1 as a template to solve equation (6.5.2) with $E(t) = E_0 \cos(\gamma t)$.

3. Use variation of parameters to find the particular solution of $x'' - 9x = 2t + 3$.

4. Use variation of parameters to find the particular solution of $x'' + x = \sin(t)$.

5. Use MATLAB codes lrc.m and yplrc.m to experiment with different $C = (0.1)10^{-6}$, $(1.0)10^{-6}$ and $(10.0)10^{-6}$ and keep $L = 2$.

6. Consider equation (6.5.2) with $E(t) = E_0 \sin(63t) \cos(1500t)$.

   (a). Use the formula $\gamma = 1/\sqrt{LC}$ and $C = 10^{-6}$ to estimate the L that will give the maximum amplitude.

   (b). Confirm this by executing the MATLAB code file lrc.m with a modified yplrc.m file.

7.   Use a variation of the MATLAB codes lrc.m and yplrc.m to solve and graph the solutions of $x'' + bx' + x = \cos(t)$ with $x(0) = 0$ and $x'(0) = 0$ for $b = 0, 1, 2$ and 4.

Chapter 7

Eigenvalues and Differential Equations

This chapter contains the eigenvalue and eigenvector approach to solving linear systems of differential equations. This is a generalization of the simple single differential equation $x' = ax + f$ to a system $\mathbf{x}' = A\mathbf{x} + \mathbf{f}$ where A is a matrix and $\mathbf{x}$ a column vector of unknown functions of time. Applications will include time dependent mixing tanks and heat conduction.

7.1 Solution of $\mathbf{x}' = A\mathbf{x}$ by Elimination

In Section 6.5 a single second order linear differential equation was expressed as a system of two first order differential equations

$$\begin{bmatrix} x \\ y \end{bmatrix}' = \begin{bmatrix} 0 & 1 \\ -c/a & -b/a \end{bmatrix} \begin{bmatrix} x \\ y \end{bmatrix} + \begin{bmatrix} 0 \\ f/a \end{bmatrix} \text{ with } \begin{bmatrix} x(0) \\ y(0) \end{bmatrix} = \begin{bmatrix} x_0 \\ y_0 \end{bmatrix}. \tag{7.1.1}$$

This is a special case of the more general initial value problem

$$\begin{bmatrix} x \\ y \end{bmatrix}' = \begin{bmatrix} a & b \\ c & d \end{bmatrix} \begin{bmatrix} x \\ y \end{bmatrix} + \begin{bmatrix} f \\ g \end{bmatrix} \text{ with } \begin{bmatrix} x(0) \\ y(0) \end{bmatrix} = \begin{bmatrix} x_0 \\ y_0 \end{bmatrix}. \tag{7.1.2}$$

Furthermore, there can be more than two unknowns and two differential equations. The objective of this section is to present some examples that will indicate how one can approach these problems so that larger systems can be solved in a systematic way.

7.1.1 Overdamped Mass-Spring System

The solution of the homogeneous second order differential equation $ax'' + bx' + cx = 0$ has the exponential form e^{rt} where r is the solution of the characteristic equations $ar^2 + br + c = 0$. There are three possibilitie, either r is distinct real, repeated real or complex root. Consider the case where r has distinct real values and, hence, the general solution is $x(t) = c_1 e^{r_1 t} + c_2 e^{r_2 t}$. When converting the second order equation to a system, we defined $y = x'$. Thus, the general homogeneous solution to the system in equation (7.1.1) with $f = 0$ is

$$
\begin{bmatrix} x \\ y \end{bmatrix} = \begin{bmatrix} c_1 e^{r_1 t} + c_2 e^{r_2 t} \\ (c_1 e^{r_1 t} + c_2 e^{r_2 t})' \end{bmatrix}
$$

$$
= \begin{bmatrix} c_1 e^{r_1 t} + c_2 e^{r_2 t} \\ c_1 e^{r_1 t} r_1 + c_2 e^{r_2 t} r_2 \end{bmatrix}
$$

$$
= c_1 \begin{bmatrix} 1 \\ r_1 \end{bmatrix} e^{r_1 t} + c_2 \begin{bmatrix} 1 \\ r_2 \end{bmatrix} e^{r_2 t}.
$$

The two column vectors are examples of *eigenvectors* associated with the system matrix in (7.1.1), and r_1 and r_2 are the corresponding *eigenvalues*, which will be studied in more detail in Section 7.2. If initial conditions are required, then one must choose the coefficients c_1 and c_2 so that $x(0) = x_0$ and $y(0) = y_0$.

Example 7.1.1. Reconsider overdamped mass-spring in Example 6.3.3, where

$$
x'' + 4x' + x = 0.
$$

The equivalent system is

$$
\begin{bmatrix} x \\ y \end{bmatrix}' = \begin{bmatrix} 0 & 1 \\ -1 & -4 \end{bmatrix} \begin{bmatrix} x \\ y \end{bmatrix}
$$

whose solution is

$$
\begin{bmatrix} x \\ y \end{bmatrix} = c_1 \begin{bmatrix} 1 \\ r_1 \end{bmatrix} e^{r_1 t} + c_2 \begin{bmatrix} 1 \\ r_2 \end{bmatrix} e^{r_2 t}
$$

where $r_1 = -2 + \sqrt{3}$ and $r_2 = -2 - \sqrt{3}$. Find c_1 and c_2 so that $x(0) = 1$ and $y(0) = 2$. This requires the solution of the algebraic system

$$
\begin{bmatrix} x(0) \\ y(0) \end{bmatrix} = c_1 \begin{bmatrix} 1 \\ r_1 \end{bmatrix} e^{r_1 0} + c_2 \begin{bmatrix} 1 \\ r_2 \end{bmatrix} e^{r_2 0}
$$

$$
\begin{bmatrix} 1 \\ 2 \end{bmatrix} = \begin{bmatrix} 1 & 1 \\ r_1 & r_2 \end{bmatrix} \begin{bmatrix} c_1 \\ c_2 \end{bmatrix}.
$$

Since the roots are distinct, this has unique solution given by Cramer's rule

$$
c_1 = \frac{r_2 - 2}{r_2 - r_1} = \frac{-4 - \sqrt{3}}{-2\sqrt{3}} \quad \text{and} \quad c_2 = \frac{2 - r_1}{r_2 - r_1} = \frac{4 - \sqrt{3}}{-2\sqrt{3}}.
$$

7.1.2 Two-tank Mixing System

As a second illustration consider a two-tank mixing problem similar to mixing application in Subsection 3.7.4. Suppose the first and second tanks have fixed volumes equal to 100 (L) and have $x(t)$ (kg) and $y(t)$ (kg) amounts of salt so that the concentrations are $x(t)/100$ (L/kg) and $y(t)/100$ (L/kg), respectively. Let the first tank have incoming pure water with flow rate 6 (L/min) and concentration 0 (L/kg), and incoming from the second tank with flow rate 2 (L/min) with concentration $y(t)/100$. The first tank feeds into the second tank with flow rate 8 (L/min) with concentration $x(t)/100$, and these yield the differential equation for the first tank

$$x' = 6(0) + 2y/100 - 8x/100.$$

In order for the second tank to have fixed volume, the flow rate out and not into the first tank must be 6 (L/min) with concentration $y(t)/100$ (L/kg). This gives the differential equation for the second tank

$$y' = 8x/100 - 2y/100 - 6y/100.$$

These two differential equations may be written as a system

$$\left[\begin{array}{c} x \\ y \end{array} \right]' = \left[\begin{array}{cc} -4/50 & 1/50 \\ 4/50 & -4/50 \end{array} \right] \left[\begin{array}{c} x \\ y \end{array} \right].$$

In subsequent sections we shall derive the solutions by elimination and by eigenvalues giving

$$x = c_1 e^{(-1/25)t} + c_2 e^{(-3/25)t} \text{ and } y = c_1 2 e^{(-1/25)t} + c_2(-2)e^{(-3/25)t}.$$

The vector form of this is

$$\left[\begin{array}{c} x \\ y \end{array} \right] = c_1 \left[\begin{array}{c} 1 \\ 2 \end{array} \right] e^{(-1/25)t} + c_2 \left[\begin{array}{c} 1 \\ -2 \end{array} \right] e^{(-3/25)t}.$$

Example 7.1.2. Suppose the initial amounts in the tanks are $x(0) = 2$ and $y(0) = 3$. Find the coefficients so that

$$\left[\begin{array}{c} x(0) \\ y(0) \end{array} \right] = c_1 \left[\begin{array}{c} 1 \\ 2 \end{array} \right] e^{(-1/25)0} + c_2 \left[\begin{array}{c} 1 \\ -2 \end{array} \right] e^{(-3/25)0}$$

$$\left[\begin{array}{c} 2 \\ 3 \end{array} \right] = \left[\begin{array}{cc} 1 & 1 \\ 2 & -2 \end{array} \right] \left[\begin{array}{c} c_1 \\ c_2 \end{array} \right].$$

The algebraic solution is $c_1 = (2(-2) - 1(3))/(-4) = 7/4$ and $c_2 = (1(3) - 2(2))/(-4) = 1/4$. The solution to the homogeneous first order system is

$$\left[\begin{array}{c} x \\ y \end{array} \right] = (7/4) \left[\begin{array}{c} 1 \\ 2 \end{array} \right] e^{(-1/25)t} + (1/4) \left[\begin{array}{c} 1 \\ -2 \end{array} \right] e^{(-3/25)t}.$$

7.1.3   Homogeneous Solutions by Elimination

For small systems of differential equations this method is practical, and requires one to write the first order system as a single higher order differential equation. In the case of two first order homogeneous equations (7.1.2) with $f = g = 0$ we have

$$x' = ax + by \text{ and } y' = cx + dy.$$

If $b \neq 0$, then the first differential equation can be written as

$$y = (x' - ax)/b.$$

Put this into the second differential equation to yield a single second order differential equation

$$((x' - ax)/b)' = cx + d((x' - ax)/b)$$
$$(x'' - ax')/b = cx + d((x' - ax)/b).$$

As an illustration reconsider Example 7.1.2 and derive the solution using the elimination method.

Example 7.1.2 (continued). The two differential equations are

$$x' = (-4/50)x + (1/50)y \text{ and } y' = (4/50)x + (-4/50)y.$$

In the first equation solve for y

$$y = 50x' + 4x.$$

Insert this into the second equation

$$(50x' + 4x)' = (4/50)x + (-4/50)(50x' + 4x)$$
$$50x'' + 4x' = (4/50)x - 4x' + (-16/50)x.$$

This is a second order differential equation for x

$$50x'' + 8x' + (12/50)x = 0.$$

The solution has the exponential form e^{rt} where r is the solution of the characteristic equation

$$50r^2 + 8r + 12/50 = 0$$

whose solution is $r = -1/25$ and $r = -3/25$. This means the solution of the second order differential equation is

$$x = c_1 e^{(-1/25)t} + c_2 e^{(-3/25)t}.$$

The first order system requires the solution of both x and y. Find y by using $y = 50x' + 4x$

$$
\begin{aligned}
y &= 50(c_1 e^{(-1/25)t} + c_2 e^{(-3/25)t})' + 4(c_1 e^{(-1/25)t} + c_2 e^{(-3/25)t}) \\
&= 50(c_1 e^{(-1/25)t}(-1/25) + c_2 e^{(-3/25)t}(-3/25)) + \\
&\quad 4(c_1 e^{(-1/25)t} + c_2 e^{(-3/25)t}) \\
&= c_1 e^{(-1/25)t}(-50/25 + 4) + c_2 e^{(-3/25)t}(-150/25 + 4) \\
&= c_1 e^{(-1/25)t}2 + c_2 e^{(-3/25)t}(-2).
\end{aligned}
$$

Example 7.1.3. Use elimination to find the general homogeneous solution of

$$
\begin{bmatrix} x \\ y \end{bmatrix}' = \begin{bmatrix} 2 & -3 \\ 1 & -2 \end{bmatrix} \begin{bmatrix} x \\ y \end{bmatrix}.
$$

The first differential equation $x' = 2x - 3y$ can be solved for y in terms of x

$$
y = (x' - 2x)/(-3).
$$

Then put this into the second differential equation $y' = x - 2y$

$$
((x' - 2x)/(-3))' = x - 2((x' - 2x)/(-3)).
$$

This yields the single second order differential equation $x'' - x = 0$ whose general solution is

$$
x = c_1 e^t + c_2 e^{-t}.
$$

Use $y = (x' - 2x)/(-3)$ to find the general solution for y

$$
\begin{aligned}
y &= ((c_1 e^t + c_2 e^{-t})' - 2(c_1 e^t + c_2 e^{-t}))/(-3) \\
&= (c_1 e^t(1) + c_2 e^{-t}(-1) - 2(c_1 e^t + c_2 e^{-t}))/(-3) \\
&= c_1 e^t(1/3) + c_2 e^{-t}(1).
\end{aligned}
$$

The vector form of the system solution is

$$
\begin{bmatrix} x \\ y \end{bmatrix} = c_1 \begin{bmatrix} 1 \\ 1/3 \end{bmatrix} e^t + c_2 \begin{bmatrix} 1 \\ 1 \end{bmatrix} e^{-t}.
$$

Find the coefficients so that $x(0) = 6$ and $y(0) = 4$, that is,

$$
\begin{aligned}
\begin{bmatrix} x(0) \\ y(0) \end{bmatrix} &= c_1 \begin{bmatrix} 1 \\ 1/3 \end{bmatrix} e^0 + c_2 \begin{bmatrix} 1 \\ 1 \end{bmatrix} e^{-0} \\
\begin{bmatrix} 6 \\ 4 \end{bmatrix} &= \begin{bmatrix} 1 & 1 \\ 1/3 & 1 \end{bmatrix} \begin{bmatrix} c_1 \\ c_2 \end{bmatrix}.
\end{aligned}
$$

The algebraic solution is $c_1 = 3$ and $c_2 = 3$, and so the solution of the initial value problem is

$$
\begin{bmatrix} x \\ y \end{bmatrix} = 3 \begin{bmatrix} 1 \\ 1/3 \end{bmatrix} e^t + 3 \begin{bmatrix} 1 \\ 1 \end{bmatrix} e^{-t}.
$$

7.1.4   Application to Heat Conduction

This is an extension of the heat conduction model in a wire in Subsections 3.2.3 and 3.3.4. Let the wire be broken into four parts and let $u_i(t)$ be the approximation of the temperature at time t and at location $i\Delta x$ where $\Delta x = L/4$, L is the length of the thin wire, and u_0 and u_4 are given at the left and right ends of the wire. The objective is to find $u_1(t)$, $u_2(t)$ and $u_3(t)$ given the initial temperatures $u_1(0)$, $u_2(0)$ and $u_3(0)$. The time dependent model is

$$\rho c \begin{bmatrix} u_1 \\ u_2 \\ u_3 \end{bmatrix}' = -(K/(\Delta x)^2) \begin{bmatrix} 2 & -1 & 0 \\ -1 & 2 & -1 \\ 0 & -1 & 2 \end{bmatrix} \begin{bmatrix} u_1 \\ u_2 \\ u_3 \end{bmatrix} + \begin{bmatrix} f + (K/(\Delta x)^2)u_0 \\ f \\ f + (K/(\Delta x)^2)u_4 \end{bmatrix}.$$

This presents a larger system of differential equations as well as some additional terms on the right sides. For the moment, consider the following simplification.

Example 7.1.4.

$$\begin{bmatrix} u_1 \\ u_2 \\ u_3 \end{bmatrix}' = - \begin{bmatrix} 2 & -1 & 0 \\ -1 & 2 & -1 \\ 0 & -1 & 2 \end{bmatrix} \begin{bmatrix} u_1 \\ u_2 \\ u_3 \end{bmatrix}.$$

The elimination method still works, but one can clearly see the additional complications. Solve the first differential equation for u_2 in terms of u_1

$$u_2 = u_1' + 2u_1. \tag{7.1.3}$$

Put u_2 into the second equation and solve for u_3

$$u_3 = u_1'' + 4u_1' + 3u_1. \tag{7.1.4}$$

Insert u_3 and $u_2 = u_1' + 2u_1$ into the third equation and collect like terms to obtain a single third order differential equation

$$u_1''' + 6u_1'' + 10u_1' + 4u_1 = 0. \tag{7.1.5}$$

Assume the solution of (7.1.5) is an exponential e^{rt} to get a third order characteristic equation

$$r^3 + 6r^2 + 10r + 4 = 0$$
$$(r+2)(r^2 + 4r + 2) = 0.$$

The algebraic solutions are $r = -2$, $-2 + \sqrt{2}$ and $-2 - \sqrt{2}$. Therefore, the general solution to the third order differential equation is

$$u_1(t) = c_1 e^{-2t} + c_2 e^{(-2+\sqrt{2})t} + c_3 e^{(-2-\sqrt{2})t}.$$

Use equations (7.1.3) and (7.1.4) to find $u_2(t)$ and $u_3(t)$, and write the general solution to the homogeneous system in vector form

$$\begin{bmatrix} u_1 \\ u_2 \\ u_3 \end{bmatrix} = c_1 \begin{bmatrix} 1 \\ 0 \\ -1 \end{bmatrix} e^{-2t} + c_2 \begin{bmatrix} 1 \\ \sqrt{2} \\ 1 \end{bmatrix} e^{(-2+\sqrt{2})t} + c_3 \begin{bmatrix} 1 \\ -\sqrt{2} \\ 1 \end{bmatrix} e^{(-2-\sqrt{2})t}.$$

Find the three coefficients so that the initial temperatures are $u_1(0) = 20$, $u_2(0) = 30$ and $u_3(0) = 20$. Insert $t = 0$ into the above general solution to obtain the algebraic system

$$\begin{bmatrix} 20 \\ 30 \\ 20 \end{bmatrix} = \begin{bmatrix} 1 & 1 & 1 \\ 0 & \sqrt{2} & -\sqrt{2} \\ -1 & 1 & 1 \end{bmatrix} \begin{bmatrix} c_1 \\ c_2 \\ c_3 \end{bmatrix}.$$

The algebraic solution is $c_1 \approx 0.0000$, $c_2 \approx 20.6066$ and $c_3 \approx -0.6066$ and so the solution to the system of differential equations is

$$\begin{bmatrix} u_1 \\ u_2 \\ u_3 \end{bmatrix} \approx 20.6066 \begin{bmatrix} 1 \\ \sqrt{2} \\ 1 \end{bmatrix} e^{(-2+\sqrt{2})t} - 0.6066 \begin{bmatrix} 1 \\ -\sqrt{2} \\ 1 \end{bmatrix} e^{(-2-\sqrt{2})t}.$$

7.1.5 Solution of Systems Using MATLAB

The MATLAB command dsolve() can be used to solve simple linear systems of differential equations. The solution of the two-tank mixing in Example 7.1.2 is found as follows.

```
>> soltank = dsolve('Dx = (-4/50)*x + (1/50)*y',
                    'Dy = (4/50)*x + (-4/50)*y',
                    'x(0) = 2', 'y(0) = 3', 't')
soltank =
      x: [1x1 sym]
      y: [1x1 sym]
>> soltank.x
      1/4*exp(-3/25*t) + 7/4*exp(-1/25*t)
>> soltank.y
      7/2*exp(-1/25*t) - 1/2*exp(-3/25*t)
```

The solution to the heat conduction in Example 7.1.4 with three equations can also be easily done. But, one can observe the increased symbolic output. For more complicated problems numerical calculations such as done by MATLAB's command ode45() are a preferred method.

```
>> solheat = dsolve('Du1 = -2*u1 + u2', 'Du2 = u1 -2*u2 + u3',
                    'Du3 = u2 -2*u3', 'u1(0) = 20',
                    'u2(0) = 30', 'u3(0) = 20', 't')
solheat =
    u1: [1x1 sym]
    u2: [1x1 sym]
    u3: [1x1 sym]
>> solheat.u1
    -15/2*2^(1/2)*exp(-(2+2^(1/2))*t) +
    15/2*2^(1/2)*exp((-2+2^(1/2))*t) +
    10*exp(-(2+2^(1/2))*t) + 10*exp((-2+2^(1/2))*t)
>> solheat.u2
    15*exp(-(2+2^(1/2))*t) + 15*exp((-2+2^(1/2))*t) -
    10*2^(1/2)*exp(-(2+2^(1/2))*t) +
    10*2^(1/2)*exp((-2+2^(1/2))*t)
>> solheat.u3
    -15/2*2^(1/2)*exp(-(2+2^(1/2))*t) +
    15/2*2^(1/2)*exp((-2+2^(1/2))*t) +
    10*exp(-(2+2^(1/2))*t) + 10*exp((-2+2^(1/2))*t)
```

7.1.6   Exercises

1.   Consider the overdamped mass-spring model $x'' + 8x' + x = 0$.
   (a).   Define $y = x'$ and write this as a first order system for x and y.
   (b).   Find the vector form of the system solution.
   (c).   Find the coefficients so that $x(0) = 2$ and $y(0) = 0$.
2.   Consider differential equation $x'' - 3x' + 2x = 0$.
   (a).   Define $y = x'$ and write this as a first order system for x and y.
   (b).   Find the vector form of the system solution.
   (c).   Find the coefficients so that $x(0) = 1$ and $y(0) = -2$.
3.   Consider the two-tank mixing model where both tanks have fixed volume equal to 50 (L).
   (a).   Derive the model to show it has the form

$$\begin{bmatrix} x \\ y \end{bmatrix}' = \begin{bmatrix} -4/25 & 1/25 \\ 4/25 & -4/25 \end{bmatrix} \begin{bmatrix} x \\ y \end{bmatrix}.$$

   (b).   Use elimination to find the vector form of the general solution.
   (c).   Find the solution of the initial value problem $x(0) = 2$ and $y(0) = 3$.
4.   Use elimination to solve the initial value problem

$$\begin{bmatrix} x \\ y \end{bmatrix}' = \begin{bmatrix} 1 & -1 \\ 2 & 4 \end{bmatrix} \begin{bmatrix} x \\ y \end{bmatrix} \text{ with } \begin{bmatrix} x(0) \\ y(0) \end{bmatrix} = \begin{bmatrix} 1 \\ 4 \end{bmatrix}.$$

5. Use elimination to solve the initial value problem

$$\begin{bmatrix} x \\ y \end{bmatrix}' = \begin{bmatrix} 3 & -1 \\ -1 & 3 \end{bmatrix} \begin{bmatrix} x \\ y \end{bmatrix} \text{ with } \begin{bmatrix} x(0) \\ y(0) \end{bmatrix} = \begin{bmatrix} 2 \\ 4 \end{bmatrix}.$$

6. Use elimination to solve the initial value problem

$$\begin{bmatrix} x \\ y \end{bmatrix}' = \begin{bmatrix} 4 & 1 \\ -2 & 1 \end{bmatrix} \begin{bmatrix} x \\ y \end{bmatrix} \text{ with } \begin{bmatrix} x(0) \\ y(0) \end{bmatrix} = \begin{bmatrix} 1 \\ 3 \end{bmatrix}.$$

7. Consider the heat conduction model and Example 7.1.4.
   (a). Derive equations (7.1.4) and (7.1.5).
   (b). Derive the expressions for $u_2(t)$ and $u_3(t)$.
8. Solve the initial value problem

$$\begin{bmatrix} u_1 \\ u_2 \\ u_3 \end{bmatrix}' = \begin{bmatrix} -3 & 1 & 0 \\ 1 & -3 & 1 \\ 0 & 1 & -3 \end{bmatrix} \begin{bmatrix} u_1 \\ u_2 \\ u_3 \end{bmatrix} \text{ with } \begin{bmatrix} u_1(0) \\ u_2(0) \\ u_3(0) \end{bmatrix} = \begin{bmatrix} 10 \\ 30 \\ 5 \end{bmatrix}.$$

9. Use MATLAB to confirm any calculations in exercises 3-6 and 8.

7.2 Real Eigenvalues and Eigenvectors

The examples in the previous section suggest that the solutions of $\mathbf{x}' = A\mathbf{x}$ are linear combination vector functions of the form $\mathbf{u}e^{rt}$. The exact expression for the vector functions was derived from the elimination method, which transforms the system into a single higher order differential equation. Because the elimination method becomes complicated for larger systems, this is not a preferred method. An alternative is to find the vectors $\mathbf{u}$ and the associated numbers r, called eigenvalues, by noting that they must satisfy $A\mathbf{u} = r\mathbf{u}$. In the discussion of the eigenvalue method we restrict the examples to the cases where the eigenvalues are distinct and real. The computation of the required eigenvalues and eigenvectors for larger systems can be done by software such as MATLAB.

7.2.1 Eigenvalue and Eigenvector Definitions

Assume a solution of $\mathbf{x}' = A\mathbf{x}$ has the form $\mathbf{x} = \mathbf{u}e^{rt}$ where $\mathbf{u}$ is a nonzero constant vector with possibly complex components and r is possibly a complex number. We must choose $\mathbf{u}$ and r so that $\mathbf{x}' = A\mathbf{x}$ holds

$$\begin{aligned} (\mathbf{u}e^{rt})' &= A(\mathbf{u}e^{rt}) \\ \mathbf{u}e^{rt}r &= (A\mathbf{u})e^{rt}. \end{aligned}$$

Since e^{rt} is never zero,

$$\begin{aligned} A\mathbf{u} &= r\mathbf{u} \\ A\mathbf{u} &= rI\mathbf{u} \\ (A - rI)\mathbf{u} &= 0. \end{aligned}$$

Because $\mathbf{u}$ is not a zero vector, $A - rI$ must not have an inverse and, thus, the determinant of $A - rI$ must be zero. This gives a single polynomial equation that r must satisfy. Once r is found, solve algebraic system $(A - rI)\mathbf{u} = \mathbf{0}$ for the unknown constant vector $\mathbf{u}$ that is associated with this r.

Definition 7.2.1. Let A be an $n \times n$ matrix. An *eigenvalue* is a number (real or complex) r such that there is a nonzero vector $\mathbf{u}$ and $A\mathbf{u} = r\mathbf{u}$. The vector $\mathbf{u}$ is called the *eigenvector* of A corresponding to the eigenvalue r.

Note, any constant times $\mathbf{u}$ is also an eigenvector, because $A(c\mathbf{u}) = c(A\mathbf{u}) = c(r\mathbf{u}) = r(c\mathbf{u})$. Examples 7.1.1-7.1.4 in the previous section give the following examples of eigenvectors and eigenvectors .

Example 7.2.1. Let $A = \begin{bmatrix} 0 & 1 \\ -1 & -4 \end{bmatrix}$ and consider $\mathbf{u} = \begin{bmatrix} 1 \\ r_1 \end{bmatrix}$ and $r = r_1 = -2 + \sqrt{3}$.

$$
\begin{aligned}
A\mathbf{u} &= \begin{bmatrix} 0 & 1 \\ -1 & -4 \end{bmatrix} \begin{bmatrix} 1 \\ -2 + \sqrt{3} \end{bmatrix} \\
&= \begin{bmatrix} 0(1) + 1(-2 + \sqrt{3}) \\ -1(1) - 4(-2 + \sqrt{3}) \end{bmatrix} \\
&= (-2 + \sqrt{3}) \begin{bmatrix} 1 \\ -2 + \sqrt{3} \end{bmatrix}.
\end{aligned}
$$

Example 7.2.2. Let $A = \begin{bmatrix} -4/50 & 1/50 \\ 4/50 & -4/50 \end{bmatrix}$ and consider $\mathbf{u} = \begin{bmatrix} 1 \\ 2 \end{bmatrix}$ and $r = r_1 = -1/25$.

$$
\begin{aligned}
A\mathbf{u} &= \begin{bmatrix} -4/50 & 1/50 \\ 4/50 & -4/50 \end{bmatrix} \begin{bmatrix} 1 \\ 2 \end{bmatrix} \\
&= \begin{bmatrix} (-4/50)1 + (1/50)2 \\ (4/50)1 + (-4/50)2 \end{bmatrix} \\
&= (-1/25) \begin{bmatrix} 1 \\ 2 \end{bmatrix}.
\end{aligned}
$$

Example 7.2.3. Let $A = \begin{bmatrix} 2 & -3 \\ 1 & -2 \end{bmatrix}$ and consider $\mathbf{u} = \begin{bmatrix} 1 \\ 1 \end{bmatrix}$ and $r = r_2 = -1$.

$$
\begin{aligned}
A\mathbf{u} &= \begin{bmatrix} 2 & -3 \\ 1 & -2 \end{bmatrix} \begin{bmatrix} 1 \\ 1 \end{bmatrix} \\
&= \begin{bmatrix} 2(1) - 3(1) \\ 1(1) - 2(1) \end{bmatrix} \\
&= (-1) \begin{bmatrix} 1 \\ 1 \end{bmatrix}.
\end{aligned}
$$

Example 7.2.4. Let $A = \begin{bmatrix} -2 & 1 & 0 \\ 1 & -2 & 1 \\ 0 & 1 & -2 \end{bmatrix}$ and consider $\mathbf{u} = \begin{bmatrix} 1 \\ -\sqrt{2} \\ 1 \end{bmatrix}$ and $r = r_3 = -2 - \sqrt{2}$.

$$
\begin{aligned}
A\mathbf{u} &= \begin{bmatrix} -2 & 1 & 0 \\ 1 & -2 & 1 \\ 0 & 1 & -2 \end{bmatrix} \begin{bmatrix} 1 \\ -\sqrt{2} \\ 1 \end{bmatrix} \\
&= \begin{bmatrix} -2(1) + 1(-\sqrt{2}) \\ 1(1) - 2(-\sqrt{2}) + 1(1) \\ 1(-\sqrt{2}) - 2(1) \end{bmatrix} \\
&= (-2 - \sqrt{2}) \begin{bmatrix} 1 \\ -\sqrt{2} \\ 1 \end{bmatrix}.
\end{aligned}
$$

7.2.2 Finding Eigenvalues by Determinants

For small matrices a simple way to find eigenvalues r is to solve

$$\det(A - rI) = 0.$$

In general this is a polynomial equation of degree n and, so, there are n possible roots or eigenvalues. For each eigenvalue there is an eigenvector $\mathbf{u}$ that satisfies the algebraic system

$$(A - rI)\mathbf{u} = \mathbf{0}.$$

A simple example is the system associated with the second order differential equation $ax'' + bx' + cx = 0$ where

$$A = \begin{bmatrix} 0 & 1 \\ -c/a & -b/a \end{bmatrix}.$$

In this case the determinant is easy to evaluate and gives the characteristic algebraic equation

$$
\begin{aligned}
\det(A - rI) &= \det(\begin{bmatrix} 0 & 1 \\ -c/a & -b/a \end{bmatrix} - r \begin{bmatrix} 1 & 0 \\ 0 & 1 \end{bmatrix}) \\
&= \det(\begin{bmatrix} 0 - r & 1 \\ -c/a & -b/a - r \end{bmatrix}) \\
&= -r(-b/a - r) - 1(-c/a) \\
&= r^2 + (b/a)r + c/a = 0.
\end{aligned}
$$

The solutions may be complex numbers, but we will focus on systems where the roots of $\det(A - rI) = 0$ are real and distinct.

Example 7.2.1 (continued). Find the eigenvalues using determinants by solving

$$\det(A - rI) = \det(\begin{bmatrix} 0-r & 1 \\ -1 & -4-r \end{bmatrix})$$
$$= r^2 + 4r + 1 = 0.$$

The two eigenvalues are $r_1 = -2 + \sqrt{3}$ and $r_2 = -2 - \sqrt{3}$. Find the eigenvector associated with $r_1 = -2 + \sqrt{3}$. Solve the algebraic system

$$(A - r_1 I)\mathbf{u} = 0$$
$$\begin{bmatrix} 0-(-2+\sqrt{3}) & 1 \\ -1 & -4-(-2+\sqrt{3}) \end{bmatrix}\begin{bmatrix} u_1 \\ u_2 \end{bmatrix} = \begin{bmatrix} 0 \\ 0 \end{bmatrix}.$$

This is equivalent to two algebraic equations, which are multiples of each other,

$$-(-2+\sqrt{3})u_1 + u_2 = 0 \text{ and } -u_1 + (2-\sqrt{3})u_2 = 0.$$

This gives one degree of freedom to choose, say, $u_1 = 1$ and then $u_2 = -2 + \sqrt{3}$. This agrees with one homogeneous solution obtained using the elimination method. In order to obtain the other homogeneous solution, let $r_2 = -2 - \sqrt{3}$ and repeat the above calculations.

Example 7.2.2 (continued). Find the eigenvalues using determinants by solving

$$\det(A - rI) = \det(\begin{bmatrix} -4/50-r & 1/50 \\ 4/50 & -4/50-r \end{bmatrix})$$
$$= (-4/50-r)(-4/50-r) - (1/50)(4/50) = 0.$$

The two eigenvalues are $r_1 = -1/25$ and $r_2 = -3/25$. Find the eigenvector associated with $r_1 = -1/25$. Solve the algebraic system

$$(A - r_1 I)\mathbf{u} = 0$$
$$\begin{bmatrix} -4/50-(-1/25) & 1/50 \\ 4/50 & -4/50-(-1/25) \end{bmatrix}\begin{bmatrix} u_1 \\ u_2 \end{bmatrix} = \begin{bmatrix} 0 \\ 0 \end{bmatrix}.$$

The two algebraic equations are

$$(-2/50)u_1 + (1/50)u_2 = 0 \text{ and } (4/50)u_1 + (-2/50)u_2 = 0.$$

This gives one degree of freedom to choose, say, $u_1 = 1$ and so $u_2 = 2$. In order to find the other homogeneous solution, let $r_2 = -3/25$ and repeat the above calculations.

Example 7.2.3 (continued). Find the eigenvalues using determinants by solving

$$\det(A - rI) = \det(\begin{bmatrix} 2-r & -3 \\ 1 & -2-r \end{bmatrix})$$
$$= (2-r)(-2-r) - (-3)1 = 0.$$

The two eigenvalues are $r_1 = 1$ and $r_2 = -1$. Find the eigenvector associated with $r_2 = -1$. Solve the algebraic system

$$(A - r_1 I)\mathbf{u} = \mathbf{0}$$

$$\begin{bmatrix} 2 - (-1) & -3 \\ 1 & -2 - (-1) \end{bmatrix} \begin{bmatrix} u_1 \\ u_2 \end{bmatrix} = \begin{bmatrix} 0 \\ 0 \end{bmatrix}.$$

The two algebraic equations are

$$3u_1 - 3u_2 = 0 \text{ and } u_1 - u_2 = 0.$$

This gives one degree of freedom to choose, say, $u_1 = 1$ and then $u_2 = 1$. In order to find the other homogeneous solution, let $r_1 = 1$ and repeat the above calculations.

Example 7.2.4 (continued). Find the eigenvalues using determinants by solving

$$\det(A - rI) = \det\left(\begin{bmatrix} -2 - r & 1 & 0 \\ 1 & -2 - r & 1 \\ 0 & 1 & -2 - r \end{bmatrix} \right)$$

$$= (-2 - r)((-2 - r)(-2 - r) - 1) - 1(1(-2 - r) - 0)$$

$$= (-2 - r)[(-2 - r)(-2 - r) - 2] = 0.$$

The three eigenvalues are $r_1 = -2$, $r_2 = -2 + \sqrt{2}$ and $r_3 = -2 - \sqrt{2}$. Find the eigenvector associated with $r_1 = -2$. Solve the algebraic system

$$(A - r_1 I)\mathbf{u} = \mathbf{0}$$

$$\begin{bmatrix} -2 - (-2) & 1 & 0 \\ 1 & -2 - (-2) & 1 \\ 0 & 1 & -2 - (-2) \end{bmatrix} \begin{bmatrix} u_1 \\ u_2 \\ u_3 \end{bmatrix} = \begin{bmatrix} 0 \\ 0 \\ 0 \end{bmatrix}.$$

The three algebraic equations are

$$u_2 = 0, \ u_1 + u_3 = 0 \text{ and } u_2 = 0.$$

This gives one degree of freedom to choose, say, $u_1 = 1$ and so $u_3 = -1$. The other two eigenvectors can be found by similar calculations.

7.2.3 Application to Three-tank Mixing

This section continues the discussion of the three-tank mixing application in Subsection 3.7.4 where the steady state model was considered. Here we will study the time dependent model for the amounts $x(t)$, $y(t)$ and $z(t)$ in the three 24 (L) tanks. The flow rates are as indicated in Figure 3.7.1, and the system of differential equations is

$$\begin{bmatrix} x \\ y \\ z \end{bmatrix}' = \begin{bmatrix} -5/12 & 2/12 & 0 \\ 5/12 & -6/12 & 1/12 \\ 0 & 4/12 & -4/12 \end{bmatrix} \begin{bmatrix} x \\ y \\ z \end{bmatrix} + \begin{bmatrix} 1 \\ 2 \\ 0 \end{bmatrix}. \tag{7.2.1}$$

The solution of $\det(A - rI) = 0$ done by-hand calculations is a little messy and, so, the MATLAB commands det() and solve() are used.

```
>> syms r
>> poly = det(A - r*eye(3))          % compute the determinant
        -5/144 - 5/12*r - 5/4*r^2 - r^3
>> r = solve(poly)      % solve for the roots of this polynomial
>> r1 = eval(r(1))      % converts the symbolic expression to numbers
        -.1264
>> r2 = eval(r(2))
        -.7643
>> r3 = eval(r(3))
        -.3593
```

From here one can solve the algebraic system $(A - rI)\mathbf{u} = \mathbf{0}$ to find the eigenvectors, but an easier and more robust approach is to use another MATLAB command eig(), which computes all the eigenvalues and their associated eigenvectors.

7.2.4   Eigenvalues and Eigenvectors Using MATLAB

The MATLAB command [U R] = eig(A) computes the eigenvectors and eigenvalues of the matrix A. All three symbols are square matrices. The columns of U, indicated by $U(:, j)$, are the eigenvectors. The matrix R is a diagonal matrix whose j^{th} diagonal components are the eigenvalues associated with the jth column of U. One homogeneous solution is $U(:, j)e^{R(j,j)t}$, and the general homogeneous solution is a linear combination of $U(:, j)e^{R(j,j)t}$. The eigenvectors and eigenvalues for Example 7.2.3 are easily computed by [U R] = eig(A). The columns of U have been scaled to have length equal to one, and appear different than those done by-hand calculations. However, any eigenvector can be scaled to obtain another eigenvector associated with the same eigenvalue.

```
>> A = [2 -3;1 -2]
>> [U R] = eig(A)
        U =
            .9487      .7071
            .3162      .7071
        R =
            1     0
            0    -1
>> newu1 = U(:,1)/U(1,1)      % scale the first eigenvector
        1.0000
        0.3333
>> newu2 = U(:,2)/U(1,2)      % scale the second eigenvector
        1.0000
        1.0000
```

Return to the three-tank mixing problem in (7.2.1), and use $[U\ R] = \text{eig}(A)$ to find the eigenvectors and eigenvalues.

```
>> A = [5/12 2/12 0; 5/12 -6/12 1/12; 0 4/12 -4/12]
>> [U R] = eig(A)
   U =
```

.3546	-.2201	.2898
.7396	-.0758	.5047
-.5721	.9725	.8132

```
   R =
```

-.7643	0.0000	0.0000
0.0000	-.3593	0.0000
0.0000	0.0000	-.1264

The solution to the three-tank mixing problem (7.2.1) is the general homogeneous solution plus the steady state solution, which was calculated in Subsection 3.7.4. The steady state solution is the algebraic solution of (7.2.1) when the time derivatives on the left side are all zero. The time dependent solution is

$$
\begin{bmatrix} x(t) \\ y(t) \\ z(t) \end{bmatrix} = c_1 \begin{bmatrix} .3546 \\ .7396 \\ -.5721 \end{bmatrix} e^{-.7643t} + c_2 \begin{bmatrix} -.2201 \\ -.0758 \\ .9725 \end{bmatrix} e^{-.3593t} +
$$

$$
c_3 \begin{bmatrix} .2898 \\ .5047 \\ .8132 \end{bmatrix} e^{-.1264t} + \begin{bmatrix} 7.2 \\ 12.0 \\ 12.0 \end{bmatrix}.
$$

Since the eigenvalues are all negative, the time dependent solution converges to the steady state solution as time increases.

7.2.5   Exercises

1.   Consider Example 7.2.1.
   (a).   Verify $r = r_2 = -2 - \sqrt{3}$ and $\mathbf{u} = [1\ \ (-2 - \sqrt{3})]^T$ are the other eigenvalue and eigenvector.
   (b).   Solve the algebraic system $(A - r_2 I)\mathbf{u} = \mathbf{0}$ to find the above eigenvector.

2.   Consider Example 7.2.2.
   (a).   Verify $r = r_2 = -3/25$ and $\mathbf{u} = [1\ \ -2]^T$ are the other eigenvalue and eigenvector.
   (b).   Solve the algebraic system $(A - r_2 I)\mathbf{u} = \mathbf{0}$ to find the above eigenvector.

3.   Consider Example 7.2.3.
   (a).   Verify $r = r_1 = 1$ and $\mathbf{u} = [1\ \ 1/3]^T$ are the other eigenvalue and eigenvector.
   (b).   Solve the algebraic system $(A - r_1 I)\mathbf{u} = \mathbf{0}$ to find the above eigenvector.

4.   Consider Example 7.2.4.

(a).   Verify $r = r_2 = -2 + \sqrt{2}$ and $\mathbf{u} = [1 \quad \sqrt{2} \quad 1]^T$ are the second eigenvalue and eigenvector.

(b).   Solve the algebraic system $(A - r_2 I)\mathbf{u} = \mathbf{0}$ to find the above eigenvector.

5.   Use determinants to find both eigenvalues and eigenvectors for

$$A = \begin{bmatrix} -4/25 & 1/25 \\ 4/25 & -4/25 \end{bmatrix}.$$

6.   Use determinants to find both eigenvalues and eigenvectors for

$$A = \begin{bmatrix} 3 & -1 \\ -1 & 3 \end{bmatrix}.$$

7.   Use determinants to find both eigenvalues and eigenvectors for

$$A = \begin{bmatrix} 4 & 1 \\ -2 & 1 \end{bmatrix}.$$

8.   Use determinants to find all three eigenvalues and eigenvectors for

$$A = \begin{bmatrix} -3 & 1 & 0 \\ 1 & -3 & 1 \\ 0 & 1 & -3 \end{bmatrix}.$$

9.   Use the MATLAB command $[U \ R] = \text{eig}(A)$ to confirm any calculations in exercises 5-8.

7.3   Solution of $\mathbf{x}' = A\mathbf{x} + \mathbf{f}(t)$

Assume the $n \times n$ matrix A has n distinct real eigenvalues such as in Examples 7.2.1-7.2.4 of the previous section. Consider the solution of the linear system of differential equations with unknown functions $\mathbf{x} = [x_1(t) \quad x_2(t) \quad \cdots \quad x_n(t)]^T$

$$\mathbf{x}' = A\mathbf{x} + \mathbf{f}(t). \tag{7.3.1}$$

In this case, one can show, see Subsection 7.3.3, there are n linearly independent homogeneous solutions formed by the eigenvectors $\mathbf{u}_j$ and eigenvalues r_j

$$\mathbf{u}_1 e^{r_1 t}, \mathbf{u}_2 e^{r_2 t}, \cdots, \mathbf{u}_n e^{r_n t}.$$

Any *homogeneous* solution $\mathbf{x}_h(t)$ satisfies $\mathbf{x}_h' = A\mathbf{x}_h$ and is a linear combination of these

$$\mathbf{x}_h(t) = c_1 \mathbf{u}_1 e^{r_1 t} + c_2 \mathbf{u}_2 e^{r_2 t} + \cdots + c_n \mathbf{u}_n e^{r_n t}.$$

A *particular* solution is any $\mathbf{x}_p(t)$ that satisfies $\mathbf{x}_p' = A\mathbf{x}_p + \mathbf{f}(t)$, and they can be found for simple cases by the method of undetermined coefficients, see

Subsection 7.3.2, or by the variation of parameters method, see Subsection 7.3.3. Since

$$
\begin{aligned}
(\mathbf{x}_h(t) + \mathbf{x}_p(t))' &= \mathbf{x}'_h + \mathbf{x}'_p \\
&= A\mathbf{x}_h + A\mathbf{x}_p + \mathbf{f}(t) \\
&= A(\mathbf{x}_h + A\mathbf{x}_p) + \mathbf{f}(t),
\end{aligned}
$$

the *general* solution is

$$
\mathbf{x}(t) = \mathbf{x}_h(t) + \mathbf{x}_p(t).
$$

The assumption that the matrix has distinct real eigenvalues does limit our discussion (see exercise 1 for other cases), and the interested reader should consult [1] or [7] for more details about repeated or complex eigenvalues. When the homogeneous and particular solutions have been found, the *initial value problem*, which also requires $\mathbf{x}(t)$ to satisfy a given initial vector $\mathbf{x}(0) = \mathbf{x}_0$, can be solved. A number of examples will be given, and an application to time dependent heat conduction will be presented.

7.3.1 A Has an Inverse and f a Constant Vector

The application to the three-tank mixing problem presented in the previous section has a differential equation for the amount in each tank. This gives a linear system where $n = 3$, $\mathbf{f}(t) = \mathbf{f}$ is a constant vector, and the eigenvalues of the matrix are distinct, real and negative. The matrix A has an inverse, and the steady state solution is defined by the solution of the algebraic system

$$
\mathbf{0} = A\mathbf{x} + \mathbf{f} \text{ so that } \mathbf{x} = -A^{-1}\mathbf{f}.
$$

The steady state solution is a constant vector, and it is also an example of a particular solution. The general solution is a linear combination of the homogeneous solutions plus the steady state solution

$$
\mathbf{x}(t) = c_1 \mathbf{u}_1 e^{r_1 t} + c_2 \mathbf{u}_2 e^{r_2 t} + c_3 \mathbf{u}_3 e^{r_3 t} + (-A^{-1}\mathbf{f}).
$$

Because the eigenvalues r_1, r_2 and r_3 are negative, the general solution must converge to the steady state solution.

Example 7.3.1.   Find the general homogeneous, steady state and general solution for

$$
\mathbf{x}' = A\mathbf{x} + \mathbf{f}
$$

$$
\begin{bmatrix} x_1 \\ x_2 \end{bmatrix}' = \begin{bmatrix} 4 & 1 \\ 10 & 1 \end{bmatrix} \begin{bmatrix} x_1 \\ x_2 \end{bmatrix} + \begin{bmatrix} 6 \\ 12 \end{bmatrix}.
$$

The homogeneous solutions are found by first finding both eigenvalues and their corresponding eigenvectors. Solve

$$
\begin{aligned}
\det(A - rI) &= 0 \\
\det\left(\begin{bmatrix} 4 - r & 1 \\ 10 & 1 - r \end{bmatrix}\right) &= 0 \\
(4 - r)(1 - r) - 10 &= 0
\end{aligned}
$$

to obtain $r_1 = 6$ and $r_2 = -1$. Let $r = 6$ and find the corresponding eigenvector by solving the algebraic systems $(A - rI)\mathbf{u} = \mathbf{0}$:

$$\begin{bmatrix} 4-6 & 1 \\ 10 & 1-6 \end{bmatrix} \begin{bmatrix} u_1 \\ u_2 \end{bmatrix} = \begin{bmatrix} 0 \\ 0 \end{bmatrix} \quad \text{or}$$

$$\begin{aligned} -2u_1 + u_2 &= 0 \text{ and} \\ 10u_1 + (-5)u_2 &= 0. \end{aligned}$$

So, for $r = r_1 = 6$ the corresponding eigenvector is $\mathbf{u}_1 = [1 \ \ 2]^T$. In a similar calculation for $r = r_2 = -1$ we obtain the eigenvector $\mathbf{u}_2 = [1 \ \ -5]^T$. The general homogeneous solution is

$$\begin{aligned} \mathbf{x}_h(t) &= c_1 \mathbf{u}_1 e^{r_1 t} + c_2 \mathbf{u}_2 e^{r_2 t} \\ &= c_1 \begin{bmatrix} 1 \\ 2 \end{bmatrix} e^{6t} + c_2 \begin{bmatrix} 1 \\ -5 \end{bmatrix} e^{-t}. \end{aligned}$$

The steady state solution is the solution of $\mathbf{0} = A\mathbf{x} + \mathbf{f}$

$$\begin{aligned} \mathbf{x}_p &= -A^{-1}\mathbf{f} \\ &= -\begin{bmatrix} 4 & 1 \\ 10 & 1 \end{bmatrix}^{-1} \begin{bmatrix} 6 \\ 12 \end{bmatrix} \\ &= -\begin{bmatrix} 1/(-6) & -1/(-6) \\ -10/(-6) & 4/(-6) \end{bmatrix} \begin{bmatrix} 6 \\ 12 \end{bmatrix} = \begin{bmatrix} -1 \\ -2 \end{bmatrix}. \end{aligned}$$

The general solution is

$$\begin{aligned} \mathbf{x}(t) &= c_1 \mathbf{u}_1 e^{r_1 t} + c_2 \mathbf{u}_2 e^{r_2 t} + \mathbf{x}_p \\ &= c_1 \begin{bmatrix} 1 \\ 2 \end{bmatrix} e^{6t} + c_2 \begin{bmatrix} 1 \\ -5 \end{bmatrix} e^{-t} + \begin{bmatrix} -1 \\ -2 \end{bmatrix}. \end{aligned}$$

Since the first eigenvalue $r_1 = 6$ is positive, the general solution may not converge to the steady state solution!

7.3.2   Particular Solution by Undetermined Coefficients

Consider particular solutions when $\mathbf{f}(t)$ is a simple vector function of time such as $\mathbf{g}t + \mathbf{h}$ or $\mathbf{g}e^{ct}$ or $\mathbf{g}\cos(ct) + \mathbf{h}\sin(ct)$ where $\mathbf{g}$ and $\mathbf{h}$ are given constant vectors. Reasonable speculations are that any particular solution should have the forms $\mathbf{x}_p(t) = \mathbf{a}t + \mathbf{b}$ or $\mathbf{a}e^{ct}$ or $\mathbf{b}\cos(ct) + \mathbf{c}\sin(ct)$, respectively. The $\mathbf{a}$ and $\mathbf{b}$ constant vectors are to be found so that $\mathbf{x}'_p = A\mathbf{x}_p + \mathbf{f}(t)$. This is the vector version of undetermined coefficients that was used in Sections 6.1 and 6.4. Here we will equate vector coefficients of the like functions of time and solve the resulting algebraic systems.

Example 7.3.2. Find the particular solution of

$$\mathbf{x}' = A\mathbf{x} + \mathbf{g}t + \mathbf{h}$$

$$\begin{bmatrix} x_1 \\ x_2 \end{bmatrix}' = \begin{bmatrix} 4 & 1 \\ 10 & 1 \end{bmatrix} \begin{bmatrix} x_1 \\ x_2 \end{bmatrix} + \begin{bmatrix} 2 \\ 6 \end{bmatrix} t + \begin{bmatrix} 6 \\ 12 \end{bmatrix}.$$

Assume the particular solution has the form

$$\mathbf{x}_p(t) = \mathbf{a}t + \mathbf{b}$$

and choose **a** and **b** so that

$$
\begin{aligned}
(\mathbf{a}t + \mathbf{b})' &= A(\mathbf{a}t + \mathbf{b}) + \mathbf{g}t + \mathbf{h} \\
\mathbf{a}1 + \mathbf{0} &= (A\mathbf{a})t + A\mathbf{b} + \mathbf{g}t + \mathbf{h} \\
\mathbf{0}t + \mathbf{a} &= (A\mathbf{a} + \mathbf{g})t + A\mathbf{b} + \mathbf{h}.
\end{aligned}
$$

Match the vector coefficients of t and 1 to get

$$\mathbf{0} = A\mathbf{a} + \mathbf{g} \text{ and } \mathbf{a} = A\mathbf{b} + \mathbf{h}.$$

The solution of the first algebraic system is

$$
\begin{aligned}
\mathbf{a} &= -A^{-1}\mathbf{g} \\
&= -\begin{bmatrix} 1/(-6) & -1/(-6) \\ -10/(-6) & 4/(-6) \end{bmatrix} \begin{bmatrix} 2 \\ 6 \end{bmatrix} = \begin{bmatrix} -2/3 \\ 2/3 \end{bmatrix}.
\end{aligned}
$$

Use **a** in the second equation to solve for **b**

$$
\begin{aligned}
\mathbf{b} &= A^{-1}(\mathbf{a} - \mathbf{h}) \\
&= \begin{bmatrix} 1/(-6) & -1/(-6) \\ -10/(-6) & 4/(-6) \end{bmatrix} \left(\begin{bmatrix} -2/3 \\ 2/3 \end{bmatrix} - \begin{bmatrix} 6 \\ 12 \end{bmatrix} \right) = \begin{bmatrix} -7/9 \\ -32/9 \end{bmatrix}.
\end{aligned}
$$

Thus, a particular solution is

$$
\begin{aligned}
\mathbf{x}_p(t) &= \mathbf{a}t + \mathbf{b} \\
&= \begin{bmatrix} -2/3 \\ 2/3 \end{bmatrix} t + \begin{bmatrix} -7/9 \\ -32/9 \end{bmatrix}.
\end{aligned}
$$

Use the homogeneous solution from Example 7.3.1 to find the general solution

$$
\begin{aligned}
\mathbf{x}(t) &= c_1 \mathbf{u}_1 e^{r_1 t} + c_2 \mathbf{u}_2 e^{r_2 t} + \mathbf{a}t + \mathbf{b} \\
&= c_1 \begin{bmatrix} 1 \\ 2 \end{bmatrix} e^{6t} + c_2 \begin{bmatrix} 1 \\ -5 \end{bmatrix} e^{-t} + \begin{bmatrix} -2/3 \\ 2/3 \end{bmatrix} t + \begin{bmatrix} -7/9 \\ -32/9 \end{bmatrix}.
\end{aligned}
$$

Example 7.3.3. Find the particular solution of

$$\mathbf{x}' = A\mathbf{x} + \mathbf{g}e^{2t}$$

$$\begin{bmatrix} x_1 \\ x_2 \end{bmatrix}' = \begin{bmatrix} 4 & 1 \\ 10 & 1 \end{bmatrix} \begin{bmatrix} x_1 \\ x_2 \end{bmatrix} + \begin{bmatrix} 4 \\ 2 \end{bmatrix} e^{2t}.$$

Assume the particular solution has the form

$$\mathbf{x}_p(t) = \mathbf{a}e^{2t},$$

and choose $\mathbf{a}$ so that

$$(\mathbf{a}e^{2t})' = A(\mathbf{a}e^{2t}) + \mathbf{g}e^{2t}$$
$$\mathbf{a}e^{2t}2 = (A\mathbf{a})e^{2t} + \mathbf{g}e^{2t}.$$

Matching the vector coefficients of e^{2t} we have

$$2\mathbf{a} = A\mathbf{a} + \mathbf{g}$$
$$2I\mathbf{a} = A\mathbf{a} + \mathbf{g}$$
$$(A - 2I)\mathbf{a} = -\mathbf{g}.$$

Since 2 is not an eigenvalue of A, the matrix $A - 2I$ has an inverse and we may solve for $\mathbf{a}$

$$
\begin{aligned}
\mathbf{a} &= -(A - 2I)^{-1}\mathbf{g} \\
&= -\begin{bmatrix} 4-2 & 1 \\ 10 & 1-2 \end{bmatrix}^{-1} \begin{bmatrix} 4 \\ 2 \end{bmatrix} \\
&= -\begin{bmatrix} -1/(-12) & -1/(-12) \\ -10/(-12) & 2/(-12) \end{bmatrix} \begin{bmatrix} 4 \\ 2 \end{bmatrix} = \begin{bmatrix} -1/2 \\ -3 \end{bmatrix}.
\end{aligned}
$$

Hence, a particular solution is

$$\mathbf{x}_p(t) = \mathbf{a}e^{2t} = \begin{bmatrix} -1/2 \\ -3 \end{bmatrix} e^{2t},$$

and the general solution is

$$
\begin{aligned}
\mathbf{x}(t) &= c_1 \mathbf{u}_1 e^{r_1 t} + c_2 \mathbf{u}_2 e^{r_2 t} + \mathbf{a}e^{2t} \\
&= c_1 \begin{bmatrix} 1 \\ 2 \end{bmatrix} e^{6t} + c_2 \begin{bmatrix} 1 \\ -5 \end{bmatrix} e^{-t} + \begin{bmatrix} -1/2 \\ -3 \end{bmatrix} e^{2t}.
\end{aligned}
$$

Example 7.3.4. Find the particular solution of

$$\mathbf{x}' = A\mathbf{x} + \mathbf{g}\cos(3t) + \mathbf{h}\sin(3t)$$
$$\begin{bmatrix} x_1 \\ x_2 \end{bmatrix}' = \begin{bmatrix} 4 & 1 \\ 10 & 1 \end{bmatrix} \begin{bmatrix} x_1 \\ x_2 \end{bmatrix} + \begin{bmatrix} 1 \\ 2 \end{bmatrix} \cos(3t) + \begin{bmatrix} 3 \\ 4 \end{bmatrix} \sin(3t).$$

Assume the particular solution has the form

$$\mathbf{x}_p = \mathbf{a}\cos(3t) + \mathbf{b}\sin(3t),$$

and choose the vector coefficients so that the differential equation holds

$$
\begin{aligned}
(\mathbf{a}\cos(3t) + \mathbf{b}\sin(3t))' \;&=\; A(\mathbf{a}\cos(3t) + \mathbf{b}\sin(3t)) + \\
&\quad\; \mathbf{g}\cos(3t) + \mathbf{h}\sin(3t) \\
\mathbf{a}\sin(3t)(-3) + \mathbf{b}\cos(3t)(3) \;&=\; (A\mathbf{a})\cos(3t) + (A\mathbf{b})\sin(3t) + \\
&\quad\; \mathbf{g}\cos(3t) + \mathbf{h}\sin(3t) \\
&=\; (A\mathbf{a} + \mathbf{g})\cos(3t) + (A\mathbf{b} + \mathbf{h})\sin(3t).
\end{aligned}
$$

Match the vector coefficients of the sine and cosine function to obtain

$$
\mathbf{a}(-3) = A\mathbf{b} + \mathbf{h} \text{ and } \mathbf{b}3 = A\mathbf{a} + \mathbf{g}.
$$

The first equation can be solved for $\mathbf{a} = (A\mathbf{b}+\mathbf{h})/(-3)$ and put into the second equation

$$
\begin{aligned}
\mathbf{b}3 \;&=\; A((A\mathbf{b} + \mathbf{h})/(-3)) + \mathbf{g} \\
\mathbf{b}3I \;&=\; A^2\mathbf{b}/(-3) + A\mathbf{h}/(-3) + \mathbf{g} \\
(3I - A^2/(-3))\mathbf{b} \;&=\; A\mathbf{h}/(-3) + \mathbf{g}.
\end{aligned}
$$

Solve this for $\mathbf{b}$ and then find $\mathbf{a} = (A\mathbf{b} + \mathbf{h})/(-3)$. Another way to solve this system is to write it as a block 2×2 algebraic system

$$
\begin{bmatrix} A & -3I \\ 3I & A \end{bmatrix}
\begin{bmatrix} \mathbf{a} \\ \mathbf{b} \end{bmatrix}
=
\begin{bmatrix} -\mathbf{g} \\ -\mathbf{h} \end{bmatrix}.
$$

This is easily solved using MATLAB

$$
\mathbf{a} = \begin{bmatrix} -4/10 \\ -2/10 \end{bmatrix} \text{ and } \mathbf{b} = \begin{bmatrix} -8/30 \\ -22/30 \end{bmatrix}.
$$

The general solution is

$$
\begin{aligned}
\mathbf{x}(t) \;&=\; c_1\mathbf{u}_1 e^{r_1 t} + c_2\mathbf{u}_2 e^{r_2 t} + \mathbf{a}\cos(3t) + \mathbf{b}\sin(3t) \\
&=\; c_1 \begin{bmatrix} 1 \\ 2 \end{bmatrix} e^{6t} + c_2 \begin{bmatrix} 1 \\ -5 \end{bmatrix} e^{-t} + \\
&\quad\; \begin{bmatrix} -4/10 \\ -2/10 \end{bmatrix} \cos(3t) + \begin{bmatrix} -8/30 \\ -22/30 \end{bmatrix} \sin(3t).
\end{aligned}
$$

The method of undetermined coefficients does not always work or is too complicated to implement. An alternative is the variation of parameters method, which can also be generalized to systems of differential equations. This method gives a closed formula for the particular solution, but it may involve some difficult integrations. A third way to solve systems is to use numerical approximations such as implemented in the MATLAB command ode45().

7.3.3   Particular Solution by Variation of Parameters

The vector version of variation of parameters is a generalization of those discussed in Sections 6.1 and 6.5. Again, assume there are n distinct real eigenvalues of the matrix A so that the homogeneous solution of (7.3.1) is a linear combination of

$$\mathbf{x}_h(t) = c_1 \mathbf{u}_1 e^{r_1 t} + c_2 \mathbf{u}_2 e^{r_2 t} + \cdots + c_n \mathbf{u}_n e^{r_n t}.$$

Let $X(t)$ be an $n \times n$ matrix with column j equal to $\mathbf{x}_j(t) \equiv \mathbf{u}_j e^{r_j t}$

$$X(t) \equiv \begin{bmatrix} \mathbf{x}_1(t) & \mathbf{x}_2(t) & \cdots & \mathbf{x}_n(t) \end{bmatrix}. \tag{7.3.2}$$

Then the homogeneous solution can be written as the matrix $X(t)$ times the coefficient vector $\mathbf{c} = \begin{bmatrix} c_1 & c_2 & \cdots & c_n \end{bmatrix}^T$

$$\mathbf{x}_h(t) = X(t)\mathbf{c}. \tag{7.3.3}$$

As in Section 6.5, the variation of parameters method assumes the particular solution can be expressed as

$$\mathbf{x}_p(t) = X(t)\mathbf{v}(t).$$

Here the column vector $\mathbf{v}(t) = \begin{bmatrix} v_1(t) & v_2(t) & \cdots & v_n(t) \end{bmatrix}^T$ is to be found so that $\mathbf{x}_p(t) = X(t)\mathbf{v}(t)$ satisfies $\mathbf{x}_p' = A\mathbf{x}_p + \mathbf{f}(t)$.
    In order to find $v_j(t)$, expand the matrix-vector product by columns

$$\begin{aligned}
\mathbf{x}_p(t) &= X(t)\mathbf{v}(t) \\
&= v_1(t)\mathbf{x}_1(t) + v_2(t)\mathbf{x}_2(t) + \cdots + v_n(t)\mathbf{x}_n(t).
\end{aligned}$$

The derivative with respect to time may be computed by columns using the product rule

$$\begin{aligned}
\mathbf{x}_p'(t) &= (v_1(t)\mathbf{x}_1(t) + v_2(t)\mathbf{x}_2(t) + \cdots + v_n(t)\mathbf{x}_n(t))' \\
&= (v_1(t)\mathbf{x}_1(t))' + (v_2(t)\mathbf{x}_2(t))' + \cdots + (v_n(t)\mathbf{x}_n(t))' \\
&= v_1\mathbf{x}_1' + v_1'\mathbf{x}_1 + v_2\mathbf{x}_2' + v_2'\mathbf{x}_2 + \cdots + v_n\mathbf{x}_n' + v_n'\mathbf{x}_n.
\end{aligned}$$

Because each $\mathbf{x}_j(t) \equiv \mathbf{u}_j e^{r_j t}$ is a homogeneous solution, $\mathbf{x}_j' = A\mathbf{x}_j$, the above can be rewritten using the column version of matrix-vector products

$$\begin{aligned}
\mathbf{x}_p'(t) &= v_1 A\mathbf{x}_1 + v_1'\mathbf{x}_1 + v_2 A\mathbf{x}_2 + v_2'\mathbf{x}_2 + \cdots + v_n A\mathbf{x}_n + v_n'\mathbf{x}_n \\
&= (v_1 A\mathbf{x}_1 + v_2 A\mathbf{x}_2 + \cdots + v_n A\mathbf{x}_n) + (v_1'\mathbf{x}_1 + v_2'\mathbf{x}_2 + \cdots + v_n'\mathbf{x}_n) \\
&= (AX)\mathbf{v} + X\mathbf{v}'.
\end{aligned}$$

Choose $\mathbf{v}$ so that $\mathbf{x}_p = X\mathbf{v}$ satisfies $\mathbf{x}_p' = A\mathbf{x}_p + \mathbf{f}(t)$, which means

$$\begin{aligned}
(AX)\mathbf{v} + X\mathbf{v}' &= A(X\mathbf{v}) + \mathbf{f}(t) \\
&= (AX)\mathbf{v} + \mathbf{f}(t).
\end{aligned}$$

If $X = X(t)$ has an inverse, then $\mathbf{v} = \mathbf{v}(t)$ should be chosen so that

$$X\mathbf{v'} = \mathbf{f}(t) \text{ or } \mathbf{v'} = X^{-1}\mathbf{f}(t).$$

The column vector $X^{-1}\mathbf{f}(t)$ can be integrated component by component to obtain the desired vector $\mathbf{v} = \mathbf{v}(t)$. A *variation of parameters formula* for a particular solution of (7.3.1) is

$$\mathbf{x}_p = X(t) \int^t X^{-1}\mathbf{f}. \qquad (7.3.4)$$

The general solution is the sum of the homogeneous solution in (7.3.3) and the particular solution in (7.3.4)

$$
\begin{aligned}
\mathbf{x}(t) &= \mathbf{x}_h(t) + \mathbf{x}_p(t) \\
&= X(t)\mathbf{c} + X(t) \int^t X^{-1}\mathbf{f}, \qquad (7.3.5)
\end{aligned}
$$

which is a valid formula provided $X(t)$ has an inverse and the components of $X^{-1}\mathbf{f}$ can be integrated.

A matrix $X(t)$ in (7.3.2) formed by the columns of n linearly independent homogeneous solutions is called a *fundamental matrix* of the system (7.3.1). The fundamental matrix and its properties for different classes of eigenvalues are more completely discussed in Chapter 9 in [7]. Here we will assume the eigenvalues of A are distinct and real and then, with some theoretical effort, one can show the eigenvectors are linearly independent. By using Theorem 3.7.3 and the properties of determinants

$$
\begin{aligned}
\det(X(t)) &= \det([\mathbf{u}_1 e^{r_1 t} \quad \mathbf{u}_2 e^{r_2 t} \quad \cdots \quad \mathbf{u}_n e^{r_n t}]) \\
&= e^{r_1 t} e^{r_2 t} \cdots e^{r_2 t} \det([\mathbf{u}_1 \quad \mathbf{u}_2 \quad \cdots \quad \mathbf{u}_n]) \\
&= e^{r_1 t} e^{r_2 t} \cdots e^{r_2 t} \det(X(0)) \neq 0.
\end{aligned}
$$

Thus, $X(t)$ has an inverse.

Example 7.3.5. Consider Example 7.3.2 and find a particular solution by the variation of parameters formula in (7.3.4). The matrix $X(t)$ and its inverse are

$$X(t) = \begin{bmatrix} e^{6t} & e^{-t} \\ 2e^{6t} & -5e^{-t} \end{bmatrix} \text{ and } X^{-1}(t) = \begin{bmatrix} -5e^{-t} & -e^{-t} \\ -2e^{6t} & e^{6t} \end{bmatrix} /(-7e^{5t}).$$

The matrix-vector product is

$$X^{-1}\mathbf{f} = \begin{bmatrix} (5/7)e^{-6t} & (1/7)e^{-6t} \\ (2/7)e^{t} & (-1/7)e^{t} \end{bmatrix} \begin{bmatrix} 2t+6 \\ 6t+12 \end{bmatrix} = \begin{bmatrix} e^{-6t}(16t/7 + 6) \\ e^{t}(-2t/7 + 0) \end{bmatrix}.$$

Use integration by parts to compute the integrals in the vector

$$\int^t X^{-1}\mathbf{f} = \begin{bmatrix} e^{-6t}(-8t/21 - 67/63) \\ e^{t}(-2t/7 + 2/7) \end{bmatrix}.$$

The particular solution is

$$\mathbf{x}_p(t) = X(t) \int^t X^{-1}\mathbf{f} = \left[\begin{array}{c} -2t/3 - 7/9 \\ 2t/3 - 32/9 \end{array} \right],$$

which agrees with the particular solution in Example 7.3.2 that was found by undetermined coefficients.

The by-hand calculations in this and similar examples using the variation of parameters formula (7.3.4) can be complicated, and one should consider using a symbolic computing tool such as Maple or the symbolic toolbox in MATLAB. The following MATLAB code will execute the formula in (7.3.4) for Example 7.3.5.

MATLAB code varpar.m

```
% This is a symbolic calculation to find
% a particular solution using variation
% of parameters for a 2D system of DEs.
syms t
f = [2*t + 6;6*t + 12];
x1 = [exp(6*t); 2*exp(6*t)];
x2 = [exp(-t); -5*exp(-t)];
X = [x1 x2];
invX = inv(X)
vprime = invX*f
v = int(vprime,'t')
xp = X*v
newxp = simple(xp)
```

Everyone should do at least one variation of parameter by-hand calculation, and then confirm the calculation using a symbolic computation tool.

7.3.4   Initial Value Problem $\mathbf{x}' = A\mathbf{x} + \mathbf{f}(t)$ and $\mathbf{x}(0) = \mathbf{x}_0$

Assume one has found n linearly independent homogeneous solutions and a particular solution to (7.3.1). Form the fundamental matrix $X(t)$ by using the homogeneous solutions, and write the general solution as

$$\begin{aligned} \mathbf{x}(t) &= X(t)\mathbf{c} + \mathbf{x}_p(t) \\ &= c_1\mathbf{x}_1(t) + c_2\mathbf{x}_2(t) + \cdots + c_n\mathbf{x}_n(t) + \mathbf{x}_p(t). \end{aligned}$$

The *initial value problem* is to find $\mathbf{x}(t)$ that satisfies

$$\mathbf{x}' = A\mathbf{x} + \mathbf{f}(t) \text{and } \mathbf{x}(0) = \mathbf{x}_0.$$

The coefficients must be chosen so that the initial condition holds

$$\mathbf{x}(0) = X(0)\mathbf{c} + \mathbf{x}_p(0) = \mathbf{x}_0.$$

This requires the solution of the algebraic system

$$X(0)\mathbf{c} = \mathbf{x}_0 - \mathbf{x}_p(0). \tag{7.3.6}$$

When the columns in the fundamental matrix are linearly independent for all $0 \le t \le T$, which is the case for n distinct real eigenvalues, then $X(t)$ will have an inverse and one can solve for the coefficients in **c**.

Outline for x′ = Ax+f(t) and x(0) = x₀ with Distinct Real Eigenvalues.
1.   Find the real eigenvalues r_j and eigenvectors $\mathbf{u}_j$.
2.   Find the fundamental matrix

$$X(t) = [\ \mathbf{x}_1(t) \quad \mathbf{x}_2(t) \quad \cdots \quad \mathbf{x}_n(t)\ ] \text{ where } \mathbf{x}_j(t) \equiv \mathbf{u}_j e^{r_j t}.$$

3.   Find $\mathbf{x}_p(t)$ by undetermined coefficients or variation of parameters.
4.   Find the general solution $\mathbf{x}(t) = X(t)\mathbf{c} + \mathbf{x}_p(t)$.
5.   Find **c** by solving $X(0)\mathbf{c} = \mathbf{x}_0 - \mathbf{x}_p(0)$.

Examples 7.3.1, 7.3.2 and 7.3.5 (continued). Find the solution of the initial value problem

$$\left[\begin{array}{c} x_1 \\ x_2 \end{array}\right]' = \left[\begin{array}{cc} 4 & 1 \\ 10 & 1 \end{array}\right]\left[\begin{array}{c} x_1 \\ x_2 \end{array}\right] + \left[\begin{array}{c} 2 \\ 6 \end{array}\right]t + \left[\begin{array}{c} 6 \\ 12 \end{array}\right] \text{ and}$$

$$\left[\begin{array}{c} x_1(0) \\ x_2(0) \end{array}\right] = \left[\begin{array}{c} 2 \\ 1 \end{array}\right].$$

Step 1. Both eigenvalues and eigenvectors were found in Example 7.3.1 where determinants were used. One could also do this by using the MATLAB command eig().

Step 2. The fundamental matrix $X(t)$ is a 2×2 matrix formed by the eigenvalues and eigenvectors

$$\begin{aligned} X(t) &= [\ \mathbf{x}_1(t) \quad \mathbf{x}_2(t)\ ] \\ &= [\ \mathbf{u}_1 e^{r_1 t} \quad \mathbf{u}_2 e^{r_2 t}\ ] \\ &= \left[\begin{array}{cc} 1e^{6t} & 1e^{-t} \\ 2e^{6t} & -5e^{-t} \end{array}\right]. \end{aligned}$$

Step 3. A particular solution was found in Example 7.3.2 using undetermined coefficients, and it was also found in Example 7.3.5 using variation of parameters

$$\mathbf{x}_p(t) = \left[\begin{array}{c} -2t/3 - 7/9 \\ 2t/3 - 32/9 \end{array}\right].$$

Step 4. The general solution is the homogeneous solution plus the particular solution

$$\begin{aligned} \mathbf{x}(t) &= X(t)\mathbf{c} + \mathbf{x}_p(t) \\ &= \left[\begin{array}{cc} 1e^{6t} & 1e^{-t} \\ 2e^{6t} & -5e^{-t} \end{array}\right]\left[\begin{array}{c} c_1 \\ c_2 \end{array}\right] + \left[\begin{array}{c} -2t/3 - 7/9 \\ 2t/3 - 32/9 \end{array}\right] \\ &= c_1\left[\begin{array}{c} 1e^{6t} \\ 2e^{6t} \end{array}\right] + c_2\left[\begin{array}{c} 1e^{-t} \\ -5e^{-t} \end{array}\right] + \left[\begin{array}{c} -2t/3 - 7/9 \\ 2t/3 - 32/9 \end{array}\right]. \end{aligned}$$

Step 5. Find c_1 and c_2 by solving

$$X(0)\mathbf{c} + \mathbf{x}_p(0) \;=\; \mathbf{x}_0$$

$$\begin{bmatrix} 1 & 1 \\ 2 & -5 \end{bmatrix} \begin{bmatrix} c_1 \\ c_2 \end{bmatrix} + \begin{bmatrix} 0 - 7/9 \\ 0 - 32/9 \end{bmatrix} \;=\; \begin{bmatrix} 2 \\ 1 \end{bmatrix}.$$

$$\begin{bmatrix} c_1 \\ c_2 \end{bmatrix} = \begin{bmatrix} 1 & 1 \\ 2 & -5 \end{bmatrix}^{-1} \left(\begin{bmatrix} 2 \\ 1 \end{bmatrix} - \begin{bmatrix} 0 - 7/9 \\ 0 - 32/9 \end{bmatrix} \right)$$

$$= \begin{bmatrix} 5/7 & 1/7 \\ 2/7 & -1/7 \end{bmatrix} \begin{bmatrix} 25/9 \\ 41/9 \end{bmatrix} = \begin{bmatrix} 166/63 \\ 9/63 \end{bmatrix}.$$

The solution to the initial value problem is

$$\begin{bmatrix} x_1(t) \\ x_2(t) \end{bmatrix} = 166/63 \begin{bmatrix} 1e^{6t} \\ 2e^{6t} \end{bmatrix} + 9/63 \begin{bmatrix} 1e^{-t} \\ -5e^{-t} \end{bmatrix} + \begin{bmatrix} -2t/3 - 7/9 \\ 2t/3 - 32/9 \end{bmatrix}.$$

7.3.5 Application to Time Dependent Heat Conduction

Heat conduction models have been discussed in Sections 3.2, 3.3 and 7.1 (Example 7.1.4). The last version considered a thin wire with three interior cells and attempted to find the approximate temperature in the cells as a function of time. As a simplification we assumed no internal heat source and zero temperature at the left and right ends of the wire. The model had the form $\mathbf{u}' = A\mathbf{u}$ where A is 3×3 and $\mathbf{u}$ is a column vector with three temperatures.

The present model will be for four unknown temperatures and have nonzero internal heat source such as from electrical resistance. The unknowns $u_i(t)$ are approximations of the temperature in the wire at position $x_i = i\Delta x$ where $\Delta x = L/(n+1)$ and $n = 4$. By following the time dependent model in Subsection 7.1.4, the system of differential equation is

$$\rho c \begin{bmatrix} u_1 \\ u_2 \\ u_3 \\ u_4 \end{bmatrix}' = -(K/(\Delta x)^2) \begin{bmatrix} 2 & -1 & 0 & 0 \\ -1 & 2 & -1 & 0 \\ 0 & -1 & 2 & -1 \\ 0 & 0 & -1 & 2 \end{bmatrix} \begin{bmatrix} u_1 \\ u_2 \\ u_3 \\ u_4 \end{bmatrix} +$$

$$\begin{bmatrix} f + (K/(\Delta x)^2)u_0 \\ f \\ f \\ f + (K/(\Delta x)^2)u_5 \end{bmatrix}.$$

The left and right temperatures are u_0 and u_5, and the initial temperatures for the interior cells will be given.

Let the thermal properties be $\rho = 1$, $c = 1$ and $K = 0.001$ with $L = 10$ so that $\Delta x = 2$. Assume the left, right and initial temperature are always zero. Model the increasing heat source in the wire by $f = 0.1 - 0.1e^{-0.1t}$. Then the initial value problem is

$$\mathbf{u}' = A\mathbf{u} + \mathbf{g} + \mathbf{h}e^{-0.1t} \text{ and } \mathbf{u}(0) = \mathbf{0}$$

$$\begin{bmatrix} u_1 \\ u_2 \\ u_3 \\ u_4 \end{bmatrix}' = -.00025 \begin{bmatrix} 2 & -1 & 0 & 0 \\ -1 & 2 & -1 & 0 \\ 0 & -1 & 2 & -1 \\ 0 & 0 & -1 & 2 \end{bmatrix} \begin{bmatrix} u_1 \\ u_2 \\ u_3 \\ u_4 \end{bmatrix} +$$

$$\begin{bmatrix} 0.1 \\ 0.1 \\ 0.1 \\ 0.1 \end{bmatrix} + \begin{bmatrix} -0.1 \\ -0.1 \\ -0.1 \\ -0.1 \end{bmatrix} e^{-0.1t}.$$

The eigenvalues and eigenvectors can be found by the MATLAB command [U R] = eig(A), which generates the four eigenvectors in the four columns of U and the four eigenvalues in the diagonal positions of R. This means the homogeneous solutions, for the distinct real eigenvalue case, are $U(:,j)e^{R(j,j)t}$. The fundamental matrix is

$$X(t) = \begin{bmatrix} U(:,1)e^{R(1,1)t} & U(:,2)e^{R(2,2)t} & U(:,3)e^{R(3,3)t} & U(:,4)e^{R(4,4)t} \end{bmatrix}$$

so that $X(0) = U$. The general homogeneous solution is

$$X(t)\mathbf{c} = c_1 U(:,1)e^{R(1,1)t} + c_2 U(:,2)e^{R(2,2)t} + c_3 U(:,3)e^{R(3,3)t} + c_4 U(:,4)e^{R(4,4)t}.$$

In order to find a particular solution, use the method of undetermined coefficients by assuming the particular solution has the form $\mathbf{a} + \mathbf{b}e^{-0.1t}$. Choose the vector coefficients so that the system of differential equations holds

$$\begin{aligned} (\mathbf{a} + \mathbf{b}e^{-0.1t})' &= A(\mathbf{a} + \mathbf{b}e^{-0.1t}) + \mathbf{g} + \mathbf{h}e^{-0.1t} \\ \mathbf{0} + \mathbf{b}e^{-0.1t}(-0.1) &= A\mathbf{a} + A\mathbf{b}e^{-0.1t} + \mathbf{g} + \mathbf{h}e^{-0.1t} \\ &= (A\mathbf{a} + \mathbf{g}) + (A\mathbf{b} + \mathbf{h})e^{-0.1t}. \end{aligned}$$

Match the like terms to obtain two vector equations

$$\mathbf{0} = A\mathbf{a} + \mathbf{g} \text{ and } \mathbf{b}(-0.1) = A\mathbf{b} + \mathbf{h}.$$

The first equation is easily solved by $\mathbf{a} = -A^{-1}\mathbf{g}$. The second equation can be written as $(-.1)I\mathbf{b} = A\mathbf{b} + \mathbf{h}$ or as $(A + (0.1)I)\mathbf{b} = -\mathbf{h}$. If -0.1 is not an eigenvector, then $(A - (-0.1)I)$ has an inverse and the second equation is solved by $\mathbf{b} = -(A + (0.1)I)^{-1}\mathbf{h}$.

The initial conditions are satisfied by solving $X(0)\mathbf{c} = \mathbf{u}_0 - (\mathbf{a} + \mathbf{b}e^{-0.1(0)})$ $= \mathbf{0} - (\mathbf{a} + \mathbf{b})$. Since $X(0) = U$, the solution is $\mathbf{c} = -U^{-1}(\mathbf{a} + \mathbf{b})$. Once these computations have been done, the solution to the initial value problem is

$$\begin{aligned} \mathbf{u} &= X(t)\mathbf{c} + \mathbf{a} + \mathbf{b}e^{-0.1t} \\ &= c_1 U(:,1)e^{R(1,1)t} + c_2 U(:,2)e^{R(2,2)t} + c_3 U(:,3)e^{R(3,3)t} + c_4 U(:,4)e^{R(4,4)t} \\ &\quad + \mathbf{a} + \mathbf{b}e^{-0.1t}. \end{aligned}$$

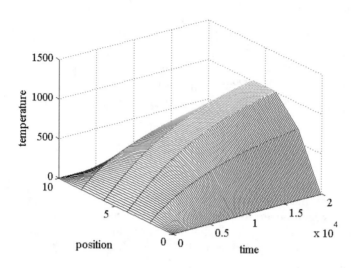

Figure 7.3.1: Heat Diffusion in Thin Wire

The calculations have been implemented in the MATLAB code file heat4.m, and the output is given in the Figure 7.3.1. Since the eigenvalues $(-.0009045, -.0006545, -.0003455$ and $-.0000955)$ are all negative, the homogeneous solution decays to zero. The particular solution decays to $\mathbf{a}$ and, so, the solution to the initial problem converges to $\mathbf{a} = [800\ 1200\ 1200\ 800]^T$. This is suggested by the right side of the graph in Figure 7.3.1.

7.3.6   Exercises

1.   Find the eigenvalues of the following matrices:

(a). $A = \begin{bmatrix} 2 & -1 \\ 1/4 & 3 \end{bmatrix}$ has repeated eigenvalues,

(b). $A = \begin{bmatrix} 2 & 1 \\ -1 & 2 \end{bmatrix}$ has complex eigenvalues and

(c). $A = \begin{bmatrix} 1 & 2 \\ 1/2 & 1 \end{bmatrix}$ has a zero eigenvalue and no inverse.

2.   Use undetermined coefficients to find the particular solution of $\mathbf{x}' = A\mathbf{x} + \mathbf{f}$ where $A = \begin{bmatrix} 2 & 3 \\ 3 & 2 \end{bmatrix}$ with

(a). $\mathbf{f} = \begin{bmatrix} 1 \\ 2 \end{bmatrix} + \begin{bmatrix} 2 \\ 3 \end{bmatrix} t$ and

(b). $\mathbf{f} = \begin{bmatrix} 1 \\ -1 \end{bmatrix} e^{2t}$.

3.   Consider the system of differential equations in exercise 2.
   (a).   Find the eigenvalues and eigenvectors.
   (b).   Find the fundamental matrix and its inverse.
   (c).   Use variation of parameters to find a particular solution in exercise
2a.
   (d).   Use variation of parameters to find a particular solution in exercise
2b.

4.   Consider the system of differential equations in exercise 2 with initial
condition
$$\mathbf{x}(0) = \begin{bmatrix} 3 \\ 5 \end{bmatrix}.$$
   (a).   Solve the initial value problem for $\mathbf{f}$ in exercise 2a.
   (a).   Solve the initial value problem for $\mathbf{f}$ in exercise 2b.

5.   Solve the initial value problem $\mathbf{x}' = A\mathbf{x} + \mathbf{f}$ where
$$A = \begin{bmatrix} 1 & -1 \\ 2 & 4 \end{bmatrix}, \mathbf{f} = \begin{bmatrix} 7 \\ -2 \end{bmatrix} + \begin{bmatrix} 1 \\ 3 \end{bmatrix} t \text{ and } \mathbf{x}(0) = \begin{bmatrix} 1 \\ 0 \end{bmatrix}.$$

6.   Solve the initial value problem $\mathbf{x}' = A\mathbf{x} + \mathbf{f}$ where
$$A = \begin{bmatrix} 3 & -1 \\ -1 & 3 \end{bmatrix}, \mathbf{f} = \begin{bmatrix} 4 \\ -2 \end{bmatrix} + \begin{bmatrix} 1 \\ 3 \end{bmatrix} e^{3t} \text{ and } \mathbf{x}(0) = \begin{bmatrix} 1 \\ -1 \end{bmatrix}.$$

7.   Solve the initial value problem $\mathbf{x}' = A\mathbf{x} + \mathbf{f}$ where
$$A = \begin{bmatrix} 4 & 1 \\ -2 & 1 \end{bmatrix}, \mathbf{f} = \begin{bmatrix} 4 \\ -2 \end{bmatrix} \cos(2t) + \begin{bmatrix} 1 \\ 3 \end{bmatrix} \sin(2t) \text{ and } \mathbf{x}(0) = \begin{bmatrix} 1 \\ 1 \end{bmatrix}.$$

8.   Solve the initial value problem $\mathbf{x}' = A\mathbf{x} + \mathbf{f}$ where
$$A = \begin{bmatrix} -3 & 1 & 0 \\ 1 & -3 & 1 \\ 0 & 1 & -3 \end{bmatrix}, \mathbf{f} = \begin{bmatrix} 4 \\ 5 \\ 4 \end{bmatrix} t \text{ and } \mathbf{x}(0) = \begin{bmatrix} 1 \\ 2 \\ 3 \end{bmatrix}.$$

9.   In exercise 3 confirm your by-hand calculations using a symbolic computation tool.

10.   Use MATLAB's command dsolve() to confirm any by-hand calculations
in exercises 4-8.

11.   Consider the application to heat conduction in a wire as implemented in
the MATLAB code heat4.m. Modify this code so that there are $n = 5$ unknown
temperatures. The matrix now will be 5×5, there will be five differential
equations and the column vectors must have five components.
   (a).   Compare the graphical output with the $n = 4$ output in Figure
7.3.1.
   (b).   Experiment with the length of the time interval by adjusting the
increment size in the time loop from $t = (k - 1) * 200$ to $t = (k - 1) * 500$.
Observe the convergence of the approximate temperatures as time increases.

Chapter 8

Image Processing in Space Domain

Chapters eight and nine use matrices to represent images. Basic matrix operations can be used to alter the image contrast, light-dark and blurred-sharpened qualities of the image. In this chapter the matrix operations will be on the original matrix, which is often called the space domain. An interesting variety of images will be enhanced. A very nice general reference to image processing is [2], where the indicated www site links to many interesting images. The pollen, aerial, moon and micro chip images were used in [2], and the Mars Rover image was taken from NASA [11].

8.1   Matrices and Images

Most images have some initial distortion or equipment noise associated with them. Because one can some times determine the nature of the distortion, it may be possible to eliminate or reduce the undesirable attributes in the image. As indicated in Subsection 2.5.6, digital images can be associated with matrices of integers. The general plan is to convert the image file to a matrix, manipulate the matrix and then convert the manipulated matrix into an enhanced image.

8.1.1   Enhanced Images

The following pairs of images have been enhanced by simple operations on the image matrix, which will be discussed in this and the next chapter. Figures 8.1.1 and 8.1.2 are images of pollen where the first figure lacks contrast and is dark. The pollen image in Figure 8.1.1 is provided courtesy of Roger Heady, Australian National University. Figure 8.1.2 was created by using a piecewise linear transformation and sharpening.

Figure 8.1.1:   Pollen Image

Figure 8.1.2:   Enhanced Pollen Image

Images in Figures 8.1.3 and 8.1.4 are an aerail photo, which initially is too bright in some areas. The aerail image in Figure 8.1.3 is provided courtesy of NASA. The enhanced version was obtained by a power array transformation and gives a clearer image of the water channel.

Figure 8.1.3: Aerial Photo

Figure 8.1.4: Enhanced Aerial Photo

The Mars Rover image in Figure 8.1.6 has been enhanced to see the darker regions in Figure 8.1.5. The Mars image in Figure 8.1.5 is provided courtesy of NASA.

Figure 8.1.5: Mars Rover Photo

Figure 8.1.6: Enhanced Mars Rover Photo

The second image of the moon in Figure 8.1.8 has sharper craters than the original moon image in Figure 8.1.7. The moon image in Figure 8.1.7 is provided courtesy of NASA.

Figure 8.1.7: Moon

Figure 8.1.8: Sharper Moon Image

8.1.2   Grayscale Images

Grayscale images are associated with $m \times n$ matrices whose components are integers. For 8-bit images the integers range from 0 to $255 = 2^8 - 1$, and for 16-bit images they range from 0 to $65535 = 2^{16} - 1$. The black image pixel is associated with 0, and the white image pixel is associated with 255 (8-bit) or 65535 (16-bit). In MATLAB one can "view" the image in several ways. First, just inspect the matrix components. Second, use the MATLAB command mesh() to generate a surface of the image where the indices of the matrix are on the xy-plane and the intensity of the image is on the z-axis. Third, one can map the matrix into a standard image file such as a *.jpg file. This can be done by the MATLAB command imwrite(). The inverse of the imwrite() is imread(), which generates a 8-bit integer matrix from an image file.

Example 8.1.1. We will create alphabetical letters from 50×40 matrices. The letter "C" can be created by defining a 50×40 matrix to initially be zero and then nonzero for some components to form the letter. This will produce a letter with black background and with lighter regions to form the letter. The following MATLAB function defines the letter "C" where the lighter region has an input value equal to g.

MATLAB code letterc.m
```
function letc = letterc(g)
letc = zeros(50,40);
letc(10:40,4:8) = g;
letc(10:14,8:36) = g;
letc(36:40,8:36) = g;
```

Figure 8.1.9 is the mesh plot of the matrix C, which was generated by the MATLAB function letterc(200). The MATLAB command imwrite(newC, 'C.jpg') maps the matrix newC (C with 8-bit integers) into the image file C.jpg, which is viewed in Figure 8.1.10. Figure 8.1.11 is the "negative" of this letter.

```
>> C = letterc(200);
>> newC =uint8(C);
>> newC(5:20,1:12)
>> ans =
     0 0 0 0 0 0 0 0 0 0 0 0
     0 0 0 0 0 0 0 0 0 0 0 0
     0 0 0 0 0 0 0 0 0 0 0 0
     0 0 0 0 0 0 0 0 0 0 0 0
     0 0 0 0 0 0 0 0 0 0 0 0
     0 0 0 200 200 200 200 200 200 200 200 200
     0 0 0 200 200 200 200 200 200 200 200 200
     0 0 0 200 200 200 200 200 200 200 200 200
     0 0 0 200 200 200 200 200 200 200 200 200
     0 0 0 200 200 200 200 200 200 200 200 200
     0 0 0 200 200 200 200 200 0 0 0 0
```

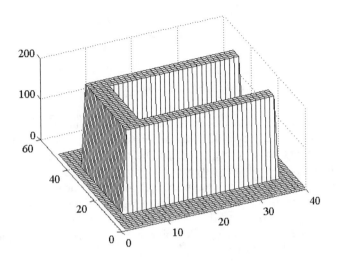

Figure 8.1.9: Plot of the Matrix C

0 0 0 200 200 200 200 200 0 0 0 0

0 0 0 200 200 200 200 200 0 0 0 0

0 0 0 200 200 200 200 200 0 0 0 0

0 0 0 200 200 200 200 200 0 0 0 0

0 0 0 200 200 200 200 200 0 0 0 0

>> mesh(C)

>> imwrite(newC, 'C.jpg')

Figure 8.1.10:   Image of Letter C

>> negC = 255 - letterc(200);

>> newnegC = uint8(negC)

>> imwrite(newnegC,'mac8111.jpg');

Figure 8.1.11:   Negative Image

Example 8.1.2. One can augment the matrices of letters to form words or other objects. In this example four letters are created by functions similar to the letter function letterc(g) to form the matrix *ncsu* and then the image NCSU. The mesh plot of the matrix is given in Figure 8.1.12, the image NCSU with black background is in Figure 8.1.13 and the "negative" of the image NCSU with white background is in Figure 8.1.14. The "negative" is created by replacing the components in the matrix ncsu by $255 - ncsu(i, j)$, that is, $negncsu(i, j) = 255 - ncsu(i, j)$. The following MATLAB code imagncsu.m was used to generate these figures.

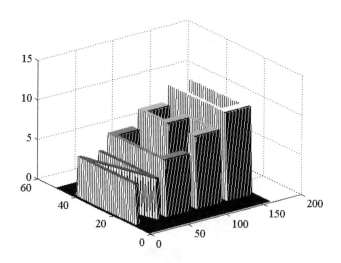

Figure 8.1.12:   Matrix NCSU

MATLAB code imagncsu.m
```
ncsu = [lettern(5) letterc(7) letters(9) letteru(11)];
mesh(ncsu);
newncsu = 20*ncsu;
newncsu = newncsu(50:-1:1,:);
negncsu = 255*ones(size(newncsu))-newncsu;
newncsu1 = uint8(newncsu);
```

```
imwrite(newncsu1, 'ncsu.jpg');
negncsu1 = uint8(negncsu);
imwrite(negncsu1, 'negncsu.jpg');
```

Figure 8.1.13:   Image of NCSU

Figure 8.1.14:   Negative Image of NCSU

8.1.3   Color Images

Color images are decomposed into red, green and blue. Three $m \times n$ matrices, or a 3D array whose dimensions are m, n and 3, are used to represent the color image. In MATLAB this is given by $array(:, :, 1)$ for the red matrix, $array(:, :, 2)$ for the green matrix and $array(:, :, 3)$ for the blue matrix. The MATLAB command imwrite() maps to the image array into the array.jpg file that is associated with color image. The following MATLAB code file color_imagncsu.m can be used to generate color versions of the image NCSU.

MATLAB code color_imagncsu.m
```
red = 10; % 220
green = 10; % 100
blue = 250; % 10
ncsu = [lettern(1) letterc(1) letters(1) letteru(1)];
newncsu = ncsu(50:-1:1,:);
redncsu = red*newncsu;
greenncsu = green*newncsu;
bluencsu = blue*newncsu;
colorncsu(:,:,1) = redncsu;
colorncsu(:,:,2) = greenncsu;
colorncsu(:,:,3) = bluencsu;
colorncsu1 = uint8(colorncsu);
imwrite(colorncsu1, 'colorncsu.jpg');
```

Figure 8.1.15:   Center Grain in Pollen

8.1.4   Cropped Images

In order to cut or crop unwanted parts of an image, one needs to first find the image matrix from the image file. This can be done by using the MATLAB command imread(). Once the image matrix is found, the appropriate part of the image matrix can be selected. Then map this back into a smaller image file. For example, if one wants to examine the pollen grain in the center of the image in Figure 8.1.1, use the following MATLAB commands.

```
>> pollenmat = imread('pollen.jpg')
>> size(pollenmat)
   ans =
        500 500
>> imwrite(pollenmat(200:400,100:400),'mac8115.jpg')
```

Figure 8.1.15 contains this cropped image. In order to distinguish the dark and white areas, there needs to be more separation between the dark and light components in the image matrix. This is known as "contrast" in the image matrix and will be more carefully discussed in the next section.

8.1.5   Exercises

1.   Consider the letter function letterc(g).
   (a).   Experiment with different input values in the letter function letterc(g).
   (b).   Modify the shape of the letter "C" generated by letterc(g).
   (c).   Modify the background shade of the letter.
2.   Consider the image of NCSU.
   (a).   Define additional letter functions for the letters in your first name.
   (b).   Modify the MATLAB code imagncsu.m to generate your first name image.
   (c).   Modify the background shade for your first name image.
3.   Experiment with different colors in the image NCSU as generated in color_imagncsu.m.

4. Modify color_imagncsu.m to generate a color image of your first name.
5. Crop the aerial photo given in aerial.jpg so that the landing strip is the main part of a smaller image file.
6. Crop the Mars photo given in mars.jpg so that the rock formation in the center right is the central part of a smaller image file.

8.2 Contrast and Histograms

Histograms of an image matrix will be introduced to quantify brightness and contrast of an image. The shift, piecewise linear and power matrix transformations of the image matrix will be used to adjust the brightness or contrast of an image. These will be applied to enhance the images of pollen, aerial photo and Mars.

8.2.1 Histogram of an Image

Images often are said to be light or dark or to lack contrast. In an attempt to be more precise about this, we use the matrix representation of an image. Each pixel's image intensity is recorded by an entry of its matrix. If the entry is small (large), then the pixel is dark (light). If there are a large number of dark (light) pixels, then one may say the image is dark (light). If most of the entries have similar values, then the image lacks contrast.

Definition 8.2.1. Let $k = u(i, j)$ be an integer from 0 to 255 and represent a shade of a pixel at the location (i, j) in an $m \times n$ image matrix, u. Define a *histogram* of the 8-bit image u to be the function from integers $0 : 1 : 255$ to the integers $0 : 1 : mn$ given by

$$h(k) = \text{number of pixels whose shade equals } k, \text{ that is, } u(i, j) = k.$$

If the image is dark (light), the graph of the histogram will be shifted to the left (right). If the image lacks contrast, the graph of the histogram will be clustered. In order to find the histogram of an image matrix, one must total the number of components that have a particular shade. This can be done by using three nested loops where the two outside loops range over the pixels (indices of the image matrix) and the inner most loop ranges over all possible shades. The first example illustrates this for the pollen image in Figure 8.1.1.

Example 8.2.1. Find the histogram for the pollen image in pollen.jpg. First, we find the 8-bit image matrix *pollenmat* by using the MATLAB command imread(), and second, the three nested loops are used to construct the histogram function. The following MATLAB code pollenhist.m does this and then generates the graph of the histogram in Figure 8.2.1.

MATLAB code pollenhist.m
```
% This code creates a histogram for pollen.jpg and
% plots its graph.
```

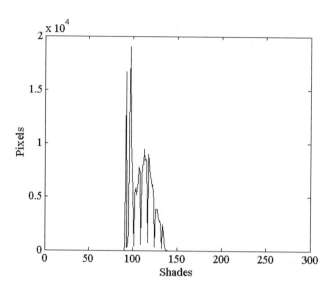

Figure 8.2.1:   Histogram of Pollen Image

```
clear;
% Creates an image matrix for pollen.jpg.
pollenmat = imread('pollen.jpg');
u = double(pollenmat);
[nx ny] = size(u)
nshades = 256;
hist = zeros(nshades,1);
for i = 1:nx        % creates the histogram, hist
     for j = 1:ny
          for k = 0:nshades-1
               if u(i,j) == k
                    hist(k+1) = hist(k+1)+1;
               end
          end
     end
end
plot(hist);
```

The histogram in Figure 8.2.1 indicates low contrast and a darker image because the pixels are clumped together and have lower shades. In subsequent subsections we will transform the image matrix so that the histogram is more evenly distributed.

8.2.2 Shift Transformation

The shift transform either increases or decreases all of the pixels (components of the image matrix) by the same amount. The histogram in Figure 8.2.1 for the pollen image indicates that most of the components have shades between 90 and 150, and so a shift to the right equal to 100 would tend to brighten the image.

Definition 8.2.2. Let u be an 8-bit image matrix and let $shift$ be a positive or negative integer. The shift transformation of the image matrix is a mapping to another 8-bit image matrix Tu whose components are

$$(Tu)(i,j) \equiv \begin{cases} 0, & \text{if } u(i,j) + shift < 0 \\ u(i,j) + shift, & \text{if } 0 \leq u(i,j) + shift \leq 255 \\ 255, & \text{if } u(i,j) + shift > 255 \end{cases}.$$

The truncation below 0 or above 255 can seriously distort the associated image. The MATLAB command uint8() does this truncation as illustrated in the following trivial calculations:

```
>> u = [-100.3   45.0   257;   23.7   150.0   301.1]
        -100.3000   045.0000   257.0000
         023.7000   150.0000   301.1000
>> shift_u = u + 100
         -00.3000   145.0000   357.0000
          123.7000   250.0000   401.1000
>> new_u = uint8(shift_u)
          000   145   255
          124   250   255
```

In terms of the MATLAB command uint8(), the shift transformation can be written as

$$Tu = \text{uint8}(u + shift) \text{ or in component form}$$
$$(Tu)(i,j) = \text{uint8}(u(i,j) + shift).$$

Example 8.2.2. This example illustrates the shift transformation applied to the image matrix of pollen.jpg. The following MATLAB code pollenlight.m is a shift to a brighter image than in Figure 8.1.1, generates a new histogram in Figure 8.2.2 and a new image file in Figure 8.2.3.

MATLAB code pollenlight.m

```
% This code uses the shift transformation to either
% brighten or darken all of the image.
clear;
pollenmat = imread('pollen.jpg');
u = double(pollenmat);
[nx ny] = size(u)
```

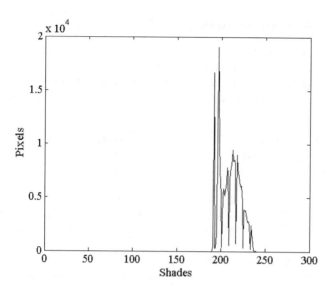

Figure 8.2.2:   Histogram of Lighter Pollen Image

```
nshades = 256;
shift = 100; % Transform by shifting to light.
% shift = -50; % Transform by shifting to dark.
ulight = u + ones(nx,ny)*shift;
hist = zeros(nshades,1);
for i=1:nx
       for j=1:ny
              for k=0:nshades-1
                    if ulight(i,j)==k
                           hist(k+1)=hist(k+1)+1;
                    end
              end
       end
end
plot(hist);
pollen_light = uint8(ulight);
imwrite(pollen_light, 'pollenlight.jpg');
```

8.2.3   Piecewise Linear Transformation

In order to spread the shades (increase the contrast) in the pollen image in
Figure 8.1.1 so that its histogram in Figure 8.2.1 is less clumped in the center,
we can use a linear map applied to the pixels in the center region. If the slope
of the linear map is larger than one, then the shades in the original image

Figure 8.2.3: Lighter Pollen Image

will be spread over a larger range of shades. In order to have the transformed image matrix have components between 0 and 255, we apply a piecewise linear transformation to each component of the image matrix. An illustration of this is in Figure 8.2.4, where one can choose different values for the points (r_1, s_1) and (r_2, s_2). The MATLAB function file for a piecewise linear function is in spread.m.

MATLAB function spread.m

```
function uspread = spread(r1,s1,r2,s2,u)
    if (u< r1)
        uspread = ((s1-0)/(r1-0))*u;
    end
    if ((u>=r1) & (u<= r2))
        uspread = ((s2 - s1)/(r2 - r1))*(u - r1)+ s1;
    end
    if (u>r2)
        uspread = ((255 - s2)/(255 - r2))*(u - r2) + s2;
    end
```

Since the values of the piecewise linear function may not be 8-bit integers, they must be rounded or truncated to be integers between 0 and 255.

Definition 8.2.3. The *piecewise linear* transformation of an image matrix u is a 8-bit image matrix $T_{pw}u$ whose (i, j) components are given by the composition

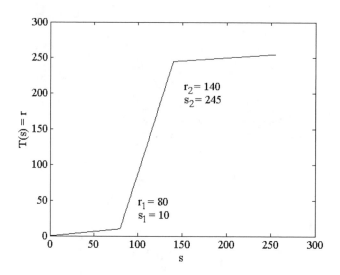

Figure 8.2.4:   Piecewise Linear Function

of the piecewise linear function $T(s)$ and the 8-bit valued MATLAB function
uint8()

$$(T_{pw}u)(i,j) \equiv \text{uint8}(T(u(i,j))).$$

A simple example of this is given by the following MATLAB calculations:

```
>> u = [60 80 150; 135 150 250]
>> newu = [spread(80,10,140,245,60) spread(80,10,140,245,80)
              spread(80,10,140,245,150);
           spread(80,10,140,245,135) spread(80,10,140,245,150)
              spread(80,10,140,245,250)]
    newu =
           007.5000   010.0000   245.8696
           225.4167   245.8696   254.5652
>> TpwU = uint8(newu)
    TpwU =
           008   010   246
           225   246   255
```

Example 8.2.3. A more interesting illustration of this transformation is given
in the MATLAB code pollenspread.m where the pollen image is transformed to
have much better contrast. This code uses the function file spread.m to do a
piecewise linear transform of the image, creates a histogram in Figure 8.2.5 and
generates the enhanced image in Figure 8.2.6.

MATLAB code pollenspread.m
```
% This file enhances the image pollen.jpg by using the
% piecewise linear transformation as given in the function
% file spread().
clear;
pollenmat = imread('pollen.jpg');
u = double(pollenmat);
[nx ny] = size(u)
nshades = 256;
r1 = 80; s1 = 10; % Transformation by piecewise linear function.
r2 = 140; s2 = 245;
for i = 1:nx
    for j = 1:ny
        uspread(i,j) = spread(r1,s1,r2,s2,u(i,j));
    end
end
pollen_spread = uint8(uspread);
hist = zeros(nshades,1);
for i = 1:nx
    for j = 1:ny
        for k = 0:nshades-1
            if pollen_spread(i,j) == k
                hist(k+1) = hist(k+1) + 1;
            end
        end
    end
end
plot(hist);
imwrite(pollen_spread, 'pollenspread.jpg');
```

8.2.4   Power Transformation

On the unit interval x^2 is concave upwards and also has values on the unit interval. The square root function $x^{1/2}$ is concave downwards and has values on the unit interval when x is in the unit interval. By replacing the x by a normalized image component $u(i,j)/255$, one can adjust the relative brightness and darkness of the image. In order to ensure the image is in the 8-bit integer range, one must use a scaled power $255(u(i,j)/255)^p$ and the uint8() function.

Definition 8.2.4. The *power* transformation of an image matrix u is a 8-bit image matrix $T_p u$ whose (i,j) components are given by the composition of the power function x^p with positive p and the 8-bit valued MATLAB function uint8()

$$(T_p u)(i,j) \equiv \text{uint8}(255(u(i,j)/255)^p).$$

A trivial illustration is done using the array operations in MATLAB:

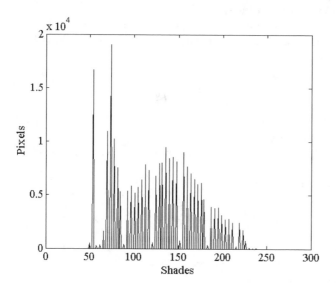

Figure 8.2.5: Histogram for Enhanced Pollen Image

Figure 8.2.6: Higher Contrast Pollen Image

```
>> u = [ 16 64 144; 50 225 40]
   u =
      16   064   144
      50   225   040
>> newu = (u/255).^.5
   newu =
      0.2505   0.5010   0.7515
      0.4428   0.9393   0.3961
>> TpU = uint8(255*newu)
   TpU =
      064   128   192
      113   240   101
```

Example 8.2.4. The following MATLAB code aerialpower.m with $p = 2$ was used to enhance the aerial image in Figures 8.1.3 and 8.1.4.

MATLAB code aerialpower.m

```
% This code uses the power transformation to enhance
% the photo in aerial.jpg.
clear;
aerialmat = imread('aerial.jpg');      % aerial photo
u = double(aerialmat);
[nx ny] = size(u)
c = 1.;
power = 2;
newu =255*c*(u/255).^power;
aeriall = uint8(newu);
imwrite(aeriall, 'aeriall.jpg');
```

The MATLAB code aerialpower.m can be easily modified to use the power transformation on other images. Figures 8.2.7 and 8.2.8 were generated by the MATLAB code marspower.m, which transforms the Mars Rover image mars3.jpg by the power method with $p = 1/2$ and $p = 2$, respectively. The image in Figure 8.2.7 is much brighter because the square root function is concave down and on the unit interval it is larger than the square function.

8.2.5 Exercises

1. Consider the MATLAB code file pollenhist.m. Alter it to find the histogram of the Mars Rover image in mars3.jpg.
2. Consider the MATLAB code file pollenhist.m. Alter it to find the histogram of the aerial image in aerail.jpg.
3. Apply the shift transformation with $shift = -40$ and

$$u = \begin{bmatrix} 275 & 240.7 & 210 \\ 195.2 & 57 & 37.1 \end{bmatrix}.$$

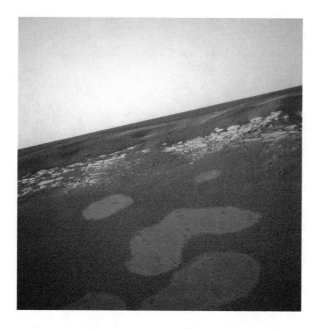

Figure 8.2.7: Mars Rover Image Using Power 1/2

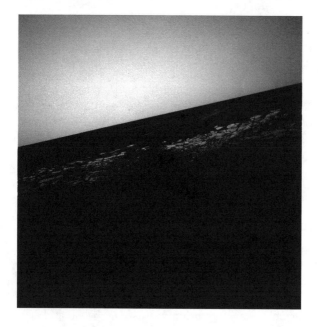

Figure 8.2.8: Mars Rover Image Using Power 2

4.   Consider the MATLAB code file pollenlight.m. Experiment with different values of $shift = -100, -50, 150$ and 200. Note the histograms and the corresponding images.

5.   Apply the piecewise linear transformation with $(r_1, s_1) = (20, 90)$, $(r_2, s_2) = (180, 200)$ and

$$u = \begin{bmatrix} 275 & 240.7 & 210 \\ 195.2 & 57 & 37.1 \end{bmatrix}.$$

6.   Consider the MATLAB code file pollenspread.m. Experiment with different values of (r_1, s_1) and (r_2, s_2) in the function spread.m. Note the histograms and the corresponding images.

7.   Consider the MATLAB code file pollenspread.m. Alter this to apply the piecewise linear transformation to the image file aerial.jpg. Experiment with different values of (r_1, s_1) and (r_2, s_2) in the function spread.m. Note the histograms and the corresponding images.

8.   Consider the MATLAB code file pollenspread.m. Alter this to apply the piecewise linear transformation to the image file mar3.jpg. Experiment with different values of (r_1, s_1) and (r_2, s_2) in the function spread.m. Note the histograms and the corresponding images.

9.   Apply the power transformation with $p = 0.5$, 1.5 and

$$u = \begin{bmatrix} 275 & 240.7 & 210 \\ 195.2 & 57 & 37.1 \end{bmatrix}.$$

10.   Consider the MATLAB code file aerialpower.m and experiment with additional values of $p = 0.3$, 0.5, 0.7 and 0.9. Note the histograms and the corresponding images.

8.3   Blurring and Sharpening

The distortion of an image may be a result of several factors such as deflection of light rays, defective equipment

and electronic noise. One model is related to heat diffusion where the heat energy spreads from high values to lower values. This is analogous to high intensity light spreading to lower intensities, which results in a sharp image becoming less sharp or blurred. By reversing the "diffusion" process, that is, use backward or negative time steps, one can deblur or sharpen images. Several examples will be given and include sharpening the pollen and moon images.

8.3.1   Images in One Dimension

Blurring of a 1D image can be modeled by the explicit finite difference model of heat where the 1D image is viewed as the initial temperature within a long thin wire. A simple explicit time discrete model for the approximate temperature

$u_i^k \approx u(i\Delta x, k\Delta t)$ was formulated in Section 3.2

$$u_i^{k+1} = u_i^k + \alpha(u_{i-1}^k - 2u_i^k + u_{i+1}^k) \text{ with } \alpha \equiv \frac{K}{\rho c}\frac{\Delta t}{\Delta x^2}.$$

The reverse of heat diffusion is to use negative time steps and to start with the temperature distribution in the future. This suggests that deblurring an image may be modeled by starting with a blurred image as a future "temperature" and apply the explicit finite difference model with negative time steps. In the blurring of the image, one must be careful to choose the time step small enough so that $1 - 2\alpha$ is positive. The backward iterations eventually start to oscillate and give badly distorted images. The first example illustrates this for a simple 1D image.

Example 8.3.1. The image is given by a simple piecewise constant function, which could be viewed as profiles of two rectangular buildings. Blurring could be caused by moisture in the air, but here it will be simulated by the forward heat diffusion model of blurring. Once this is done, negative time steps will be used in an attempt to deblur the distorted image. The MATLAB code blur1d.m is an implementation of this.

MATLAB code blur1d.m

```
clear;
clf;
% This code illustrates possible blurring and deblurring in 1D.
% It is motivated by heat diffusion in 1D.
% Blurring is modeled by diffusion of heat in positive time steps.
% Deblurring is done by diffusion of "heat" in negative time steps.
% In order to see the changes, push a key.
u = zeros(101,1);
newu = u;
u(30:50) = 10;
u(70:73) = 15;
dt = .00001;
dx = 1/100;
%
% Forward in "time" or "blurring"
%
subplot(1,3,1)
plot(u)
axis([0 101 0 16]);
title('Image')
figure(1)
pause
subplot(1,3,2)
for k = 1:10
        for i = 2:100
```

```
            newu(i) = u(i) + (dt/(dx*dx))*(u(i-1) -2*u(i) + u(i+1));
        end
        plot(newu);
        axis([0 101 0 16]);
        title('Blurred Image')
        k
        pause;
        u = newu;
    end
    %
    % Backward in "time" or "deblurring"
    %
    subplot(1,3,3)
    plot(u)
    for k = 1:10
        for i = 2:100
            newu(i) = u(i) - (dt/(dx*dx))*(u(i-1) -2*u(i) + u(i+1));
        end
        plot(newu);
        axis([0 101 0 16]);
        title('Deblurred Image')
        k
        pause;
        u = newu;
    end
```

The graphical output is given in Figure 8.3.1. The left graph is the unblurred image and the center graph is the blurred image after $k = 10$ "time" steps with $\Delta t = 0.00001$ and $\Delta x = 0.01$. These values have been chosen so that the blurring and deblurring process gives reasonable results. The graph in the right side of Figure 8.3.1 is the deblurred image after $k = 10$ steps. Note the additional "peaks" in the deblurred image. A more accurate deblurred image is found at $k = 8$ steps, and the reader should run the code to see this. So, the deblurring process can be too aggressive and insert nonexistent image components.

8.3.2   Images in Two Dimensions

Blurring of a 2D image can also be modeled by the explicit finite difference model of heat diffusion where the 2D image is viewed as the initial temperature within a thin plate. Heat is allowed to diffuse in both the x and y directions so that a simple explicit time discrete model for the approximate temperature

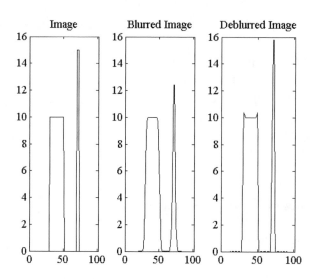

Figure 8.3.1:   Deblurred 1D Image

$u_{ij}^k \approx u(i\Delta x, j\Delta y, k\Delta t)$ is

$$
\begin{aligned}
u_{ij}^{k+1} &= u_{ij}^k + \alpha_x(u_{i-1,j}^k - 2u_{ij}^k + u_{i+1,j}^k) \\
&\quad + \alpha_y(u_{i,j-1}^k - 2u_{ij}^k + u_{i,j+1}^k) \\
\text{with } \alpha_x &\equiv \frac{K}{\rho c}\frac{\Delta t}{\Delta x^2} \text{ and } \alpha_y \equiv \frac{K}{\rho c}\frac{\Delta t}{\Delta y^2}.
\end{aligned}
$$

The reverse of heat diffusion is to use negative time steps and to start with the temperature distribution in the future. As in the 1D case deblurring an image may be modeled by starting with a blurred image as a future "temperature" and apply the explicit finite difference model with negative time steps. In the blurring of the image, one must choose the time step small enough so that $1 - 2\alpha_x - 2\alpha_y$ is positive. Like the 1D example backward iterations eventually start to oscillate and give badly distorted images. The second example illustrates the blurring and deblurring of the image NCSU.

Example 8.3.2. Consider the image NCSU that was created in Subsection 8.1.2. The following MATLAB code, blur2d_ncsu.m, illustrates deblurring in 2D using negative time steps. The reader will find it informative to experiment with the time step sizes and the length of the outer time loops. The mesh() plots are also interesting to observe, but are not printed out in this text. Figures 8.3.2, 8.3.3 and 8.3.4 are original, blurred and deblurred image files, respectively. The deblurring is still a little fuzzy relative to the original image.

MATLAB code blur2d_ncsu.m

```
clear;
clf;
% This code illustrates possible blurring and deblurring in 2D.
% It is motivated by heat diffusion in 2D.
% Blurring is modeled by diffusion of heat in positive time steps.
% Deblurring is done by diffusion of "heat" in negative time steps.
ncsu = [lettern(5) letterc(7) letters(9) letteru(11)];
u = ncsu;
newu = u;
[nx ny] = size(ncsu)
lightncsu = 20*u;
lightncsu = lightncsu(nx:-1:1,:);
lightncsu1 = uint8(lightncsu);
imwrite(lightncsu1, 'mac832.jpg');
figure(1)
subplot(1,3,1)
mesh(u)
axis([0 ny 0 nx 0 15]);
title('Image NCSU')
dt = .00006;
dx = 1/(nx-1);
dy = 3.2/(ny-1);
subplot(1,3,2)
%
% Forward in "time" or "blurring"
%
for k = 1:20
    for j = 2:ny-1
        for i = 2:nx-1
            newu(i,j) = u(i,j) + ...
                (dt/(dx*dx))*(u(i-1,j) -2*u(i,j) + u(i+1,j))+...
                (dt/(dy*dy))*(u(i,j-1) -2*u(i,j) + u(i,j+1));
        end
    end
    mesh(newu);
    axis([0 ny 0 nx 0 15]);
    title('Blurred NCSU')
    k
    pause;
    u = newu;
end
blurncsu = 20*u;
blurncsu = blurncsu(nx:-1:1,:);
blurncsu1 = uint8(blurncsu);
```

Figure 8.3.2:   Original NCSU

Figure 8.3.3:   Blurred NCSU

```
imwrite(blurncsu1, 'mac833.jpg');
subplot(1,3,3)
%
% Backward in "time" or "deblurring"
%
for k = 1:16
    for j = 2:ny-1
        for i = 2:nx-1
            newu(i,j) = u(i,j) - ...
                (dt/(dx*dx))*(u(i-1,j) -2*u(i,j) + u(i+1,j))-...
                (dt/(dy*dy))*(u(i,j-1) -2*u(i,j) + u(i,j+1));
        end
    end
    mesh(newu);
    axis([0 ny 0 nx 0 15]);
    title('Deblurred NCSU')
    k
    pause;
    u = newu;
end
deblurncsu = 20*u;
deblurncsu = deblurncsu(nx:-1:1,:);
deblurncsu1 = uint8(deblurncsu);
imwrite(deblurncsu1, 'mac834.jpg');
```

8.3.3   Sharpen the Pollen Image

The original pollen image pollen.jpg in Figure 8.1.1 was enhanced by increasing the contrast via a piecewise linear transformation to the image in Figure 8.2.6. The image in Figure 8.2.6 will be deblurred or sharpened by using negative time

Figure 8.3.4: Deblurred NCSU

steps and the 2D heat diffusion method as implemented in the Example 8.3.2. The MATLAB code pollensharpen.m does both the contrast enhancement in Figure 8.3.5 and subsequent sharpening in Figure 8.3.6. The reader will find it interesting to compare the last image with the original image in Figure 8.1.1 and to experiment with different values of $intensity = 1.05$ and $coeff = 0.05$ and the length of the k-loop.

MATLAB code pollensharpen.m

```
clear;
% This code is a composition of the piecewise linear
% transformation and the negative time step diffusion
% operation. The pollen image will have an increased contrast,
% be brighter and be sharper.
pollenmat = imread('pollen.jpg');
u = double(pollenmat);
[nx ny] = size(u)
nshades = 256;
%
% Transformation by piecewise linear function.
%
r1 = 80; s1 = 10;
r2 = 140; s2 = 245;
for i = 1:nx
    for j = 1:ny
        if (u(i,j)< r1)
            uspread(i,j) = ((s1-0)/(r1-0))*u(i,j)
        end
        if ((u(i,j)>=r1) & (u(i,j)<= r2))
            uspread(i,j) = ((s2 - s1)/(r2 - r1))*(u(i,j) - r1)+ s1;
        end
        if (u(i,j)>r2)
            uspread(i,j) = ((255 - s2)/(255 - r2))*(u(i,j) - r2) + s2;
        end
    end
end
pollenspread = uint8(uspread);
imwrite(pollenspread, 'mac835.jpg');
%
```

Figure 8.3.5: Increased Contrast Pollen

```
% Brighten and sharpen the image.
%
u = double(pollenspread);
newu = u;
coeff = .06;
intensity = 1.05;
for k = 1:5
     for j = 2:ny-1
          for i = 2:nx-1
               newu(i,j) = intensity*u(i,j) ...
                         - coeff*(u(i-1,j) -2*u(i,j) + u(i+1,j))...
                         - coeff*(u(i,j-1) -2*u(i,j) + u(i,j+1));
          end
     end
     u = newu;
end
pollen_sharpen = uint8(u);
imwrite(pollen_sharpen, 'mac836.jpg');
```

Figure 8.3.6: Brighter and Sharper Pollen

8.3.4 Sharpen the Moon Image

The MATLAB code moonsharpen.m is a variation of the pollensharpen.m code, and also it is used to both brighten and sharpen the moon image moon.jpg. The original moon image is brightened by 20% by using *intensity* $= 1.2$, and then one negative time step is used to sharpen the image with $coeff = 1$. Again, the reader will find it interesting to experiment with different values of *intensity*, *coeff* and the length of the k-loop. The original moon image is in Figure 8.3.7, and the brightened and sharpened moon image is in Figure 8.3.8.

MATLAB code moonsharpen.m

```
clear;
% The original moon image is brighten and sharpened. The
% brightening is controlled by intensity, and the sharpening
% is controlled by coeff and the k-loop.
moonmat = imread('moon.jpg');
u = double(moonmat);
[nx ny] = size(u)
nshades = 256;
%
% Sharpen the image.
%
coeff = 1;
```

Figure 8.3.7: Original Moon Image

```
intensity = 1.2;
for k = 1:1
      for j = 2:ny-1
            for i = 2:nx-1
                  newu(i,j) = intensity*u(i,j) - ...
                              coeff*(u(i-1,j) -2*u(i,j) + u(i+1,j))-...
                              coeff*(u(i,j-1) -2*u(i,j) + u(i,j+1));
            end
      end
      u = newu;
end
moon_sharpen = uint8(u);
imwrite(moon_sharpen, 'mac838.jpg');
```

8.3.5 Exercises

1.   Experiment with blur1d.m and vary the parameters for the time step and the k-loop.
   (a).   Vary $dt = 0.00005, 0.00010$ and 0.00020.
   (b).   Vary $k = 1 : 5, 1 : 10$ and $1 : 20$.
2.   Experiment with blur2d_ncsu.m and vary the parameters for the time step and the k-loop.

Figure 8.3.8: Brightened and Sharpened

    (a).   Vary $dt = 0.0003$, 0.0006 and 0.00012.

    (b).   Vary $k = 1 : 10, 1 : 20$ and $1 : 30$.

    (c).   Can you do a better job of deblurring the image in Figure 8.3.3?

3.   Experiment with pollensharpen.m and vary the parameters *intensity*, *coeff* and the k-loop.

    (a).   Vary *intensity* $= 1.00$, 1.04 and 1.08.

    (b).   Vary *coeff* $= 0.02$, 0.04 and 0.06.

    (c).   Vary $k = 1 : 5, 1 : 10$ and $1 : 20$.

4.   Experiment with moonsharpen.m and vary the parameters *intensity*, *coeff* and the k-loop.

    (a).   Vary *intensity* $= 1.1$, 1.2 and 1.4.

    (b).   Vary *coeff* $= 0.5$, 1.0 and 1.5.

    (c).   Vary $k = 1 : 5, 1 : 10$ and $1 : 20$.

5.   Consider the sharpened image for pollen in Figure 8.3.6. Use a variation of *intensity* in moonsharpen.m to modify pollensharpen.m so that the sharpened image will be brighter than in Figure 8.3.6.

Chapter 9

Image Processing in Frequency Domain

The discrete Fourier transform is introduced to filter unwanted attributes of an image or signal such as from electronic noise or from atmospheric distortion. The discrete Fourier transform is represented by the Fourier matrix whose components are complex numbers. By using properties of complex numbers and matrices a number of important properties of the discrete Fourier and fast Fourier transform are derived. These are then used to filter in the frequency domain a number of images.

9.1 Laplace and Fourier Transforms

Laplace and Fourier transformations can be used to solve differential equations and to reduce periodic noise in signals and images. Basically, they convert the derivative operations into multiplication, differential equations into algebraic equations, and periodic functions are associated with their frequencies. This latter property is very useful in image enhancement where one wants to discard or filter unwanted periodic components. For example, in the Section 9.2 the discrete version of the Fourier transform will be used to filter a "fuzzy" sine function.

9.1.1 Laplace Transform

The Laplace transform is traditionally introduced because it is useful in solving initial value problems for ordinary differential equations. The Laplace transformation maps a function $f(t)$ to another function of a parameter s. This transformation converts derivatives into multiplication by the parameter s, which allows one to solve initial value problems by solving algebraic systems in terms of the parameter s.

Definition 9.1.1. The *Laplace transform* of a function $f(t)$ is a function of a parameter $s > 0$

$$L(f) \equiv \int_0^\infty e^{-st} f(t) dt.$$

Because the improper integral is a linear operation, the Laplace transform is also linear, that is,

$$L(cf) = cL(f) \text{ and } L(f + g) = L(f) + L(g).$$

The Laplace transform of an exponential function $f(t) = e^{at}$ with $s > a$ is easy to compute as follows

$$
\begin{aligned}
L(e^{at}) &= \int_0^\infty e^{-st} e^{at} dt \\
&= \lim_{N \to \infty} \int_0^N e^{-st} e^{at} dt \\
&= \lim_{N \to \infty} \int_0^N e^{-(s-a)t} dt \\
&= \lim_{N \to \infty} [\frac{1}{-(s-a)} e^{-(s-a)N} - \frac{1}{-(s-a)}] \\
&= \frac{1}{s-a}.
\end{aligned}
$$

A special case is $a = 0$ so that $L(1) = 1/s$. There are a number of other rules for computing Laplace transforms, but here we list only four important rules.

Basic Rules for Laplace Transform. Assume all functions are such that the improper integrals in the definition exist.

1. $L(cf) = cL(f)$,
2. $L(f + g) = L(f) + L(g)$,
3. $L(\frac{df}{dt}) = sL(f) - f(0)$ and
4. $L(f * g) = L(f)L(g)$ where $f * g$ is called the *convolution* of f and g

$$(f * g)(t) \equiv \int_0^t f(\tau) g(t - \tau) d\tau.$$

The third rule is important because it converts the derivative operation into multiplication by the parameter s. The third property is established by using the integration by parts formula

$$\int v \, du = uv - \int u \, dv$$

with $du = \frac{df}{dt} dt$ and $v = e^{-st}$. Assume $f(t)$ and the parameter s are such that

$\lim_{N \to \infty} f(N)e^{-sN} = 0$ and, thus,

$$
\begin{aligned}
L(\frac{df}{dt}) &= \int_0^\infty e^{-st}\frac{df}{dt}dt \\
&= \lim_{N \to \infty} \int_0^N e^{-st}\frac{df}{dt}dt \\
&= \lim_{N \to \infty} [f(N)e^{-sN} - f(0)e^{-s0} - \int_0^N (-s)e^{-st}f(t)dt] \\
&= 0 - f(0) + sL(f).
\end{aligned}
$$

The proof of the convolution formula is more complicated, but it will be used in the next subsection.

9.1.2   Differential Equations and Laplace Transform

The solution process using Laplace transforms to solve initial value problems can be outlined in the following three steps: use the Laplace transform to convert the initial value problem (often associated with the "time" domain) into an algebraic problem (often associated with "frequency" domain), solve the algebraic problem, and convert back to the "time" domain for the solution of the initial value problem.

Consider the initial value problem

$$
\frac{du}{dt} = au + f(t) \text{ with } u(0) \text{ given.}
$$

Compute the Laplace transform of both sides of the differential equation and use the first three basic rules

$$
\begin{aligned}
L(\frac{du}{dt}) &= L(au + f(t)) \\
sL(u) - u(0) &= aL(u) + L(f).
\end{aligned}
$$

Solve for the Laplace transform $L(u)$

$$
\begin{aligned}
(s-a)L(u) &= L(f) + u(0) \\
L(u) &= L(f)\frac{1}{s-a} + u(0)\frac{1}{s-a} \\
&= L(f)L(e^{at}) + u(0)L(e^{at}).
\end{aligned}
$$

Use the convolution property with $g(t) = e^{at}$ to obtain

$$
\begin{aligned}
L(u) &= L(f*g) + u(0)L(e^{at}) \\
&= L(f*g + u(0)e^{at}).
\end{aligned}
$$

Evidently,

$$
\begin{aligned}
u(t) &= (f * g)(t) + u(0)e^{at} \\
&= \int_0^t f(\tau)g(t-\tau)d\tau + u(0)e^{at} \\
&= \int_0^t f(\tau)e^{a(t-\tau)}d\tau + u(0)e^{at} \\
&= e^{at}\int_0^t f(\tau)e^{-a\tau}d\tau + u(0)e^{at}.
\end{aligned}
$$

This agrees with the solution in Section 6.1 where the homogeneous solution is e^{at} and the particular solution is from the variation of parameters formula.

9.1.3   Fourier Transform

The Fourier transform can be viewed as a variation of the Laplace transform where the parameter s is replaced by an imaginary parameter iw and the domain of $f(t)$ is expanded to the real line. The Fourier transform has similar properties, which we will only state.

Definition 9.1.2. The *Fourier transform* of a function $f(t)$ is a function of a parameter $w > 0$

$$
F(f) \equiv \int_{-\infty}^{\infty} e^{-iwt} f(t)dt.
$$

Four Basic Properties of Fourier Transform. Assume all functions are such that the improper integrals in the definition exist.

1. $F(cf) = cF(f)$,
2. $F(f+g) = F(f) + F(g)$,
3. $F(\frac{df}{dt}) = -iwF(f)$,
4. $F(f * g) = F(f)F(g)$ where $f * g$ is called the *convolution* of f and g

$$
(f * g)(t) \equiv \int_{-\infty}^{\infty} f(\tau)g(t-\tau)d\tau.
$$

The solution of a differential equation can be found using the Fourier transformation. For example, find a solution of

$$
-\frac{d^2u}{dt^2} + cu = f(t).
$$

As in the use of the Laplace transform, compute the Fourier transform of both sides of the differential equation and use the above rules

$$
\begin{aligned}
F(-\frac{d^2u}{dt^2} + cu) &= F(f(t)) \\
-(-iwF(\frac{du}{dt})) + cF(u) &= F(f(t)) \\
-((-iw)^2 F(u) + cF(u) &= F(f(t)).
\end{aligned}
$$

Solve for $F(u)$

$$F(u) = F(f) \frac{1}{\omega^2 + c}.$$

Let $g = g(t)$ be such that $F(g) = \frac{1}{\omega^2 + c}$ and use the fourth rule with the convolution to conclude

$$u(t) = (f * g)(t).$$

Some functions and their transforms can become complicated, but they can be computed by using tables or symbolic computation tools.

9.1.4 Transforms Using MATLAB

The symbolic toolbox in MATLAB has commands to compute the Laplace and Fourier transforms and their inverses. Some simple illustrations are given below.

```
>> laplace(t)                      % computes the Laplace transform
   1/s^2
>> laplace(t^2)
   2/s^3
>> ilaplace(1/(s*(s+1)))           % the inverse Laplace transform
   1-exp(-t)
>> ilaplace(2/(s-3)^2)
   2*t*exp(3*t)
```

The Fourier transforms of many functions give Dirac delta functionals, which can be viewed as a limit of impulse functions.

Definitions 9.1.3. The *unit step* function, $H(t - t_0)$, or *Heaviside* function is defined as

$$H(t - t_0) \equiv \begin{cases} 0, & \text{for } t < t_0 \\ 1, & \text{for } t \geq t_0 \end{cases}.$$

The *impulse* function $I_\epsilon(t - t_0)$ can be written as

$$I_\epsilon(t - t_0) \equiv \frac{1}{2\epsilon} (H(t - (t_0 - \epsilon)) - H(t - (t_0 + \epsilon)))$$

$$= \begin{cases} \frac{1}{2\epsilon}, & \text{for } t_0 - \epsilon \leq t \leq t_0 + \epsilon \\ 0, & t < t_0 - \epsilon \text{ or } t_0 + \epsilon < t \end{cases}.$$

The *Dirac delta functional* is given by the limit

$$\delta(t - t_0) = \lim_{\epsilon \to 0} I_\epsilon(t - t_0).$$

Another perspective on the limit $I_\epsilon(t - t_0)$ is how it operates as an integral of the product with a continuous function $f(t)$

$$\lim_{\epsilon \to 0} \int f(t) I_\epsilon(t - t_0) dt = \lim_{\epsilon \to 0} \int_{t_0 - \epsilon}^{t_0 + \epsilon} f(t) \frac{1}{2\epsilon} dt = f(t_0).$$

The limit of $I_e(t - t_0)$ can be thought of as an "ideal" impulse.

```
>> fourier(sin(t))                % computes the Fourier transform
   -i*pi*Dirac(w-1) + i*pi*Dirac(w+1)    % gives delta functionals
>> fourier(cos(t))
   pi*Dirac(w-1) + pi*Dirac(w+1)
>> ifourier(w,'t')                % the inverse Fourier transform
   -i*Dirac(1,t)
>> ifourier(w+3*w^2,'t')
   -i*Dirac(1,t) - 3*Dirac(2,t).
```

The Fourier transform identifies the frequency associated with the $\sin(2\pi ft)$ and $\cos(2\pi ft)$ according to the location of the delta functional

```
>> syms f
>> fourier(sin(2*pi*f*t))
   -i*pi*Dirac(w-2*pi*f) + i*pi*Dirac(w+2*pi*f)
>> fourier(cos(2*pi*f*t))
   pi*Dirac(w-2*pi*f) + pi*Dirac(w+2*pi*f)
>> fourier(8*cos(2*pi*100*t) + 4*sin(2*pi*40*t))
   8*pi*Dirac(w-200*pi) + 8*pi*Dirac(w+200*pi)
   -4*i*pi*Dirac(w-80*pi) + 4*i*pi*Dirac(w+80*pi).
```

The discrete version of the Fourier transform is similar to this.

9.1.5   Discrete Fourier Transform (DFT)

The discrete Fourier transform is derived from the Fourier transform of $f(t)$ by assuming $f(t)$ is periodic on the unit interval. Let $\omega = 2\pi k$ where k are integers from 0 to $n - 1$. In this case the improper integral may be approximated as follows by letting $[0 \ \infty)$ be approximated by $[0 \ n]$

$$F(f) = \int_{-\infty}^{\infty} e^{-iwt} f(t)dt \approx 2\int_0^n e^{-iwt} f(t)dt = 2n\int_0^1 e^{-iwt} f(t)dt.$$

Approximate the integral over the unit interval by a Riemann sum

$$F(f) \approx 2n\sum_{j=0}^{n-1} e^{-i2\pi k(j/n)} f(j/n)(1/n)$$

$$= 2\sum_{j=0}^{n-1} (e^{-i2\pi/n})^{kj} f(j/n).$$

Definition 9.1.4. Let the complex number $z \equiv e^{-i2\pi/n}$ and let f_j be complex numbers where j are integers from 0 to $n - 1$. The *discrete Fourier transform*, DFT, is a mapping from $\mathbb{C}^n$ into $\mathbb{C}^n$ whose k^{th} component is

$$\sum_{j=0}^{n-1} z^{kj} f_j.$$

One can view this as a matrix-vector product

$$DFT(\mathbf{f}) = \mathbb{F}\mathbf{f}$$

where the kj-components of the $n \times n$ matrix $\mathbb{F}$ are z^{kj}. The matrix $\mathbb{F}$ is called the *Fourier* matrix. Another useful perspective is that the kth component equals the evaluation of a polynomial at a $x = z^k$

$$[\mathbb{F}\mathbf{f}]_k = p_\mathbf{f}(z^k) = \sum_{j=0}^{n-1} f_j \, (z^k)^j \text{ where}$$

$$p_\mathbf{f}(x) \equiv \sum_{j=0}^{n-1} f_j \, x^j.$$

Example 9.1.1. Let $n = 3$ so that $z = e^{-i2\pi/3} = \cos(2\pi/3) - i\sin(2\pi/3) = -1/2 - i\sqrt{3}/2$. The Fourier matrix is 3×3 with complex numbers as components

$$\mathbb{F} = \begin{bmatrix} z^0 & z^0 & z^0 \\ z^0 & z^1 & z^2 \\ z^0 & z^2 & z^4 \end{bmatrix} = \begin{bmatrix} 1 & 1 & 1 \\ 1 & z^1 & z^2 \\ 1 & z^2 & z^1 \end{bmatrix}.$$

If $\mathbf{f} = \begin{bmatrix} 1 & 7 & 2 \end{bmatrix}^T$ so that $p_\mathbf{f}(x) \equiv 1 + 7x + 2x^2$ and $DFT(\mathbf{f})$ is

$$\begin{aligned} \mathbb{F}\mathbf{f} &= \begin{bmatrix} 1 & 1 & 1 \\ 1 & z^1 & z^2 \\ 1 & z^2 & z^1 \end{bmatrix} \begin{bmatrix} 1 \\ 7 \\ 2 \end{bmatrix} = \begin{bmatrix} 1+7+2 \\ 1+7z+2z^2 \\ 1+7z^2+2z \end{bmatrix} \\ &= \begin{bmatrix} p_\mathbf{f}(z^0) \\ p_\mathbf{f}(z^1) \\ p_\mathbf{f}(z^2) \end{bmatrix} = \begin{bmatrix} 10 \\ -3.50 - 4.33i \\ -3.50 + 4.33i \end{bmatrix}. \end{aligned}$$

Example 9.1.2. Let $n = 4$ so that $z = e^{-i2\pi/4} = \cos(2\pi/4) - i\sin(2\pi/4) = -i$. The Fourier matrix is 4×4 with complex numbers as components and here the subscript denotes $n = 4$

$$\mathbb{F}_4 = \begin{bmatrix} z^0 & z^0 & z^0 & z^0 \\ z^0 & z^1 & z^2 & z^3 \\ z^0 & z^2 & z^4 & z^6 \\ z^0 & z^3 & z^6 & z^9 \end{bmatrix} = \begin{bmatrix} 1 & 1 & 1 & 1 \\ 1 & z^1 & z^2 & z^3 \\ 1 & z^2 & 1 & z^2 \\ 1 & z^3 & z^2 & z^1 \end{bmatrix}.$$

The DFT of $\mathbf{f} = \begin{bmatrix} f_0 & f_1 & f_2 & f_3 \end{bmatrix}$ is

$$\mathbb{F}_4\mathbf{f} = \begin{bmatrix} 1 & 1 & 1 & 1 \\ 1 & z^1 & z^2 & z^3 \\ 1 & z^2 & 1 & z^2 \\ 1 & z^3 & z^2 & z^1 \end{bmatrix} \begin{bmatrix} f_0 \\ f_1 \\ f_2 \\ f_3 \end{bmatrix}.$$

Regroup the four components of $\mathbf{f}$ into even and odd subscripts by using a permutation matrix P and note

$$\mathbb{F}_4\mathbf{f} = (\mathbb{F}_4 P)(P\mathbf{f}) = \begin{bmatrix} 1 & 1 & 1 & 1 \\ 1 & z^2 & z & z^3 \\ 1 & 1 & z^2 & z^2 \\ 1 & z^2 & z^3 & z \end{bmatrix} \begin{bmatrix} f_0 \\ f_2 \\ f_1 \\ f_3 \end{bmatrix}$$

$$= \begin{bmatrix} \mathbb{F}_2 & D\mathbb{F}_2 \\ \mathbb{F}_2 & -D\mathbb{F}_2 \end{bmatrix} \begin{bmatrix} \mathbf{f}_{even} \\ \mathbf{f}_{odd} \end{bmatrix} \quad \text{where}$$

$$\mathbb{F}_2 = \begin{bmatrix} 1 & 1 \\ 1 & z^2 \end{bmatrix} = \begin{bmatrix} 1 & 1 \\ 1 & \widehat{z} \end{bmatrix} \quad \text{with } \widehat{z} = e^{-i2\pi/4}$$

$$D = \begin{bmatrix} 1 & 0 \\ 0 & z \end{bmatrix}, \ \mathbf{f}_{even} = \begin{bmatrix} f_0 \\ f_2 \end{bmatrix} \quad \text{and } \mathbf{f}_{odd} = \begin{bmatrix} f_1 \\ f_3 \end{bmatrix}.$$

Thus, the 4x4 matrix product reduces to two 2x2 matrix-vector products. A similar reduction occurs when $n = 8 = 2^3$ and in general for $n = 2^d$ or $d = log_2(n)$. This generates a very fast implementation of the Fourier transform called the fast Fourier transform. It requires $(n/2)log_2(n)$ operations, which is much less than the $2n^2$ operations for the general matrix-vector product.

In applications n is typically very large and an efficient method to do these computations is required. In MATLAB the command fft() is an implementation of the DFT called the *fast Fourier* transform. The computation in Example 9.1.1 is easily done

```
>> fft([1 7 2])     % n = 3
    10.0    -3.5000 - 4.3301i     -3.5000 + 4.3301i.
```

We will make frequent use of this command.

9.1.6   Exercises

1.   Use the Laplace transform to solve the initial value problem

$$\frac{du}{dt} = c(80 - u) \text{ with } u(0) = 210.$$

2.   Use Laplace transform rule three twice to find the Laplace transform of

$$\frac{d^2u}{dt^2} \text{ with } u(0) = 2 \text{ and } \frac{du}{dt}(0) = 3.$$

3.   Let $n = 5$ and consider the DFT.
  (a).   Find the Fourier matrix $\mathbb{F}$.
  (b).   Use the Fourier matrix and by-hand computations find the DFT of
          $\mathbf{f} = [2 \ 8 \ 9 \ 1 \ 3]^T$.
  (c).   Use the MATLAB command fft() to confirm the above calculation.

4.   Let $n = 8$ and consider the DFT.
    (a).   Find the Fourier matrix $\mathbb{F}$.
    (b).   Use the Fourier matrix and by-hand computations find the DFT of
$$\mathbf{f} = [2\ 8\ 9\ 1\ 3\ 2\ -1\ 3]^T.$$
    (c).   Use the MATLAB command fft() to confirm the above calculation.

9.2   Properties of DFT

The Fourier transform of a function $f(t)$ may be discretized, and many of its properties have analogues to the discrete case. The discrete Fourier transform, DFT, will be defined and three important properties will be discussed. These include the DFT of the discrete sine and cosine functions, inverse DFT and DFT of convolutions. Like the Fourier transform of functions, the DFT will be able to identify frequencies of periodic functions. This allows one to "filter" unwanted parts of signals or images.

9.2.1   Definition of DFT

The discrete Fourier transform is derived from the Fourier transform. The *discrete Fourier transform*, DFT, is a mapping from $\mathbb{C}^n$ into $\mathbb{C}^n$ whose k^{th} component is

$$\sum_{j=0}^{n-1} z^{kj} f_j.$$

This is a matrix-vector product

$$DFT(\mathbf{f}) = \mathbb{F}\mathbf{f}$$

where the kj-components of the $n \times n$ Fourier matrix $\mathbb{F}$ are z^{kj}. Or, the kth component of $DFT(\mathbf{f})$ is the evaluation of a polynomial $p_\mathbf{f}(x) \equiv \sum_{j=0}^{n-1} f_j\, x^j$ at a $x = z^k$

$$[\mathbb{F}\mathbf{f}]_k = p_\mathbf{f}(z^k) = \sum_{j=0}^{n-1} f_j\, (z^k)^j.$$

Example 9.2.1. Let $n = 5$ so that $z = e^{-i2\pi/5} = \cos(2\pi/5) - i\sin(2\pi/5) = 0.3090 - i0.9511$. The Fourier matrix is 5×5 with complex numbers as components

$$\mathbb{F} = \begin{bmatrix} z^0 & z^0 & z^0 & z^0 & z^0 \\ z^0 & z^1 & z^2 & z^3 & z^4 \\ z^0 & z^2 & z^4 & z^6 & z^8 \\ z^0 & z^3 & z^6 & z^9 & z^{12} \\ z^0 & z^4 & z^8 & z^{12} & z^{16} \end{bmatrix} = \begin{bmatrix} 1 & 1 & 1 & 1 & 1 \\ 1 & z^1 & z^2 & z^3 & z^4 \\ 1 & z^2 & z^4 & z & z^3 \\ 1 & z^3 & z & z^4 & z^2 \\ 1 & z^4 & z^3 & z^2 & z \end{bmatrix}.$$

If $\mathbf{f} = [1\ 7\ 2\ -2\ 3]^T$, then $p_{\mathbf{f}}(x) \equiv 1 + 7x + 2x^2 - 2x^3 + 3x^4$ and $DFT(\mathbf{f})$ is

$$
\mathbb{F}\mathbf{f} =
\begin{bmatrix}
1 & 1 & 1 & 1 & 1 \\
1 & z^1 & z^2 & z^3 & z^4 \\
1 & z^2 & z^4 & z & z^3 \\
1 & z^3 & z & z^4 & z^2 \\
1 & z^4 & z^3 & z^2 & z
\end{bmatrix}
\begin{bmatrix}
1 \\ 7 \\ 2 \\ -2 \\ 3
\end{bmatrix}
$$

$$
=
\begin{bmatrix}
p_{\mathbf{f}}(z^0) \\
p_{\mathbf{f}}(z^1) \\
p_{\mathbf{f}}(z^2) \\
p_{\mathbf{f}}(z^3) \\
p_{\mathbf{f}}(z^5)
\end{bmatrix}
=
\begin{bmatrix}
11.0000 \\
4.0902 + 6.1554i \\
-7.0902 - 1.4531i \\
-7.0902 + 1.4531i \\
4.0902 - 6.1554i
\end{bmatrix} .
$$

MATLAB easily does this computation

```
>> fft([1 7 2 -2 3])'
   ans =
      11.0000
       4.0902 + 6.1554i
      -7.0902 - 1.4531i
      -7.0902 + 1.4531i
       4.0902 - 6.1554i
```

9.2.2   DFT of Sine and Cosine

Let $\mathbf{t} = 0 : 1/n : 1-1/n$ be the vector starting at 0 and increasing in increments of $1/n$ to $1-1/n$. In order to compute the DFT of the sine and cosine functions, first compute the DFT of the exponential vector $e^{i2\pi kt}$ and then use Euler's formula. We shall make use of three properties of $z \equiv e^{-i2\pi/n}$:

$$
z^n = 1,\ z^{-k} = z^{n-k} \text{ and } 1 + z + z^2 + \cdots + z^{n-1} = 0.
$$

The first two properties follow from Euler's formula. The third property follows from

$$
\begin{aligned}
z(1 + z + z^2 + \cdots + z^{n-1}) &= z + z^2 + \cdots + z^{n-1} + z^n \\
&= z + z^2 + \cdots + z^{n-1} + 1.
\end{aligned}
$$

Move the terms on the right side to the left side and use the distribution property to get

$$
(z - 1)(1 + z + z^2 + \cdots + z^{n-1}) = 0.
$$

Since $z - 1$ is not zero, the second factor must be zero. Example 9.2.2 illustrates the discrete Fourier transform of $e^{i2\pi kt}$ for $n = 3$ and $\mathbf{t} = [0/3\ 1/3\ 2/3]^T$.

Example 9.2.2. If $n = 3$, then $DFT(e^{i2\pi ft})$ is a 3×1 vector given by the Fourier matrix with $z = e^{-i2\pi/3}$

$$\mathbb{F}e^{i2\pi ft} = \begin{bmatrix} 1 & 1 & 1 \\ 1 & z^1 & z^2 \\ 1 & z^2 & z^1 \end{bmatrix} \begin{bmatrix} e^{i2\pi f0/3} \\ e^{i2\pi f1/3} \\ e^{i2\pi f2/3} \end{bmatrix}$$

$$= \begin{bmatrix} 1 & 1 & 1 \\ 1 & z^1 & z^2 \\ 1 & z^2 & z^1 \end{bmatrix} \begin{bmatrix} 1 \\ z^{-f} \\ z^{-2f} \end{bmatrix}.$$

If $f = 0$, then

$$\mathbb{F}e^{i2\pi ft} = \begin{bmatrix} 1 & 1 & 1 \\ 1 & z^1 & z^2 \\ 1 & z^2 & z^1 \end{bmatrix} \begin{bmatrix} 1 \\ 1 \\ 1 \end{bmatrix} = \begin{bmatrix} 1+1+1 \\ 1+z+z^2 \\ 1+z^2+z \end{bmatrix} = \begin{bmatrix} 3 \\ 0 \\ 0 \end{bmatrix}.$$

If $f = 1$, then

$$\mathbb{F}e^{i2\pi ft} = \begin{bmatrix} 1 & 1 & 1 \\ 1 & z^1 & z^2 \\ 1 & z^2 & z^1 \end{bmatrix} \begin{bmatrix} 1 \\ z^{-1} \\ z^{-2} \end{bmatrix} = \begin{bmatrix} 1+z^2+z \\ 1+1+1 \\ 1+z+z^2 \end{bmatrix} = \begin{bmatrix} 0 \\ 3 \\ 0 \end{bmatrix}.$$

If $f = 2$, then

$$\mathbb{F}e^{i2\pi ft} = \begin{bmatrix} 1 & 1 & 1 \\ 1 & z^1 & z^2 \\ 1 & z^2 & z^1 \end{bmatrix} \begin{bmatrix} 1 \\ z^{-2} \\ z^{-1} \end{bmatrix} = \begin{bmatrix} 1+z+z^2 \\ 1+z^2+z \\ 1+1+1 \end{bmatrix} = \begin{bmatrix} 0 \\ 0 \\ 3 \end{bmatrix}.$$

DFT of Sine and Cosine Functions. Let $[\mathbb{F}\mathbf{f}]_k$ denote the kth component of the vector $\mathbb{F}\mathbf{f}$. Then for $\mathbf{f} = e^{i2\pi ft}$

$$[\mathbb{F}e^{i2\pi ft}]_k = \begin{cases} 0, & \text{for } k \neq f \\ n, & \text{for } k = f \end{cases} \quad \text{and}$$

$$[\mathbb{F}e^{-i2\pi ft}]_k = \begin{cases} 0, & \text{for } k \neq n - f \\ n, & \text{for } k = n - f \end{cases}.$$

Use Euler's formula to obtain for $\mathbf{f} = \cos(2\pi ft)$ and $\sin(2\pi ft)$

$$[\mathbb{F}\cos(2\pi ft)]_k = \begin{cases} 0, & \text{for } k \neq n - f \text{ or } f \\ n/2, & \text{for } k = n - f \text{ or } f \end{cases} \quad \text{and}$$

$$[\mathbb{F}\sin(2\pi ft)]_k = \begin{cases} 0, & \text{for } k \neq n - f \text{ or } f \\ (-n/2)i, & \text{for } k = f \\ (+n/2)i, & \text{for } k = n - f \end{cases}.$$

The MATLAB code ffttrig.m illustrates these rules, which are analogous to the rules for the Fourier transform of the sine and cosine functions giving Dirac delta functionals.

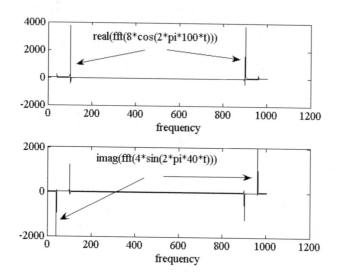

Figure 9.2.1:   DFT of Sine and Cosine

MATLAB code ffttrig.m

```
% This code uses fft() to compute the discrete
% Fourier transform of the sine and cosine.
% The real and imaginary parts of the transformed
% vectors are graphed.
t = 0:.001:1;
freq = 1:1:1001;
fftsin = fft(4*sin(2*pi*40*t));
fftcos = fft(8*cos(2*pi*100*t));
subplot(2,1,1);
plot(freq,real(fftsin),freq,real(fftcos));
subplot(2,1,2);
plot(freq,imag(fftsin),freq,imag(fftcos));
```

Figure 9.2.1 has the two graphs of the transformed functions. The reader should note the locations and magnitudes of the transformed vectors, which confirms the above important rules.

9.2.3   Inverse DFT

Let the complex number $z \equiv e^{-i2\pi/n}$ and let g_j be complex numbers where j are integers from 0 to $n - 1$. The *inverse discrete Fourier transform* is a mapping from $\mathbb{C}^n$ into $\mathbb{C}^n$ whose k^{th} component is

$$\frac{1}{n}\sum_{j=0}^{n-1} z^{-kj} g_j.$$

Inverse Fourier Matrix Property. Let $\mathbb{F}$ be the Fourier matrix. Then $\mathbb{F}$ has an inverse matrix whose kj-components are $\frac{1}{n} z^{-kj}$.

Example 9.2.3. Use the Fourier matrix and show $\mathbb{F}\mathbb{F}^{-1}$ is the 3×3 identity matrix

$$
\begin{aligned}
\mathbb{F}\mathbb{F}^{-1} &= \begin{bmatrix} 1 & 1 & 1 \\ 1 & z^1 & z^2 \\ 1 & z^2 & z^1 \end{bmatrix} \frac{1}{3}\begin{bmatrix} 1 & 1 & 1 \\ 1 & z^{-1} & z^{-2} \\ 1 & z^{-2} & z^{-1} \end{bmatrix} \\
&= \frac{1}{3}\begin{bmatrix} 1+1+1 & 1+z^2+z & 1+z+z^2 \\ 1+z+z^2 & 1+1+1 & 1+z^2+z \\ 1+z^2+z & 1+z+z^2 & 1+1+1 \end{bmatrix} \\
&= \frac{1}{3}\begin{bmatrix} 3 & 0 & 0 \\ 0 & 3 & 0 \\ 0 & 0 & 3 \end{bmatrix} = I.
\end{aligned}
$$

The inverse Fourier transform can be efficiently computed by using the same technique as in the fast Fourier transform. The following illustrates the MATLAB command ifft() for computing the inverse Fourier transform:

```
>> f = [1 7 2]
>> g = fft(f)
  g = 10.0000    -3.5000 - 4.3301i    -3.5000 + 4.3301i
>> ifft(g)
  1.0000    7.0000    2.0000.
```

9.2.4   DFT and Convolution

The objective is to define a discrete convolution so that an analogue of the convolution property can be established. Use the polynomial representation of the DFT for two vectors **a** and **b** with n components

$$[\mathbb{F}\mathbf{a}]_k = p_\mathbf{a}(z^k) \equiv \sum_{j=0}^{n-1} a_j (z^k)^j \text{ and } [\mathbb{F}\mathbf{b}]_k = p_\mathbf{b}(z^k) \equiv \sum_{j=0}^{n-1} b_j (z^k)^j.$$

Note, the product of these two polynomials is a polynomial of degree $2n - 2$ with $2n - 1$ coefficients. This suggests the convolution of **a** and **b** should be a vector with $2n - 1$ components chosen to be the components of the product of the polynomials. For example, for $n = 3$ and two vectors $\mathbf{a} = [a_0 \ a_1 \ a_2]^T$ and

$\mathbf{b} = [b_0 \ b_1 \ b_2]^T$

$$
\begin{aligned}
p_{\mathbf{a}}(x)p_{\mathbf{b}}(x) &= (a_0 + a_1 x + a_2 x^2)(b_0 + b_1 x + b_2 x^2) \\
&= (a_0 b_0) + (a_0 b_1 + a_1 b_0)x + (a_0 b_2 + a_1 b_1 + a_2 b_0)x^2 \\
&\quad + (a_1 b_2 + a_2 b_1)x^3 + (a_2 b_2)x^4.
\end{aligned}
$$

Define the convolution of $\mathbf{a}$ and $\mathbf{b}$ to be the vector of coefficients in the product polynomial

$$
\begin{bmatrix} a_0 b_0 & a_0 b_1 + a_1 b_0 & a_0 b_2 + a_1 b_1 + a_2 b_0 & a_1 b_2 + a_2 b_1 & a_2 b_2 \end{bmatrix}.
$$

Definition 9.2.1. Let $\mathbf{a}$ and $\mathbf{b}$ have n components and let $p_{\mathbf{a}}(x)$ and $p_{\mathbf{b}}(x)$ be polynomials associated with these vectors. The *convolution of two vectors*, $conv(\mathbf{a}, \mathbf{b})$, is a $2n - 1$ vector such that

$$
p_{conv(\mathbf{a},\mathbf{b})}(x) = p_{\mathbf{a}}(x)p_{\mathbf{b}}(x).
$$

The analogue of the convolution property for the discrete Fourier transforms requires the use of "padded" discrete Fourier transforms. This is a technical difficulty caused by the fact that the dimension of vectors increases during the convolution of two vectors. Let $\mathbf{a}$ be an n vector. A "padded" vector of $\mathbf{a}$ is a $2n - 1$ vector, $\mathbf{A}$, with $\mathbf{a}$ in the first n components and zeros in the last $n - 1$ components. For example, let $n = 3$ and $\mathbf{a} = [a_0 \ a_1 \ a_2]^T$ so that

$$
\mathbf{A} = [a_0 \ a_1 \ a_2 \ 0 \ 0]^T.
$$

Let $z = e^{-i2\pi/5}$ and compute the DFT of the padded vector $\mathbf{A}$

$$
\mathbb{F}\mathbf{A} =
\begin{bmatrix}
1 & 1 & 1 & 1 & 1 \\
1 & z & z^2 & z^3 & z^4 \\
1 & z^2 & z^4 & z & z^3 \\
1 & z^3 & z & z^4 & z^2 \\
1 & z^4 & z^3 & z^2 & z
\end{bmatrix}
\begin{bmatrix}
a_0 \\ a_1 \\ a_2 \\ 0 \\ 0
\end{bmatrix}
=
\begin{bmatrix}
1 \\ 1 \\ 1 \\ 1 \\ 1
\end{bmatrix} a_0 +
\begin{bmatrix}
1 \\ z \\ z^2 \\ z^3 \\ z^4
\end{bmatrix} a_1 +
\begin{bmatrix}
1 \\ z^2 \\ z^4 \\ z \\ z^3
\end{bmatrix} a_2.
$$

The component-wise or array product of $\mathbb{F}\mathbf{A}$ and $\mathbb{F}\mathbf{B}$ is

$$
\mathbb{F}\mathbf{A}. * \mathbb{F}\mathbf{B} = \left(
\begin{bmatrix} 1 \\ 1 \\ 1 \\ 1 \\ 1 \end{bmatrix} a_0 +
\begin{bmatrix} 1 \\ z \\ z^2 \\ z^3 \\ z^4 \end{bmatrix} a_1 +
\begin{bmatrix} 1 \\ z^2 \\ z^4 \\ z \\ z^3 \end{bmatrix} a_2 \right). *
$$

$$
\left(
\begin{bmatrix} 1 \\ 1 \\ 1 \\ 1 \\ 1 \end{bmatrix} b_0 +
\begin{bmatrix} 1 \\ z \\ z^2 \\ z^3 \\ z^4 \end{bmatrix} b_1 +
\begin{bmatrix} 1 \\ z^2 \\ z^4 \\ z \\ z^3 \end{bmatrix} b_2 \right)
$$

$$
= \begin{bmatrix} 1 \\ 1 \\ 1 \\ 1 \\ 1 \end{bmatrix} a_0 b_0 + \begin{bmatrix} 1 \\ z \\ z^2 \\ z^3 \\ z^4 \end{bmatrix} (a_0 b_1 + a_1 b_0)
$$

$$
+ \begin{bmatrix} 1 \\ z^2 \\ z^4 \\ z \\ z^3 \end{bmatrix} (a_0 b_2 + a_1 b_1 + a_2 b_0)
$$

$$
+ \begin{bmatrix} 1 \\ z^3 \\ z \\ z^4 \\ z^2 \end{bmatrix} (a_1 b_2 + a_2 b_1) + \begin{bmatrix} 1 \\ z^4 \\ z^3 \\ z^2 \\ z \end{bmatrix} a_2 b_2
$$

$$
= \mathbb{F} \, conv(\mathbf{a}, \mathbf{b}).
$$

This is true for general vectors and gives the desired analogue of the convolution of Fourier transform for functions.

Convolution Property of DFT. Let $\mathbf{A}$ and $\mathbf{B}$ be the "padded" vectors of two n vectors $\mathbf{a}$ and $\mathbf{b}$. The $2n - 1$ dimensional DFT of the convolution of $\mathbf{a}$ and $\mathbf{b}$ is the component-wise or array product of the DFT of $\mathbf{A}$ and $\mathbf{B}$

$$
\mathbb{F} \, conv(\mathbf{a}, \mathbf{b}) = \mathbb{F}\mathbf{A}. * \mathbb{F}\mathbf{B}.
$$

Example 9.2.4. The reader should confirm by-hand calculations the following MATLAB computations, which verify the convolution property. Let $n = 3$ and $\mathbf{a} = [2 \ 3 \ 6]$ and $\mathbf{b} = [8 \ 6 \ 3]$.

```
>> a = [2 3 6]
>> b = [8 6 3]
>> conv(a,b)
    16     36     72     45     18
>> A = [a  0  0]
>> B = [b  0  0]
>> fftA = fft(A)
    fftA =
      11.0000      -1.9271 - 6.3799i      1.4271 + 3.9430i
                    1.4271 - 3.9430i     -1.9271 + 6.3799i
>> fftB = fft(B)
    fftB =
      17.0000       7.4271 - 7.4697i      4.0729 - 0.6735i
                    4.0729 + 0.6735i      7.4271 + 7.4697i
```

```
>> lhs = fft(conv(a,b))
    lhs =
       1.0e+002 *
       1.8700        -0.6197 - 0.3299i        0.0847 + 0.1510i
                      0.0847 - 0.1510i        -0.6197 + 0.3299i
>> rhs = fftA.*fftB
    rhs =
       1.0e+002 *
       1.8700        -0.6197 - 0.3299i        0.0847 + 0.1510i
                      0.0847 - 0.1510i        -0.6197 + 0.3299i.
```

9.2.5   Application to Signal or Image Filtering

The convolution property can be restated as

$$conv(\mathbf{a}, \mathbf{b}) = \mathbb{F}^{-1}(\mathbb{F}\mathbf{A}. * \mathbb{F}\mathbf{B}).$$

If **B** is a signal or image, then its DFT will reflect the frequencies of the periodic parts of the signal or image. By masking or filtering out the unwanted frequencies one can obtain a new signal or image by applying the inverse Fourier transformation. A *filter* is a $(2n - 1) \times 1$ matrix, a $(2n - 1) \times 1$ column vector for a 1D signal or image, whose components vary from 0 to 1 and with the same dimension as the DFT of the padded vector. If the component is 1, then the frequency is allowed to pass; if the component is 0, then the frequency is tossed out.

Definition 9.2.2. Let *Filter* represent a $(2n-1) \times 1$ matrix whose components are in the unit interval. The *filtered signal or image* is

$$New\mathbf{B} = \mathbb{F}^{-1}(Filter. * \mathbb{F}\mathbf{B}).$$

As an illustration consider following MATLAB code fftsine.m, which filters a low frequency sine function with an unwanted high frequency noise. Figure 9.2.2 has the noisy sine curve with the time domain in the top graph and its shifted DFT with the frequency domain in the bottom graph. The unwanted frequencies are 200 and 210 and are associated with the small spikes away from the center of the shifted frequency domain. The filter matrix is zero on the left and right parts of the shifted frequency domain and one in the center region corresponding to the low frequencies. The top graph in Figure 9.2.3 is the filtered DFT of the noisy sine function, which has removed the higher frequencies, and the lower graph is the filtered image in the time domain.

MATLAB code fftsine.m
```
% This code is a low-pass filter for a noisy sine curve.
% It uses the DFT to transform the image to the frequency domain.
% The filter matrix is applied to eliminate the unwanted high
% frequencies. Then the convolution property and inverse DFT
```

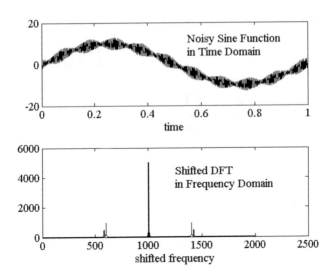

Figure 9.2.2: Noisy Sine Function

% are used to transform the filtered image back to the time domain.
clear;
t = 0:.001:1; % Define 10 sin(2 pi t) with high frequency variations.
x = 10*sin(2*pi*t) + 2*sin(2*pi*200*t) + 1*cos(2*pi*210*t);
%
fftx = fft(x,2002);   % Padded fft.
fftx = fftshift(fftx);
%
figure(1)
subplot(2,1,1);
plot(t,x);
subplot(2,1,2);
plot(abs(fftx));
%
freqfilter = zeros(1,2002);   % Low pass filter.
freqfilter(1,700:1300) = 1;
newfftx = freqfilter.*fftx;
newx = ifftshift(newfftx);
%
newx = ifft(newx,2002);   % Padded inverse fft.
%
figure(2)
subplot(2,1,1);
plot(abs(newfftx));

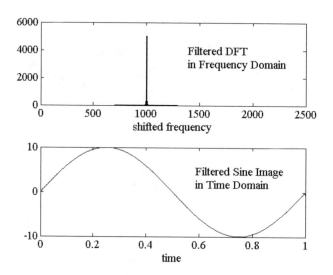

Figure 9.2.3: Filtered Sine Image

```
subplot(2,1,2);
plot(t,real(newx(1,1:1001)));
```

9.2.6   Exercises

1.   Let $n = 3$, $\mathbf{t} = [0 \ \ 1/3 \ \ 2/3]^T$ and $f = 0, 1$ and 2.
(a).   By-hand calculations find the DFT of $\cos(2\pi f \mathbf{t})$.
(b).   By-hand calculations find the DFT of $\sin(2\pi f \mathbf{t})$.
(c).   Use the MATLAB command fft() to confirm the above calculations.
2.   Consider the MATLAB code ffttrig.m.
(a).   Use a variation of this code to find the DFT of $3\cos(2\pi 30t)$.
(b).   Use it to find the DFT of $3\cos(2\pi 100t) + 4\sin(2\pi 120t) + 20\cos(2\pi t)$.
3.   Let $n = 5$ and consider the inverse DFT.
(a).   Find the inverse Fourier matrix $\mathbb{F}^{-1}$.
(b).   Verify it is the inverse of $\mathbb{F}$, that is, show $\mathbb{F}\mathbb{F}^{-1} = I$.
4.   Consider Example 9.2.4 of the convolution property.
(a).   By-hand calculations verify the MATLAB computations.
(b).   Let $n = 4$ and use MATLAB to verify the convolution property for
$\mathbf{a} = [1 \ \ 3 \ \ 4 \ \ -1]^T$ and $\mathbf{b} = [8 \ \ 9 \ \ 4 \ \ 1]^T$.
5.   Consider the MATLAB code fftsine.m. Adjust the filter matrix, freqfilter, so that the following is filtered

$$10\sin(2\pi t) + 2\sin(2\pi 50t) + 1\cos(2\pi 60t).$$

6.   Consider the MATLAB code fftsine.m. Use a variation of it to filter the noisy cosine function

$$20\cos(2\pi t) + 3\cos(2\pi 100t) + 4\sin(2\pi 120t).$$

9.3   DFT in $\mathbb{R}^n \times \mathbb{R}^n$

In the previous section the DFT of a single vector was defined to be another vector with k^{th} component

$$\sum_{j=0}^{n-1}(e^{-i2\pi/n})^{kj}f_j = [\mathbb{F}\mathbf{f}]_k = p_\mathbf{f}(z^k) \text{ where } z \equiv e^{-i2\pi/n}.$$

Three important properties were the DFT of sine and cosine functions, the inverse DFT and the convolution identity. These were used to illustrate a low-pass filter for a noisy 1D image or signal. The objective of this section is to extend the DFT to 2D images, which are represented by matrices and not a single vector. This will allow us to filter unwanted attributes of 2D images.

9.3.1   Definition of DFT in 2D

Let an image be represented by an $n_x \times n_y$ matrix f whose components are f_{j_x,j_y} where $0 \leq j_x < n_x$ and $0 \leq j_y < n_y$. The 2D version of DFT and its inverse are analogous to DFT of a single vector.

Definitions 9.3.1. The *2D DFT* of the matrix f is another matrix of the same dimension with k_x, k_y-component equal to

$$\sum_{j_y=0}^{n_y-1}\sum_{j_x=0}^{n_x-1}(e^{-i2\pi/n_y})^{k_y j_y}(e^{-i2\pi/n_x})^{k_x j_x}f_{j_x,j_y} = p_f(z_x^{k_x}, z_y^{k_y}) \text{ where}$$

$$p_f(x,y) \equiv \sum_{j_y=0}^{n_y-1}\sum_{j_x=0}^{n_x-1}f_{j_x,j_y}x^{j_x}y^{j_y}, \ z_x \equiv e^{-i2\pi/n_x} \text{ and } z_y \equiv e^{-i2\pi/n_y}.$$

Let $DFT(f) = g = p_f(z_x^{k_x}, z_y^{k_y})$. The *inverse 2D DFT* can be shown to have j_x, j_y-component equal to

$$f_{j_x,j_y} = \frac{1}{n_x n_y}\sum_{k_y=0}^{n_y-1}\sum_{k_x=0}^{n_x-1}z_x^{-k_y j_y} z_y^{-k_x j_x} g_{k_x,k_y}.$$

Both the 2D DFT and its inverse can easily be computed by using the MATLAB commands fft2() and ifft2(). For example, for $n_x = 2$ and $n_y = 3$:

```
>> a = [1  2  5; 6  7  11]
        a =      1    2    5
                 6    7    11
>> ffta = fft2(a)
        ffta =        32.0000      -5.5000 + 6.0622i
                                   -5.5000 - 6.0622i
                      -16.0000      0.5000 - 0.8660i
                                    0.5000 + 0.8660i
>> ifft2(ffta)
        =      1    2    5
               6    7    11.
```

9.3.2 2D DFT of Sine and Cosine

The following MATLAB code in ffttrig2da.m illustrates a more general rule for
2D DFT of sine and cosine functions of two variables

MATLAB code in ffttrig2da.m

```
clear;
nx = 4;
ny = 3;
fy = 2;
fx = 1;
for i = 1:nx
    for j=1:ny
        u(i,j) = sin(2*pi*(i-1)/nx*fx+2*pi*(j-1)/ny*fy);
    end
end
u
fftu = fft2(u)

>> ffttrig2da
    u =
              0.0000      -0.8660       0.8660
              1.0000      -0.5000      -0.5000
              0.0000       0.8660      -0.8660
             -1.0000       0.5000       0.5000
    fftu =
             -0.0000                   0.0000 + 0.0000i
                                       0.0000 - 0.0000i
              0.0000 - 0.0000i         0.0000 + 0.0000i
                                      -0.0000 - 6.0000i
              0.0000                   0.0000 - 0.0000i
                                      -0.0000 + 0.0000i
              0.0000 + 0.0000i        -0.0000 + 6.0000i
                                       0.0000 - 0.0000i
```

As in the 1D DFT of the sine and cosine functions, the 2D DFT transforms will identify the frequencies as "spikes" in the matrix of the Fourier transforms. Let $x = 0 : 1/n_x : 1 - 1/n_x$ and $y = 0 : 1/n_y : 1 - 1/n_y$.

The k_x, k_y-component of DFT($\cos(2\pi x f_x + 2\pi y f_y)$) is

$$\begin{cases} \frac{n_x n_y}{2}, & \text{for } (k_x, k_y) = (f_x, f_y) \\ \frac{n_x n_y}{2}, & \text{for } (k_x, k_y) = (n_x - f_x, n_y - f_y) \\ 0, & \text{for other cases.} \end{cases}$$

The k_x, k_y-component of DFT($\sin(2\pi x f_x + 2\pi y f_y)$) is

$$\begin{cases} -\frac{n_x n_y}{2}i, & \text{for } (k_x, k_y) = (f_x, f_y) \\ \frac{n_x n_y}{2}i, & \text{for } (k_x, k_y) = (n_x - f_x, n_y - f_y) \\ 0, & \text{for other cases.} \end{cases}$$

The following MATLAB code ffttrig2d.m also illustrates these rules. Figure 9.3.1 is the graph of the 2D DFT of the "padded" image of the indicated sine and cosine functions of two variables. The frequency domain has been shifted so that the center frequencies are the smaller frequencies. The "spikes" are located near a circle of radius 400 from the center. Since the "spikes" are very large, the vertical axis in the mesh plot is the $log(1 + abs(fftu))$.

MATLAB code ffttrig2d.m

```
% Computes the 2D DFT using fft2().
% Notice the locations of the "spikes" in
% the mesh plot, which indicates the frequencies.
nx = 500;
ny = 500;
for i = 1:nx
    for j=1:ny
    u(i,j) = 1 + ...
                2^0*(1+sin(2*pi*((i-1)/nx)*200))+...
                2^2.*(1+sin(2*pi*((j-1)/ny)*200))+...
                2^4.*(1+cos(2*pi*((i-1)/nx+(j-1)/ny)*141))+...
                2^6.*(1+sin(2*pi*((i-1)/nx-(j-1)/ny)*141));
    end
end
fftu = fft2(u,2*nx-1,2*ny-1);
fftu = fftshift(fftu);
mesh(log(1+abs(fftu)));
```

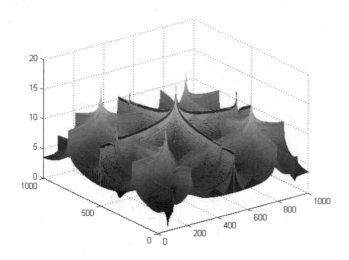

Figure 9.3.1: 2D DFT of Sine and Cosine

9.3.3 Convolution for Matrices

The convolution identity also generalizes to the 2D DFT. Let a be an $n_x \times n_y$ matrix with components $a(j_x, j_y)$ where the indices start at zero. The polynomial of two variables x and y associated with this matrix is defined to be

$$p_a(x, y) \equiv \sum_{j_y=0}^{n_y-1} \sum_{j_x=0}^{n_x-1} a(j_x, j_y) x^{j_x} y^{j_y}.$$

The *convolution* of two $n_x \times n_y$ matrices, $conv2(a, b)$, is a $(2n_x - 1) \times (2n_y - 1)$ matrix such that

$$p_{conv2(a,b)}(x, y) = p_a(x, y) \, p_b(x, y).$$

For example, consider $nx = 2$ and $ny = 3$ where we have again used MATLAB

```
>> a = [1 2 5; 6 7 11]
      1      2      5
      6      7     11
>> b = [2 5 9; 1 2 8]
      2      5      9
      1      2      8
>> conv2(a,b)
      2      9     29     43     45
     13     48    128    144    139
      6     19     73     78     88.
```

$$p_a(x, y) = 1 + 2y + 5y^2 + 6x + 7xy + 11xy^2$$
$$p_b(x, y) = 2 + 5y + 9y^2 + 1x + 2xy + 8xy^2$$

The coefficient in $p_{conv2(a,b)}(x, y)$ of x^2y^4 is
$$88 = 11 * 8 = a(1, 2) * b(1, 2).$$
The coefficient in $p_{conv2(a,b)}(x, y)$ of x^2y^3 is
$$78 = 7 * 8 + 11 * 2 = a(1, 1) * b(1, 2) + a(1, 2) * b(1, 1).$$

The convolution identity in 2D also requires the matrices to be "padded" by zeros in both the rows and columns. The justification of this identity is similar to that given for the 1D DFT. Here we have used the MATLAB notation, $fft2()$, for the fast Fourier transform in 2D.

Convolution Property for 2D DFT. Let a and b be $n_x \times n_y$ image matrices.

$$fft2(conv2(a, b)) = fft2(a, 2*n_x - 1, 2*n_y - 1). * fft2(b, 2*n_x - 1, 2*n_y - 1).$$

The following MATLAB computation verifies this for the above 2×3 matrices with $2 * n_x - 1 = 3$ and $2 * n_y - 1 = 5$:

```
>> fft2(conv2(a,b))
      1.0e+002 *
      Columns 1 through 3
         8.6400              -2.7193 + 2.0698i      -1.0757 + 0.8192i
        -2.4000 - 1.8013i     1.0034 + 0.1752i       0.2677 - 0.1158i
        -2.4000 + 1.8013i     0.5290 - 0.9709i       0.2249 - 0.4680i
      Columns 4 and 5
        -1.0757 - 0.8192i     -2.7193 - 2.0698i
         0.2249 + 0.4680i      0.5290 + 0.9709i
         0.2677 + 0.1158i      1.0034 - 0.1752i
>> norm(fft2(conv2(a,b)) - fft2(a,3,5).*fft2(b,3,5))
      4.8969e-014.
```

9.3.4 Application to Image Filtering

The convolution identity can be restated by taking the inverse DFT of both sides

$$conv2(a, b) = ifft2(fft2(a, 2*n_x - 1, 2*n_y - 1). * fft2(b, 2*n_x - 1, 2*n_y - 1)).$$

If b is an image, then the DFT of b will reflect the frequencies of the periodic parts of the image. By masking or filtering out the unwanted frequencies one can obtain a new image by applying the inverse DFT. A *filter* is a matrix with the same dimension as the DFT of the padded image. The components of the filter usually vary from 0 to 1. If the component is 1, then the frequency is allowed to pass; if the component is 0, then the frequency is deleted.

Definition 9.3.2. Let b be an $n_x \times n_y$ image matrix. Let *Filter* represent a $(2\,n_x - 1) \times (2\,n_y - 1)$ matrix whose components are in the unit interval. The filtered image is $Newb$

$$Newb = ifft2(Filter. * fft2(b, 2 * n_x - 1, 2 * n_y - 1)).$$

The following MATLAB code fftsine2d.m illustrates this for a low frequency sine wave with higher frequency sine "noise."

MATLAB code fftsine2d.m

```
% This code creates a noisy 2D sine wave.
% The 2D DFT is used to filter the unwanted
% higher frequencies.
clear;
nx = 200;
ny = 600;
for i = 1:nx       % Define the big wave with periodic noise.
     for j = 1:ny
          u(i,j) = 0+100*(1+sin(2*pi*((j-1)/ny)*5)) + ...
                   15.*(1+sin(2*pi*((i-1)/nx)*80))+...
                   15.*(1+sin(2*pi*((j-1)/ny)*80));
     end
end
sine = uint8(u);
imwrite(sine, 'sine.jpg');
fftu = fft2(u,2*nx-1,2*ny-1);      % Padded fft.
fftu = fftshift(fftu);
figure(1);
mesh(u');
figure(2)
mesh(log(1+(abs(fftu))));
filter = ones(2*nx-1,2*ny-1);
d0 = 150;                          % Use ideal low-pass filter.
for i = 1:2*nx-1
     for j =1:2*ny-1
          dist = ((i-nx)^2 + (j-ny)^2)^.5;
          if dist > d0
               filter(i,j) = 0;
          end
     end
end
figure(3)
mesh(filter);
fil_sine = filter.*fftu;
figure(4)
mesh(log(1+abs(fil_sine)));
```

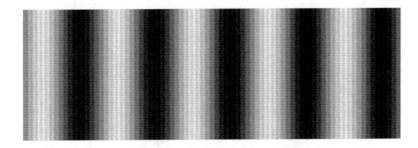

Figure 9.3.2: Noisy 2D Sine Wave

```
fil_sine = ifftshift(fil_sine);
fil_sine = ifft2(fil_sine);
fil_sine = real(fil_sine(1:nx,1:ny));
fil_sine = uint8(fil_sine);
imwrite(fil_sine, 'sine_fil.jpg');
```

Figure 9.3.2 is the noisy sine wave with unwanted higher frequencies. Figure 9.3.3 is the mesh plot of the real matrix associated with the noisy image. Figure 9.3.4 is the log plot of the magnitude of the shifted DFT for the noisy sine wave; note the spikes indicating the higher frequencies away from the center. Figure 9.3.5 is the low-pass filter used to eliminate the spikes in the previous figure. Figure 9.3.6 is the filtered DFT, and Figure 9.3.7 is the filtered image of the noisy sine wave.

9.3.5 Exercises

1.   Let $a = [1\ 3\ 5\ 6; 2\ 4\ 7\ 19]$ and $b = [2\ -3\ 4\ 7; 1\ 5\ 2\ 10]$.
    (a).   Find the fft2 of a and b and
    (b).   Verify ifft2(fft2(a)) = a.
2.   Use a variation of ffttrig2da.m to find
    (a).   fft2($\sin(2\pi(x + 3y))$) and
    (b).   fft2($\cos(2\pi(x + 3y))$).
3.   Consider a variation of ffttrig2d.m to find the DFT of

$$u = 1 + \cos(2\pi x200) + 4\cos(2\pi y200) + 16\sin(2\pi x200) + 64\cos(2\pi x200).$$

4.   Consider the vectors in exercise 1.
    (a).   Find $conv2(a, b)$ and
    (b).   Verify the convolution property for the 2D DFT.
5.   Use a variation of fftsine2d.m to find the 2D DFT of

$$u = 100(1 + \cos(2\pi y10)) + 30(1 + \cos(2\pi x120) + \sin(2\pi y120)).$$

6.   Use a variation of fftsine2d.m to find the 2D DFT of

$$u = 100(1 + \cos(2\pi y10)) + 30(1 + \cos(2\pi x120 + 2\pi y120)).$$

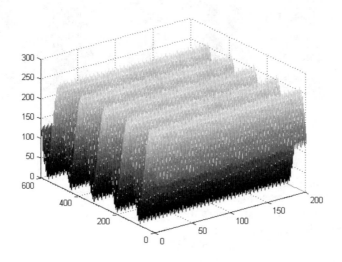

Figure 9.3.3: Mesh Plot of Noisy Sine Wave

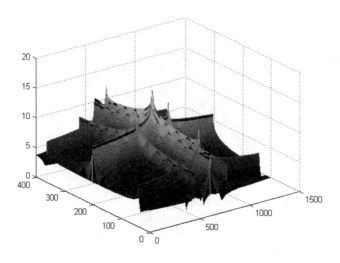

Figure 9.3.4: DFT of Noisy Sine Wave

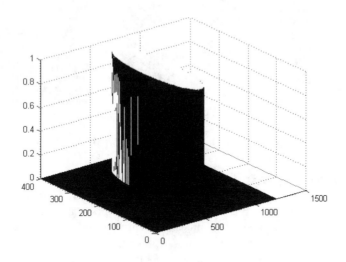

Figure 9.3.5: Low-pass Filter

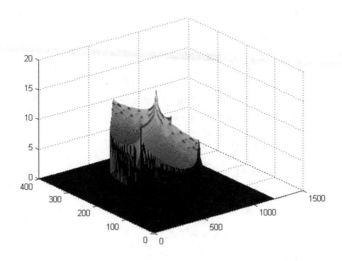

Figure 9.3.6: Filtered DFT of Sine Wave

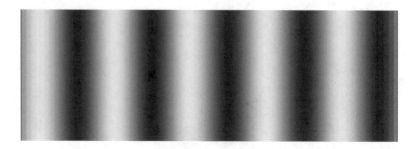

Figure 9.3.7:   Filtered Sine Wave

9.4   Frequency Filters in $\mathbb{R}^n \times \mathbb{R}^n$

The low-pass, high-pass and band-reject filters will be studied. These have various shapes including ideal (step functions), Butterworth (polynomial functions) and Gaussian (exponential functions). Applications to the NCSU image, a light and noisy aerial image and sharpening of micro chip image will be presented.

9.4.1   Classifications of Filters

A filter is a matrix with the same dimension as the 2D DFT of the padded image. The components of the filter usually vary from 0 to 1 and let *Filter* represent such a matrix. Then the filtered image is given by *Newb*

$$Newb = ifft2(Filter. * fft2(b, 2 * n_x - 1, 2 * n_y - 1)).$$

The MATLAB code fftsine2d.m in the previous section illustrated an ideal low-pass filter for a low frequency sine wave with higher frequency noise. A *high-pass* filter allows higher frequencies in the image to remain, and it can be viewed as a ones matrix minus a low-pass filter matrix. The *band-reject* filter allows higher and lower frequencies to pass and inhibits some middle range frequencies. *Ideal* filters have components that are either zeros and ones. The following are three examples of band-reject filters where the approximate frequencies to be rejected are near d_0 and we use $dist(i, j) \equiv \sqrt{(i - n_z)^2 + (j - n_y)^2}$.

Ideal Band-reject Filter.

$$Filter(i, j) = \begin{cases} 0, & \text{if } d_0 - w/2 \leq dist(i, j) \leq d_0 + w/2 \\ 1, & \text{otherwise.} \end{cases}$$

Butterworth Band-reject Filter.

$$Filter(i, j) = \frac{1}{1 + (dist(i, j)w/(dist(i, j)^2 - d_0^2)^{2n}}.$$

Gaussian Band-reject Filter.

$$Filter(i, j) = 1 - e^{\frac{-1}{2}((dist(i,j)^2 - d_0^2)/dist(i,j)w)^2}.$$

The parameters w and d_0 control the width and location of the band. In the Butterworth band-reject filter the exponent n controls the steepest of the boundaries in the band of frequencies to be rejected. There are similar versions of high and low-pass filters.

9.4.2 Ideal Low-pass for NCSU Image

The MATLAB code filter_ncsuid.m generates a noisy NCSU image, see Figure 9.4.1, and filters it via an ideal low-pass filter as indicated in Figure 9.4.2. The filtered image is in Figure 9.4.3 and has blurred edges on the letters.

MATLAB code filter_ncsuid.m

```
% This code uses an ideal low-pass filter to
% eliminate unwanted frequencies from a noisy
% NCSU image.
clear;
ncsu = [bign(5) bigc(7) bigs(9) bigu(11)];
newncsu = 20*ncsu;
[nx ny] = size(newncsu);
nx
ny
newncsu = newncsu(nx:-1:1,:);
newncsu1 = uint8(newncsu);
imwrite(newncsu1, 'ncsu.jpg');
u = newncsu;
for i = 1:nx              % This is NCSU with periodic noise.
      for j = 1:ny
            u(i,j) = u(i,j) + ...
                      15.*(1+sin(2*pi*((i-1)/nx)*80))+...
                      15.*(1+sin(2*pi*((j-1)/ny)*80));
      end
end
sinencsu = uint8(u);
imwrite(sinencsu, 'sinencsu.jpg');
fftu = fft2(u,2*nx-1,2*ny-1);
fftu = fftshift(fftu);
subplot(2,2,1);
mesh(u');
subplot(2,2,2);
mesh(log(1+(abs(fftu))));
filter = ones(2*nx-1,2*ny-1);
% Use ideal low-pass filter.
d0 = 150;
for i = 1:2*nx-1
      for j =1:2*ny-1
            dist = ((i-nx)^2 + (j-ny)^2)^.5;
```

Figure 9.4.1: Noisy NCSU Image

```
            if dist > d0
                  filter(i,j) = 0;
            end
        end
end
subplot(2,2,3);
mesh(filter);
fil_ncsu = filter.*fftu;
subplot(2,2,4);
mesh(log(1+abs(fil_ncsu)));
fil_ncsu = ifftshift(fil_ncsu);
fil_ncsu = ifft2(fil_ncsu,2*nx-1,2*ny-1);
fil_ncsu = real(fil_ncsu(1:nx,1:ny));
fil_ncsu = uint8(fil_ncsu);
imwrite(fil_ncsu, 'sinencsuid_fil.jpg');
```

9.4.3   Butterworth Band-reject for NCSU Image

The ideal low-pass filtered image in Figure 9.4.3 is not entirely satisfactory. Another approach is to use a band-reject filter where one may need to experiment with the width, w, the "slope", n, and the location, d_0, of the band. The MATLAB code filter_ncsubu.m uses the Butterworth band-reject filter and may be applied to either the noisy big sine wave or the noisy NCSU. This code generates a noisy NCSU image, see Figure 9.4.1, and filters it via the Butterworth band-reject filter as indicated in Figure 9.4.4. The filtered image is in Figure 9.4.5, and one should compare it with the ideal low-pass filtered image in Figure 9.4.3. The code is similar to the above, but the filter definition is different as indicated below:

```
% Use Butterworth band-reject filter.
d0 = 160;
n = 2;
w = 20;
```

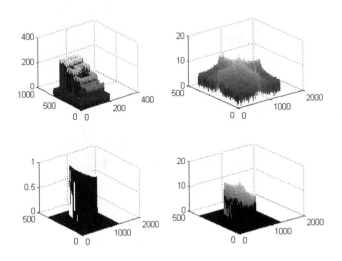

Figure 9.4.2: Low-pass Filtering of NCSU

Figure 9.4.3: Ideal Low-pass NCSU

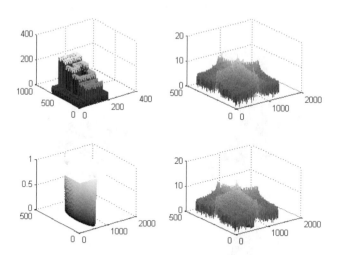

Figure 9.4.4: Band-reject Filtering of NCSU

Figure 9.4.5: Band-reject Filtered NCSU

```
for i = 1:2*nx-1
    for j =1:2*ny-1
        dist = ((i-nx)^2 + (j-ny)^2)^.5;
        if dist ~= d0
            filter(i,j)= 1/(1 + (dist*w/(dist^2 - d0^2))^(2*n));
        else
            filter(i,j) = 0;
        end
    end
end
```

9.4.4   Power/Butterworth Band-reject for Aerial Image

Another application of the band-reject filter is to the noisy aerial photograph. This image suffers from too much exposure and from banded sine and cosine noise. The light exposure is modified by use of the power transformation with the power equal to two, and then the Butterworth band-reject filter is used to reduce the noise. The MATLAB code filter_aerial.m starts with a light and noisy aerial image, see Figure 9.4.6. The image is enhanced by the power method and the Butterworth band-reject filter as is indicated in Figure 9.4.7. The filtered image is in Figure 9.4.8. The filter portion of the code is as follows:

```
c = 1.;      % Use the power transformation to darken.
gamma = 2;
f_fp =255*c*(u/255).^gamma;
u = f_fp;
fftu = fft2(u,2*nx-1,2*ny-1);
fftu = fftshift(fftu);
subplot(2,2,2)
mesh(log(1+(abs(fftu))));
filter = ones(2*nx-1,2*ny-1);
d0 = 400;      % Use Butterworth band-reject filter.
n = 2;
w = 20;
for i = 1:2*nx-1
      for j =1:2*ny-1
            dist = ((i-nx)^2 + (j-ny)^2)^.5;
            if dist ~= d0
                  filter(i,j)= 1/(1 + (dist*w/(dist^2 - d0^2))^(2*n));
            else
                  filter(i,j) = 0;
            end
      end
end
```

9.4.5   High-pass/Sharpened Micro Chip Image

This illustrates a high-pass filter to give emphasis to the higher frequencies in the image of a damaged micro chip. The micro chip image in Figure 9.4.9 is provided courtesy of Brockhouse Institute for Materials Research, McMaster University. The high-pass image is then added to the original image so as to obtain a sharper image. The reader may find it interesting to experiment with width and frequency threshold of the Butterworth or the Gaussian high-pass filters. The MATLAB code filter_micro.m uses a high-pass Gaussian filter to sharpen an image of a micro chip, see Figure 9.4.9. The high frequencies are isolated by a Gaussian filter, see Figure 9.4.10, and then these frequencies are

Figure 9.4.6: Light and Noisy Aerial Image

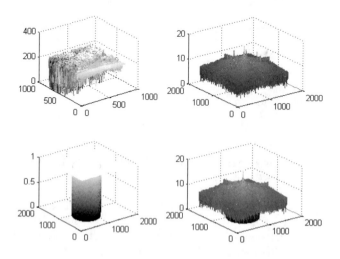

Figure 9.4.7: Filtering Aerial Image

Figure 9.4.8: Filtered Aerial Image

added to the DFT of the padded image. The sharpened image is in Figure 9.4.11. The filter portion of the code is indicated below:

```
% Use Butterworth or Gaussian high-pass filter.
filter = ones(2*nx-1,2*ny-1);
d0 = 100;
n = 4;
for i = 1:2*nx-1
      for j =1:2*ny-1
            dist = ((i-nx)^2 + (j-ny)^2)^.5;
            % Use Butterworth high-pass filter.
            % filter(i,j)= 1/(1 + (dist/d0)^(2*n));
            % filter(i,j)= 1.0 - filter(i,j);
            % Use Gaussian high-pass filter.
            filter(i,j) = exp(-dist^2/(2*d0^2));
            filter(i,j) = 1.0 - filter(i,j);
      end
end
   :
% Update image with high frequencies.
fil_micro = fftu + filter.*fftu;
```

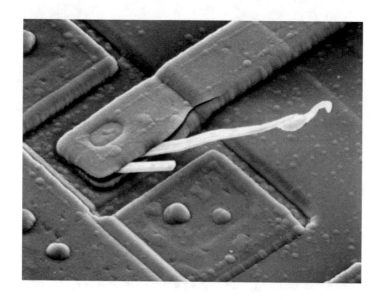

Figure 9.4.9:   Micro Chip Image

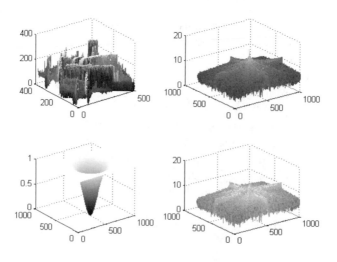

Figure 9.4.10:   Sharpening of Micro Chip Image

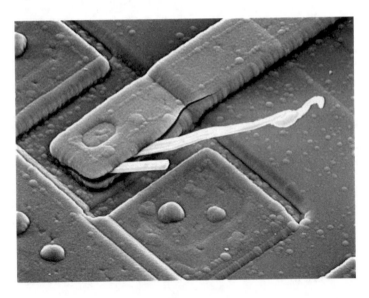

Figure 9.4.11:   Sharpened Micro Chip Image

9.4.6   Exercises

1.   Experiment with the parameter d_0 in the ideal low-pass filter in filter_ncsuid.m.

2.   Modify the filter in filter_ncsuid.m to become an ideal band-reject filter. Experiment with the location and width of the band of frequencies to be rejected.

3.   Experiment with the parameters n, w and d_0 for the Butterworth band-reject filter in filter_ncsubu.m.

4.   Modify the filter in filter_ncsubu.m to become a Gaussian band-reject filter and experiment with its parameters.

5.   Experiment with the parameters n, w and d_0 for the Butterworth band-reject filter in filter_aerail.m.

6.   Modify the filter in filter_aerial.m to become a Gaussian band-reject filter and experiment with its parameters.

7.   Experiment with the parameters in the high-pass Gaussian filter used filter_micro.m. Also, modify the sharpening step to include other linear combinations:

> % Update image with high frequencies.
> fil_micro = a*fftu + b*filter.*fftu;

8.   Repeat exercise 7 using the high-pass Butterworth filter.

Appendix A

Solutions to Odd Exercises

Chapter One
Section 1.1
1. (a). 0.333 (b). 0.111; $error = -0.000111$
3. (a). $2w + 3z = 25 + 3i$
  (b). $zw = 17 + 19i$; $z/w = (11 - 23i)/13$
5. $z^2 = -1/2 + (-\sqrt{3}/2)i$ and $1 + z + z^2 = 0$
7. $z + 0 = (3 + 4i) - (7 + 2i) = (3 - 7) + (4 - 2)i = -4 + 2i$
9. $-1/58 + (17/58)i$
11. $9/25 + (-12/25)i$
13. Prove $|z + w|^2 \le (|z| + |w|)^2$: Use the dot product as in Section 1.4.

Chapter One
Section 1.2
1. (a). $u = 2x - y + 7$ and $v = x + 2y + 1$
  (b). $z = 3/5 + (-9/5)i$
3. (a).

$$u = \frac{x^2 + y^2 + y - 2}{x^2 + (y - 1)^2} \text{ and } v = \frac{3x}{x^2 + (y - 1)^2}$$

  (b). $z = -3/2 + i$
5. $\sqrt{1 + i} = 1.0987 + 0.4551i$
7. (a). $z = -1 \pm (1.0987 + 0.4551i)$
9. $(1 + i)^{10} = (\sqrt{2})^{10} e^{i(2\pi/8)10} = 32i$
11. $3 + 4i = 5e^{i\theta}$ where $\theta = \arctan(4/3)$
  Solutions of $z^3 = 3 + 4i$ are
   $5^{1/3} e^{i\theta/3} (e^{i2\pi/3})^0 = 1.6289 + 0.5202i$
   $5^{1/3} e^{i\theta/3} (e^{i2\pi/3})^1 = -1.2650 + 1.1506i$ and
   $5^{1/3} e^{i\theta/3} (e^{i2\pi/3})^2 = -0.3640 - 1.6708i$

13. Prove $\overline{w} = w^{-1}$ where $w \equiv e^{i2\pi/n}$:

$$\begin{aligned}
\overline{w} &= \overline{e^{i2\pi/n}} \\
&= \overline{\cos(2\pi/n) + i\sin(2\pi/n)} \\
&= \cos(2\pi/n) - i\sin(2\pi/n)
\end{aligned}$$

$$\begin{aligned}
w^{-1} &= \frac{1}{\cos(2\pi/n) + i\sin(2\pi/n)} \\
&= \frac{1}{\cos(2\pi/n) + i\sin(2\pi/n)} \frac{\cos(2\pi/n) - i\sin(2\pi/n)}{\cos(2\pi/n) - i\sin(2\pi/n)} \\
&= \frac{\cos(2\pi/n) - i\sin(2\pi/n)}{1}
\end{aligned}$$

Chapter One
Section 1.3

1. $\vec{a} = 2\vec{i} + 5\vec{j}$ and $\mathbf{a} = \begin{bmatrix} 2 \\ 5 \end{bmatrix}$

3. $\vec{a} - \vec{b} = \begin{bmatrix} 3 & 2 \end{bmatrix}$

5. $\begin{bmatrix} x_1 & x_2 \end{bmatrix} = \begin{bmatrix} -1 & 3 \end{bmatrix} - \begin{bmatrix} 2 & 5 \end{bmatrix} = \begin{bmatrix} -3 & -2 \end{bmatrix}$

7. Hint: $\vec{a} = \begin{bmatrix} a_1 & a_2 \end{bmatrix}$, $\vec{b} = \begin{bmatrix} b_1 & b_2 \end{bmatrix}$ and $\vec{c} = \begin{bmatrix} c_1 & c_2 \end{bmatrix}$ and compute both sides.

9. $\|\vec{a}\| = \sqrt{1 + 49}$, $\arctan(7/1) \approx 81.87$ degrees and $\vec{a}/\|\vec{a}\| = \begin{bmatrix} 1/\sqrt{50} & 7\sqrt{50} \end{bmatrix}$

11. $13 = 29 + 10 - 2\sqrt{29}\sqrt{10}\cos(\theta)$. Then $\cos(\theta) = 13/\sqrt{290}$ and $\theta \approx 40.24$ degrees.

13. $\vec{h} \approx \begin{bmatrix} 353.55 & 473.55 \end{bmatrix}$

Chapter One
Section 1.4

1. The angle between the vectors is 30 degrees, which is also given by

$$1\sqrt{3} + \sqrt{3}(1) = 2(2)\cos(\theta).$$

3. $1(-3) + 2(1) = \sqrt{5}\sqrt{10}\cos(\theta)$ giving $\theta \approx 98.13$ degrees.

5. $\begin{bmatrix} 3 & 1 \end{bmatrix} \bullet \begin{bmatrix} x & -1 \end{bmatrix} = 3x - 1 = 0$ gives $x = 1/3$.

7. (a). Hint: $\vec{a} = \begin{bmatrix} a_1 & a_2 \end{bmatrix}$, $\vec{b} = \begin{bmatrix} b_1 & b_2 \end{bmatrix}$ and compute both sides.
   (b). Both sides are equal to 11.
   (c). Proof requires equality for all possible vectors.

9. $area^2 = (\det(\begin{bmatrix} 1 & -3 \\ 2 & 1 \end{bmatrix}))^2 = 7$ so that the $area$ is $\sqrt{7}$.

11. $\det(\begin{bmatrix} 2 & 1 \\ 1 & -1 \end{bmatrix}) = -3$ and $\det(\begin{bmatrix} 12 & 3 \\ 6 & -3 \end{bmatrix}) = -54$.

13. $work = \vec{f} \bullet \vec{d} = 3(30) + 1(6) = 96$.

Chapter One
Section 1.5
1.   (a). $y = (1/3)x - 3$ (b). $x = 3t$ and $y = t - 3$
3.   (a). $[x \ y] = [2 \ 7] + t[1 \ -1]$ (b). $x = 2 + t$ and $y = 7 - t$
5. $[0 \ 1] + (-3/5)[1 \ -2] = [-3/5 \ 11/5]$

Chapter Two
Section 2.1
3. $\vec{a} = 3\vec{i} + \vec{j} + 2\vec{k}$ or $\mathbf{a} = \begin{bmatrix} 3 \\ 1 \\ 2 \end{bmatrix}$

5.   (a). $2\vec{a} + \vec{b} = [7 \ 5 \ 5]$ (b). $\vec{x} = [-5 \ 1 \ -3]$
7.   (a). $\|\vec{a}\| = \sqrt{21}$ and $\|\vec{b}\| = \sqrt{11}$ (b). $\vec{a} \bullet \vec{b} = -3$
9. $1(1) + 2(3) - 1(-1) = \sqrt{6}\sqrt{11}\cos(\theta)$ giving $\theta \approx 10.02$ degrees
11.   (a).   Hint: $\vec{a} = [a_1 \ a_2 \ a_3]$ and $\vec{b} = [b_1 \ b_2 \ b_3]$ and compute both sides.
      (b).   Both sides are equal to 26.
      (c).   Proof requires equality for all possible vectors.
13. $work = \vec{f} \bullet \vec{d} = 10(25) + 1(20) + 2(5) = 280$

Chapter Two
Section 2.2
1. $\vec{a} \times \vec{b} = 3\vec{i} + \vec{j} - 9\vec{k}$ and $\|\vec{a} \times \vec{b}\| = \sqrt{91}$

3. $\sqrt{46}$
5.   Hint: $\vec{a} = [a_1 \ a_2 \ a_3]$ and $\vec{b} = [b_1 \ b_2 \ b_3]$ and compute both cross products.
7. $\vec{r} \times \vec{f} = [0 \ 1 \ -3]$
9.   (a). $(\vec{a} \times \vec{b}) \bullet \vec{c} = [-4 \ -3 \ 19] \bullet [2 \ 1 \ 5] = 84$
      (b).   84
11.   Hint: expand the determinants.

Chapter Two
Section 2.3
1. $[1 \ 2 \ -1] + t[3 \ 1 \ 1]$
3.   (a). $[1 \ 1 \ 2] + (7/3)[-1 \ 1 \ 1]$
      (b). $\sqrt{168/3} = \sqrt{3(56)}/3$
      (c). $\sqrt{56/3}$
5. $\sqrt{2}$
7.   (a).

$$\begin{bmatrix} 5 & -3 \\ 3 & -3 \end{bmatrix}\begin{bmatrix} t \\ \tau \end{bmatrix} = \begin{bmatrix} -4 \\ -3 \end{bmatrix}$$

      (b). $t = -1/2$ and $\tau = 1/2$

(c). $\sqrt{3/2}$

Chapter Two
Section 2.4

1.   (a). $1(2) + 2(1) + 4(2) = 12$

     (b). $[1 \; 2 \; 4]$

     (c). $[1 \; 2 \; 4] \bullet [x - 2 \; \; y - 1 \; \; z - 2] = 0$

3.   (a). $[-1 \; -2 \; -3]$

     (b). $[-1 \; -2 \; -3] \bullet [x - 1 \; \; y - 1 \; \; z - 1] = 0$

         $-x - 2y - 3y = -6$

5.   (a).

$$
\begin{aligned}
2 &= 1 + 2r + s \\
3 &= 1 + 3r + 0s \\
1 &= 1 + 0r + 0s
\end{aligned}
$$

     (b). $r = -2/(-3)$ and $s = 1/(-3)$

7. $[1 \; 1 \; 1] + (17/19) [2 \; 3 \; 1] + (60/19) [1 \; 0 \; 1] =$

         $= [113/19 \; 70/19 \; 96/19]$

9.   (a). $m = -45.00$ and $c = 1046.67$

     (b). $m4 + c = 866.67$

Chapter Two
Section 2.5

1.   (a). $\mathbf{s}^T = [5 \; 6 \; 2 \; 1 \; 5]$ and $\mathbf{p}^T = [1000 \; 850 \; 1700 \; 2220 \; 1150]$

     (b). $\mathbf{s}^T\mathbf{p} = 21470$

3.   (a).

$$
\begin{bmatrix} 1 & 2 & 4 \\ 0 & 1 & 1 \\ 1 & 1 & 4 \end{bmatrix}
\begin{bmatrix} r \\ s \\ t \end{bmatrix}
=
\begin{bmatrix} 10 \\ 1 \\ 4 \end{bmatrix}
$$

     (b). $r = 18/1$, $s = 6/1$ and $t = -5/1$

5.   (a). $[-8 \; -5 \; 7]$

     (b). $[18/7 \; 24/7 \; 0] + t[-8 \; -5 \; 7]$

Chapter Three
Section 3.1

1.   (a).   [8 19 17 23 7]

    (b).

$$\begin{bmatrix} 2 & 0 & 1 \\ 5 & 1 & 4 \\ 7 & 2 & 2 \\ 1 & 7 & 1 \\ -1 & -4 & -3 \end{bmatrix}$$

    (c).   30

    (d). $\sqrt{80}$

3.   (a).

$$\begin{bmatrix} 1 & 27 & 0 \\ 216 & 8 & 1 \\ 27 & -1 & 64 \\ 125 & 27 & -8 \end{bmatrix}$$

    (b).   [1/3 6/2 3/(−1) 5/3]

    (c).

$$\begin{bmatrix} \sin(1) & \sin(3) & \sin(0) \\ \sin(6) & \sin(2) & \sin(1) \\ \sin(3) & \sin(-1) & \sin(4) \\ \sin(5) & \sin(3) & \sin(-2) \end{bmatrix}$$

5.   (a).

$$F = \begin{bmatrix} 397 & 394 & 389 \\ 391 & 388 & 383 \end{bmatrix}$$

    (b).   [X Y] = meshgrid(1:1:3,1:1:3)
        F = 400 - 20*X.^2 - Y.^2

7.   (a). $C(x,y) = 4xy + (2x + 2y)200/(xy)$

    (b).   About $C(8.1, 8.1)$.

9.

θ	f_1	f_2
$\pi/12$	386.	-373.
$2\pi/12$	200.	-173.
$3\pi/12$	141.	-100.
$4\pi/12$	115	-57.

11. $i_1 = 0$, $i_2 = -5$ and $i_3 = -5$

Chapter Three
Section 3.2

1. $A\mathbf{x} = [35\ 37]^T$ and $A + C = \begin{bmatrix} 2 & -1 & 6 & 8 \\ 6 & -2 & 14 & 10 \end{bmatrix}$

3.   Both sides should equal $\begin{bmatrix} 2 & 21 & 2 & 12 \\ 7 & -3 & 17 & 17 \end{bmatrix}$.

5. $\begin{bmatrix} 23 & 26 & 37 & 7 \\ 1 & 0 & 9 & -7 \end{bmatrix}$

7. Both sides should equal $\begin{bmatrix} -33 & -55 \\ 37 & 135 \end{bmatrix}$.

9. $\alpha = 32/100$,
$$\mathbf{u}^1 = [64.00 \;\; 00.00 \;\; 22.40]^T,$$
$$\mathbf{u}^2 = [87.04 \;\; 27.64 \;\; 30.46]^T$$
$$\mathbf{u} = [167.50 \;\; 135.00 \;\; 102.50]^T$$

Chapter Three
Section 3.3

1. $\mathbf{x} = [2 \;\; 4 \;\; 3]^T$

3. $\mathbf{x} = [-50 \;\; 2 \;\; 5 \;\; 1]^T$

5. $\mathbf{y} = [210 \;\; 115 \;\; 156.67]^T$ and $\mathbf{x} = \mathbf{u} = [182.5 \;\; 155.0 \;\; 117.5]^T$

7. Use forward and backward substitution.

Chapter Three
Section 3.4

1. $\begin{bmatrix} 2 & -1 & 1 & 7 \\ -4 & 1 & 1 & 0 \\ 0 & 3 & 1 & -2 \end{bmatrix}$

3. $E_{43}(-3) = \begin{bmatrix} 1 & 0 & 0 & 0 \\ 0 & 1 & 0 & 0 \\ 0 & 0 & 1 & 0 \\ 0 & 0 & -7 & 1 \end{bmatrix}$, $P_{23} = \begin{bmatrix} 1 & 0 & 0 & 0 \\ 0 & 0 & 1 & 0 \\ 0 & 1 & 0 & 0 \\ 0 & 0 & 0 & 1 \end{bmatrix}$, and

$$E_4(10) = \begin{bmatrix} 1 & 0 & 0 & 0 \\ 0 & 1 & 0 & 0 \\ 0 & 0 & 1 & 0 \\ 0 & 0 & 0 & 10 \end{bmatrix}.$$

5. (a). $[A \;\; \mathbf{d}] = \begin{bmatrix} 1 & 7 & 1 & 1 \\ 2 & 0 & 7 & 2 \\ 0 & 14 & 1 & 6 \end{bmatrix}$

   (b). $E_{32}(-2) \, E_{21}(1) \, [A \;\; \mathbf{d}] = \begin{bmatrix} 1 & 7 & 1 & 1 \\ 0 & -14 & 5 & 0 \\ 0 & 0 & 6 & 6 \end{bmatrix} = [U \;\; \widehat{\mathbf{d}}]$

   (c). The solution of the equivalent system $U\mathbf{x} = \widehat{\mathbf{d}}$ is
   $\mathbf{x} = [-5/2 \;\; 5/14 \;\; 1]^T$.

7. (a). $[A \;\; \mathbf{d}] = \begin{bmatrix} 1 & -1 & 0 & 1 & 6 \\ 1 & 0 & -1 & -1 & 4 \\ 0 & 2 & 7 & 0 & 4 \\ 2 & 0 & 0 & -1 & 8 \end{bmatrix}$

   (b). $E_{43}(-2/9) \, E_{42}(-2) \, E_{32}(-2) \, E_{41}(-2) \, E_{21}(-1) \, [A \;\; \mathbf{d}] =$

$$\begin{bmatrix} 1 & -1 & 0 & 1 & 6 \\ 0 & 1 & -1 & -2 & 1 \\ 0 & 0 & 9 & 4 & 2 \\ 0 & 0 & 0 & 1/9 & -58/9 \end{bmatrix} = [U \ \widehat{\mathbf{d}}]$$

(c). The solution of the equivalent system $U\mathbf{x} = \widehat{\mathbf{d}}$ is
$\mathbf{x} = [-25 \ -89 \ 26 \ -58]^T$.

9. (a). $[A \ \mathbf{d}] = \begin{bmatrix} 1 & -1 & 1 & 0 \\ 1 & 0 & -4 & 20 \\ 0 & -2 & -4 & 10 \end{bmatrix}$

(b). $E_{32}(2) \, E_{21}(-1) \, [A \ \mathbf{d}] = \begin{bmatrix} 1 & -1 & 1 & 0 \\ 0 & 1 & -5 & 20 \\ 0 & 0 & -14 & 50 \end{bmatrix} = [U \ \widehat{\mathbf{d}}]$

(c). The solution of the equivalent system $U\mathbf{x} = \widehat{\mathbf{d}}$ is
$\mathbf{x} = \mathbf{i} = [40/7 \ 15/7 \ -25/7]^T$.

11. Hint: Do the elementary matrix products.

Chapter Three
Section 3.5

1. (a). $A^{-1} = \begin{bmatrix} 3/10 & 1/10 \\ -1//10 & 3/10 \end{bmatrix}$

(b). $E_{12}(1/3) \, E_2(3/10) \, E_1(1/3) \, E_{21}(-1/3) \, [A \ I] = [I \ A^{-1}]$

(c). $\begin{bmatrix} 3 & -1 & 0 & 0 \\ 1 & 3 & 0 & 0 \\ 0 & 0 & 3 & 0 \\ 0 & 0 & 0 & 4 \end{bmatrix}^{-1} = \begin{bmatrix} 3/10 & 1/10 & 0 & 0 \\ -1//10 & 3/10 & 0 & 0 \\ 0 & 0 & 1/3 & 0 \\ 0 & 0 & 0 & 1/4 \end{bmatrix}$

3. (a). $\begin{bmatrix} 3 & -1 & 0 \\ -1 & 3 & -1 \\ 0 & -1 & 3 \end{bmatrix}^{-1} = (1/21) \begin{bmatrix} 8 & 3 & 1 \\ 3 & 9 & 3 \\ 1 & 3 & 8 \end{bmatrix}$

(b). $(1/121) \begin{bmatrix} 8 & 3 & 1 \\ 3 & 9 & 3 \\ 1 & 3 & 8 \end{bmatrix} \begin{bmatrix} 1 \\ 2 \\ 3 \end{bmatrix} = \begin{bmatrix} 17/21 \\ 30/21 \\ 31/21 \end{bmatrix}$

5. Hint: Do the row operations.

7. (a). $A_1^{-1} = \begin{bmatrix} 1 & 0 & 0 \\ 0 & 1 & 0 \\ 4 & 0 & 1 \end{bmatrix}$ and $A_2^{-1} = \begin{bmatrix} 1 & -2 & 0 \\ 0 & 1 & 0 \\ 0 & 0 & 1 \end{bmatrix}$

(b). $(A_1 A_2)^{-1} = A_2^{-1} A_1^{-1} = \begin{bmatrix} 1 & -2 & 0 \\ 0 & 1 & 0 \\ 4 & 0 & 1 \end{bmatrix}$

$(A_2 A_1)^{-1} = A_1^{-1} A_2^{-1} = \begin{bmatrix} 1 & -2 & 0 \\ 0 & 1 & 0 \\ 4 & -8 & 1 \end{bmatrix}$

9. See text form MATLAB commands.

11.   (a).   node 1: $[-f_1 \; 0] + [0 \; -f_2] + [f_7 \; 0] + [f_{3}c \; f_{3}s] = [0 \; 0]$
              node 2: $[-f_{3}c \; -f_{3}s] + [0 \; -f_8] + [f_{4}c \; -f_{4}s] = [0 \; 0]$
              node 3: $[0 \; -f_5] + [-f_6 \; 0] + [-f_{4}c \; f_{4}s] = [0 \; 0]$
              node 4: $[f_6 \; 0] + [0 \; f_8] + [-f_7 \; 0] + [0 \; -w] = [0 \; 0]$

Chapter Three
Section 3.6

1.   (a).
$$\begin{bmatrix} 1 & 0 & 0 & 0 \\ -1/2 & 1 & 0 & 0 \\ 0 & -2/3 & 1 & 0 \\ 0 & 0 & -3/4 & 1 \end{bmatrix} \begin{bmatrix} 2 & -1 & 0 & 0 \\ 0 & 3/2 & -1 & 0 \\ 0 & 0 & 4/3 & -1 \\ 0 & 0 & 0 & 5/4 \end{bmatrix}$$

      (b). $\mathbf{y} = [470 \; 625 \; 823\frac{1}{3} \; 1087\frac{1}{2}]^T$ and $\mathbf{x} = [870 \; 1270 \; 1270 \; 870]^T$

3.   (a). $P_{13}A = \begin{bmatrix} 1 & 0 & 0 \\ 1/2 & 1 & 0 \\ 0 & 2 & 1 \end{bmatrix} \begin{bmatrix} 2 & 1 & 1 \\ 0 & 5/2 & 1/2 \\ 0 & 0 & 5 \end{bmatrix}$ and

              $P_{13}\mathbf{d} = [3 \; 2 \; 6]^T$

      (b). $\mathbf{y} = [3 \; 1/2 \; 5]^T$ and $\mathbf{x} = [1 \; 9 \; 1]^T$

5.   (a). $\widehat{A}_{22} = \begin{bmatrix} 4/3 & -1 \\ -1 & 2 \end{bmatrix}$

      (b). $\widehat{A}_{22}\mathbf{x}_2 = \widehat{\mathbf{d}}_2 = [823\frac{1}{3} \; 470]^T$
              $\mathbf{x}_2 = [1270 \; 870]^T$ and $\mathbf{x}_1 = [470 \; 1270]^T$

7-11.   See circuit3.m

Chapter Three
Section 3.7

1.   (a). $\det(A^T) = 1(-6) - (3+5) = -11$
      (b). $\det(P_{32} \; A) = 6 + 3 + 2 = 11$
3.   (a). $\det(A) = -8 - 3$
      (b). $\det(A) = -3 - 8$
5.   (a). $\det(A) = 77 - 26 = 21$
      (b). $\det(A) = \det(E_{43}(8/3) \; E_{41}(-1) \; E_{31}(-8) \; A)$
              $= 1(1)(3)(7) = 21$

7. $\begin{bmatrix} b_{12} \\ b_{22} \\ b_{32} \end{bmatrix} = \begin{bmatrix} (-1)^{2+1}M_{21}/(-11) \\ (-1)^{2+2}M_{22}/(-11) \\ (-1)^{2+3}M_{23}/(-11) \end{bmatrix} = \begin{bmatrix} 5/11 \\ 3/11 \\ -2/11 \end{bmatrix}$

11.   (a). $x' = (2 + 4(y/24)) - 10(x/24)$
              $y' = (3 + 10(x/24) + 2(z/24)) - (4(y/24) + 8(y/24))$
              $z' = 4 + 8(y/24) - (2(z/24) + 6(z/24)).$

      (b). $A \begin{bmatrix} x \\ y \\ z \end{bmatrix} = \begin{bmatrix} -2 \\ -3 \\ -4 \end{bmatrix}$

      (c). $\begin{bmatrix} x \\ y \\ z \end{bmatrix} = \begin{bmatrix} 14.1 \\ 24.0 \\ 36.0 \end{bmatrix}$

Chapter Four
Section 4.1

1.   (a). $F(m, c) = 14m^2 + 12mc + 3c^2 - 56m - 22c + 59$
    (b). $m = 18/6$ and $c = -14/7$
    (c). $y = 38/3$
3.   (a). $F(m, c) = 204m^2 + 60mc + 8c^2 - 7976m - 1634c + 85771$
    (b). $m = 7.4167$ and $c = 68.75$
    (c). $y = 180.0$
5.   (a). $u(0) = 154.2983$ and $r = 0.04419$
    (b). $u(60) = 10.8893$

Chapter Four
Section 4.2

1.   (a). $A^T = \begin{bmatrix} 2 & 7 \\ 0 & -1 \\ 1 & 2 \end{bmatrix}$ and (b). $B^T A^T = \begin{bmatrix} 5 & 12 \\ 5 & 19 \\ 10 & 27 \end{bmatrix}$

3.   Hint: Compute the transpose of $AA^{-1} = I$.

5.   (a).

$$\begin{bmatrix} 30 & 10 \\ 10 & 4 \end{bmatrix} \begin{bmatrix} m \\ c \end{bmatrix} = \begin{bmatrix} 19130 \\ 7735 \end{bmatrix}$$

    (b). $A\mathbf{z} = \begin{bmatrix} 1 & 1 \\ 2 & 1 \\ 3 & 1 \\ 4 & 1 \end{bmatrix} \begin{bmatrix} z_1 \\ z_2 \end{bmatrix} = \begin{bmatrix} 0 \\ 0 \\ 0 \\ 0 \end{bmatrix}$ implies four equations must

    hold:

$$\begin{aligned} 1z_1 + 1z_2 &= 0, \\ 2z_1 + 1z_2 &= 0, \\ 3z_1 + 1z_2 &= 0 \text{ and} \\ 4z_1 + 1z_2 &= 0. \end{aligned}$$

    Thus, $z_2 = -z_1$ from the first equation. The other equations give $z_2 = z_1 = 0$.
    (c). $m = -41.5$ and $c = 2035.5$
    (d).   Predicted price at time $t = 8$ is $(-41.5)8 + 2037.5 = 1705.50$.
7.   See the MATLAB code pop.m.

Chapter Four
Section 4.3

1.   (a).   OK
    (b).   No. The data are not "near" a plane.
    (c). $m_1 = 4.0$, $m_2 = 1.0$, $c = 92$ and $\mathbf{r}^T\mathbf{r} = 36 \gg 0.25$.

3.   (a).

$$A = \begin{bmatrix} 2 & 3 & 3 & 1 \\ 3 & 4 & 4 & 1 \\ 5 & 4 & 4 & 1 \\ 8 & 5 & 5 & 1 \\ 12 & 6 & 6 & 1 \\ 15 & 8 & 8 & 1 \\ 19 & 10 & 10 & 1 \\ 23 & 11 & 11 & 1 \\ 27 & 13 & 13 & 1 \end{bmatrix} \quad \text{and } \mathbf{d} = \begin{bmatrix} 25000 \\ 24200 \\ 23900 \\ 23200 \\ 22600 \\ 22100 \\ 21500 \\ 20500 \\ 20100 \end{bmatrix}$$

(b).   A\d gives $m_1 = -227$, $m_2 = 177$, $m_3 = -94$ and $c = 24618$

(c).   17252

5.   (a).

$$A = \begin{bmatrix} 3 & 9 & 12 \\ 5 & 8 & 13 \\ 7 & 7 & 16 \\ 9 & 5 & 18 \\ 11 & 4 & 21 \\ 12 & 3 & 23 \\ 14 & 2 & 26 \\ 15 & 1 & 24 \end{bmatrix} \quad \text{and } \widehat{\mathbf{d}} = \begin{bmatrix} (8-10)/2 \\ (7-9)/2 \\ (5-8)/2 \\ (4-7)/2 \\ (3-5)/2 \\ (2-4)/2 \\ (1-3)/2 \\ (1-2)/2 \end{bmatrix}$$

(b).   A\d gives $a = -0.2753$, $b = -0.1946$, and $c = 0.1298$

Chapter Four

Section 4.4

1.   function Fnm = Fnm(p)
      Fnm = 100 + 2*(p(1)-2)^2 + 4*(p(2)-3)^2 + 10*(p(3)-4)^2;
      >> [p Fval] = fminsearch('Fnm', [10 10 10])

3.   Use rad_fmin.m to get
$$u(t) = 155.9841e^{-0.044993t} \text{ and } u(60) = 10.4834$$

5.   Use variation of pop_us_fmin.m

Chapter Five

Section 5.1

1.   (a). $\mathbf{n} = [2 \ 2 \ -2]$ and $\widehat{\mathbf{n}} = [-1 \ -1 \ 1]$

      (b). $r = 1/2$ and $s = -1/2$

3. $\mathbf{x} = r \begin{bmatrix} -2/3 \\ 1 \\ 0 \end{bmatrix} + s \begin{bmatrix} -1/3 \\ 0 \\ 1 \end{bmatrix} + \begin{bmatrix} 2 \\ 0 \\ 0 \end{bmatrix}$

5. $\mathbf{x} = r \begin{bmatrix} -1/2 \\ 1 \\ 0 \end{bmatrix} + s \begin{bmatrix} 0 \\ 0 \\ 1 \end{bmatrix} + \begin{bmatrix} 5 \\ 0 \\ 0 \end{bmatrix}$

7. There is no particular solution because

$$\begin{bmatrix} 1 & 0 \\ -3 & 1 \end{bmatrix} \begin{bmatrix} 3 & 2 & 1 & 2 \\ 9 & 6 & 3 & 5 \end{bmatrix} = \begin{bmatrix} 3 & 2 & 1 & 2 \\ 0 & 0 & 0 & -1 \end{bmatrix}$$

implies $0x_1 + 0x_2 + 0x_3 = -1$!

9. The free variable is $x_3 = r$ and

$$\mathbf{x} = r \begin{bmatrix} -3 \\ 0 \\ 1 \end{bmatrix} + \begin{bmatrix} 3 \\ -1/2 \\ 0 \end{bmatrix}.$$

11. The free variable is $x_3 = r$ and

$$\mathbf{x} = r \begin{bmatrix} 1 \\ -1 \\ 1 \end{bmatrix} + \begin{bmatrix} -4 \\ 3 \\ 0 \end{bmatrix}.$$

Chapter Five
Section 5.2

1. (a). $\mathbf{x}_p = \begin{bmatrix} 4 \\ 0 \\ 0 \\ 0 \end{bmatrix}$

(b). $\mathbf{x}_h = r \begin{bmatrix} -2/3 \\ 1 \\ 0 \\ 0 \end{bmatrix} + s \begin{bmatrix} -1/3 \\ 0 \\ 1 \\ 0 \end{bmatrix} + t \begin{bmatrix} 0 \\ 0 \\ 0 \\ 1 \end{bmatrix}$

(c). $\mathbf{x} = \begin{bmatrix} 4 \\ 0 \\ 0 \\ 0 \end{bmatrix} + r \begin{bmatrix} -2/3 \\ 1 \\ 0 \\ 0 \end{bmatrix} + s \begin{bmatrix} -1/3 \\ 0 \\ 1 \\ 0 \end{bmatrix} + t \begin{bmatrix} 0 \\ 0 \\ 0 \\ 1 \end{bmatrix}$

3. (a). $\begin{bmatrix} 3 & 2 & 1 & 1 & 6 \\ 0 & 0 & 0 & 1 & 1 \end{bmatrix}$

(b). $\mathbf{x}_p = \begin{bmatrix} 5/3 \\ 0 \\ 0 \\ 1 \end{bmatrix}$

(c). $\mathbf{x}_h = r \begin{bmatrix} -2/3 \\ 1 \\ 0 \\ 0 \end{bmatrix} + s \begin{bmatrix} -1/3 \\ 0 \\ 1 \\ 0 \end{bmatrix}$

(d). $\mathbf{x} = \begin{bmatrix} 5/3 \\ 0 \\ 0 \\ 1 \end{bmatrix} + r \begin{bmatrix} -2/3 \\ 1 \\ 0 \\ 0 \end{bmatrix} + s \begin{bmatrix} -1/3 \\ 0 \\ 1 \\ 0 \end{bmatrix}$

5. (a). $\begin{bmatrix} 2 & 1 & 1 & 3 & 4 \\ 0 & 3 & 1 & 8 & 6 \\ 0 & 0 & 0 & 0 & 0 \end{bmatrix}$

(b).
$$\begin{bmatrix} 1 & 0 & 2/6 & 1/6 & 6/6 \\ 0 & 1 & 1/3 & 8/3 & 6/3 \\ 0 & 0 & 0 & 0 & 0 \end{bmatrix}$$

(c). $\mathbf{x} = \begin{bmatrix} 1 \\ 2 \\ 0 \\ 0 \end{bmatrix} + r \begin{bmatrix} -2/6 \\ -1/3 \\ 1 \\ 0 \end{bmatrix} + s \begin{bmatrix} -1/6 \\ -8/3 \\ 0 \\ 1 \end{bmatrix}$

7.
$$\begin{bmatrix} 1 & 0 & 0 & -1 & 0 & 0 & -1 \\ 0 & 1 & 0 & 1 & 0 & 1/2 & 1 \\ 0 & 0 & 1 & 1 & 0 & 0 & -1 \\ 0 & 0 & 0 & 0 & 1 & 1/2 & 2 \end{bmatrix}$$

Chapter Five
Section 5.3

1. $EA^{-1}E^T = [25/6]$, $\mathbf{y} = 30/29$ and $\mathbf{x} = [89/29 \ \ -72/29 \ \ 106/29]^T$

3. $N = \begin{bmatrix} -1/2 & -1/2 \\ 1 & 0 \\ 0 & 1 \end{bmatrix}$, $\mathbf{z} = [-2.4828 \ \ 3.6552]^T$ and

      $\mathbf{x} = [3.0690 \ \ -2.4828 \ \ 3.6552]^T$

Chapter Six
Section 6.1

1.   Left side $= (95/3)e^{3t} - 2/3$
      Right side $= 3[(95/9)e^{3t} - 2t/3 - 5/9] + 2t + 1$
      $x(0) = 95/9 - 5/9$

3.   (a). $x_h = Ce^{5t}$
      (b). $x_p = (-3/5)t - 13/25$
      (c). $x = (138/25)e^{5t} + (-3/5)t - 13/25$

5.   (a). $x_h = Ce^{2t}$
      (b). $x_p = 10e^{3t}$
      (c). $x = -5e^{2t} + 10e^{3t}$

7.   (a). $x_h = Ce^{5t}$
      (b). $x_p = (-25/13)\cos(t) + (5/13)\sin(t)$
      (c). $x = (38/13)e^{5t} + (-25/13)\cos(t) + (5/13)\sin(t)$

9. $x_p = e^{5t} \int e^{-5\tau}(3\tau + 2)d\tau = t(-3/5) - 13/25$

Chapter Six
Section 6.2

1.   (a).   Compute the second derivative and show $mx_2'' + kx_2 = 0$.
      (b). $x(0) = (1/2)0 + 2 = 2$
               $x'(t) = (1/2)\cos(2t)2 + 2(-\sin(2t))2$
               $x'(0) = (1/2)\cos(0)2 + 2(-\sin(0))2 = 1$

3.   Compute x' and x'' and show $x'' + 4x' + 4x = 0$.
               $x(0) = 1 + 0$
               $x'(t) = e^{-2t}(-2) + 5(1e^{-2t} + te^{-2t}(-2))$

$$x'(0) = -2 + 5(1+0) = 3$$

5.   (a). $x_h = c_1 e^{-3t} + c_2 e^{2t}$
        (b). $x = (-3/5)e^{-3t} + (-2/5)e^{2t}$
7.   (a). $x_h = c_1 e^{4t} + c_2 t e^{4t}$
        (b). $x = -2e^{4t} + 11t e^{4t}$
9.   (a).   Hint: Subtract the two equations.
        (b).   Hint: Show z must be equal to zero.

Chapter Six
Section 6.3

1.   (a). $x_1' = e^{\alpha t}\alpha\cos(\beta t) + e^{\alpha t}(-\sin(\beta t))\beta$
                $x_2' = e^{\alpha t}\alpha\sin(\beta t) + e^{\alpha t}(\cos(\beta t))\beta$
        (b). $W(0) = \begin{bmatrix} 1 & 0 \\ \alpha & \beta \end{bmatrix}$
        (c). $\det(W(0)) = \beta \neq 0$
3.   (a). $x_h = c_1\cos(3t) + c_2\sin(3t)$
        (b). $x = 2\cos(3t) + (1/3)\sin(3t)$
5.   (a). $x_h = c_1 e^{-t}\cos(\sqrt{2}t) + c_2 e^{-t}\sin(\sqrt{2}t)$
        (b). $x = 1e^{-t}\cos(\sqrt{2}t) + (3/\sqrt{2})e^{-t}\sin(\sqrt{2}t)$
7. $k = 1$: $(\sqrt{3}+3/2)e^{(-2+\sqrt{3})t} + (\sqrt{3}/2)(-2+\sqrt{3})e^{(-2-\sqrt{3})t}$
        $k = 4$: $3e^{-2t} + 6te^{-2t}$
        $k = 10$: $\sqrt{6}e^{-2t}\sin(\sqrt{6}t) + 3e^{-2t}\cos(\sqrt{6}t)$

Chapter Six
Section 6.4

1. $x = c_1 e^{-t} + c_2 e^{-4t} + (3/4)t + 1/16$
3. $x = c_1 e^{-t} + c_2 e^{-4t} + (1/6)e^{2t}$
5. $x = c_1\cos(2t) + c_2\sin(2t) + 1\cos(t) + (1/3)\sin(t)$
7. $x = 6e^{-t} + (-13/4)e^{-2t} + (3/2)t - 7/4$
9. $x = -e^{-t} + (29/48)e^{-4t} + (1/3)e^{2t} + (3/4)t + 1/16$
11.   (a). $x_p = \frac{1}{1-\omega^2}\cos(\omega t)$
        (b). $\lim_{w\to 1}((\frac{-1}{1-\omega^2})\cos(t) + \frac{1}{1-\omega^2}\cos(\omega t)) = (1/2)t\sin(t)$
        (c).   Start with $x_p = t(a\cos(t) + b\sin(t))$.

Chapter Six
Section 6.5

3. $x_p = e^{3t}v_1 + e^{-3t}v_2 = (-2/9)t - 1/3$ where
           $v_1 = (1/6)[(-2/3)e^{-3t}t + (-11/9)e^{-3t}]$ and
           $v_2 = (-1/6)[(2/3)e^{3t}t + (7/9)e^{3t}]$

Chapter Seven
Section 7.1

1.   (a). $\begin{bmatrix} x \\ y \end{bmatrix}' = \begin{bmatrix} 0 & 1 \\ -1 & -8 \end{bmatrix}\begin{bmatrix} x \\ y \end{bmatrix}$

(b). $\begin{bmatrix} x \\ y \end{bmatrix} = c_1 \begin{bmatrix} 1 \\ -4 + \sqrt{15} \end{bmatrix} e^{(-4+\sqrt{15})t} +$

$\qquad\qquad c_2 \begin{bmatrix} 1 \\ -4 - \sqrt{5} \end{bmatrix} e^{(-4-\sqrt{15})t}$

(c). $c_1 = (-4 - \sqrt{15})/(-\sqrt{15})$ and $c_2 = (-4 + \sqrt{15})/(-\sqrt{15})$

3.   (a).   Replace 100 (L) by 50 (L).

(b). $\begin{bmatrix} x \\ y \end{bmatrix} = c_1 \begin{bmatrix} 1 \\ 2 \end{bmatrix} e^{(-2/25)t} + c_2 \begin{bmatrix} 1 \\ -2 \end{bmatrix} e^{(-6/25)t}$

(c). $c_1 = 7/4$ and $c_2 = 1/4$

5. $\begin{bmatrix} x \\ y \end{bmatrix} = (-1) \begin{bmatrix} 1 \\ -1 \end{bmatrix} e^{4t} + 3 \begin{bmatrix} 1 \\ -2 \end{bmatrix} e^{2t}$

Chapter Seven
Section 7.2

1.   (a).   Show $\begin{bmatrix} 0 & 1 \\ -1 & -4 \end{bmatrix} \begin{bmatrix} 1 \\ -2 - \sqrt{3} \end{bmatrix} = (-2 - \sqrt{3}) \begin{bmatrix} 1 \\ -2 - \sqrt{3} \end{bmatrix}$

(b).   Let $r = -2 - \sqrt{3}$ and

$\begin{bmatrix} 0-r & 1 \\ -1 & -4-r \end{bmatrix} \begin{bmatrix} u_1 \\ u_2 \end{bmatrix} = \begin{bmatrix} 0 \\ 0 \end{bmatrix}$ implies

$(2 + \sqrt{3})u_1 + 1u_2 = 0$ and $-u_1 + (-2 + \sqrt{3})u_2 = 0$.

Let $u_1 = 1$ and then compute $u_2 = -(2 + \sqrt{3})$.

3.   (a).   Show $\begin{bmatrix} 2 & -3 \\ 1 & -2 \end{bmatrix} \begin{bmatrix} 1 \\ 1/3 \end{bmatrix} = (1) \begin{bmatrix} 1 \\ 1/3 \end{bmatrix}$

(b). $\begin{bmatrix} 2-(1) & -3 \\ 1 & -2-(1) \end{bmatrix} \begin{bmatrix} u_1 \\ u_2 \end{bmatrix} = \begin{bmatrix} 0 \\ 0 \end{bmatrix}$ implies

$(1)u_1 + (-3)u_2 = 0$ and $(1)u_1 + (-3)u_2 = 0$.

Let $u_1 = 1$ and then compute $u_2 = -1/3$.

5. $\det(\begin{bmatrix} -4/25 - r & 1/25 \\ 4/25 & -4/25 - r \end{bmatrix}) = (-4/25 - r)^2 - 4/25^2 = 0$ implies

$r = -6/25$ and $-2/25$.

7. $\det(\begin{bmatrix} 4-r & 1 \\ -2 & 1-r \end{bmatrix}) = (4 - r)(1 - r) + 2 = 0$ implies

$r = 2$ and 3.

Chapter Seven
Section 7.3

1.   (a). $r = -5/2$ and $-5/2$
      (b). $r = 2 + i$ and $2 - i$
      (c). $r = 0$ and 2.

3.   (a). $r_1 = -1$ and $\mathbf{u}_1 = \begin{bmatrix} 1 \\ -1 \end{bmatrix}$; $r_1 = 5$ and $\mathbf{u}_1 = \begin{bmatrix} 1 \\ 1 \end{bmatrix}$

(b). $X(t) = \begin{bmatrix} e^{-t} & e^{5t} \\ -e^{-t} & e^{5t} \end{bmatrix}$ and $X(t)^{-1} = \begin{bmatrix} e^{t}/2 & -e^{t}/2 \\ e^{-5t}/2 & e^{-5t}/2 \end{bmatrix}$

(c). $X(t)v(t) = \begin{bmatrix} e^{-t} & e^{5t} \\ -e^{-t} & e^{5t} \end{bmatrix} \begin{bmatrix} e^{t}(-t/2 + 0) \\ e^{-5t}(-t/2 - 4/10) \end{bmatrix}$

$$= \begin{bmatrix} -t - 2/5 \\ -2/5 \end{bmatrix}$$

(d). $\quad X(t)v(t) = \begin{bmatrix} e^{-t} & e^{5t} \\ -e^{-t} & e^{5t} \end{bmatrix} \begin{bmatrix} e^{3t}/3 \\ 0 \end{bmatrix} = \begin{bmatrix} e^{2t}/3 \\ -e^{2t}/3 \end{bmatrix}$

5. $\quad (37/4) \begin{bmatrix} 1 \\ -1 \end{bmatrix} e^{2t} + (-28/9) \begin{bmatrix} 1 \\ -2 \end{bmatrix} e^{3t} + \begin{bmatrix} -185/36 \\ 109/36 \end{bmatrix} + \begin{bmatrix} -7/6 \\ -1/6 \end{bmatrix} t$

7. $\quad (-7/2) \begin{bmatrix} 1 \\ -2 \end{bmatrix} e^{2t} + (67/13) \begin{bmatrix} 1 \\ -1 \end{bmatrix} e^{3t} + \begin{bmatrix} -17/26 \\ -11/13 \end{bmatrix} \cos(2t) +$

$$\begin{bmatrix} 7/26 \\ -10/13 \end{bmatrix} \sin(2t)$$

Chapter Eight
Section 8.1

1. (c). Replace letc = zeros(50,40) with
shade = 100 % for example
letc = shade*ones(50,40)

3. Vary red, green and blue in color_imagncsu.m

Chapter Eight
Section 8.2

1. In the code file pollenhist.m replace pollen.jpg by mars3.jpg.

3. $\quad unit8(u - 40) = \begin{bmatrix} 235 & 201 & 170 \\ 155 & 17 & 0 \end{bmatrix}$

5. $\quad newu = \begin{bmatrix} 255 & 245 & 222 \\ 211 & 115 & 102 \end{bmatrix}$

9. Using $p = 1.5$

$$newu = \begin{bmatrix} 255 & 234 & 191 \\ 171 & 27 & 14 \end{bmatrix}$$

Chapter Eight
Section 8.3

Have fun comparing the new images.

Chapter Nine
Section 9.1

1. $\quad sL(u) - 210 = c80(1/s) - cL(u)$
$L(u) = 210/(s + c) + c80(1/(cs) - 1/(c(s + c)))$
$\quad\quad = L(210e^{-ct} + 80 - 80e^{-ct})$
$u(t) = 130e^{-ct} + 80$

3. (a). Let $z = e^{-i2\pi/5}$ and

$$\mathbf{F}_5 = \begin{bmatrix} 1 & 1 & 1 & 1 & 1 \\ 1 & z & z^2 & z^3 & z^4 \\ 1 & z^2 & z^4 & z & z^3 \\ 1 & z^3 & z & z^4 & z^2 \\ 1 & z^4 & z^3 & z^2 & z \end{bmatrix}$$

(b). $\mathbb{F}_5 \mathbf{f} = \begin{bmatrix} 1 & 1 & 1 & 1 & 1 \\ 1 & z & z^2 & z^3 & z^4 \\ 1 & z^2 & z^4 & z & z^3 \\ 1 & z^3 & z & z^4 & z^2 \\ 1 & z^4 & z^3 & z^2 & z \end{bmatrix} \begin{bmatrix} 2 \\ 8 \\ 9 \\ 1 \\ 3 \end{bmatrix} = \begin{bmatrix} 23.0000 \\ -2.6910 + 9.4576i \\ -3.8090 - 4.6695i \\ -3.8090 + 4.6695i \\ -2.6910 - 9.4576i \end{bmatrix}$

Chapter Nine
Section 9.2

1. (a). Let $z = e^{-i2\pi/3}$ and $f = 1$.

$$\mathbb{F}_3 \sin(2\pi 1 \mathbf{t}) = \begin{bmatrix} 1 & 1 & 1 \\ 1 & z & z^2 \\ 1 & z^2 & z \end{bmatrix} \begin{bmatrix} \sin(2\pi 1(0)) \\ \sin(2\pi 1(1/3)) \\ \sin(2\pi 1(2/3)) \end{bmatrix} = \begin{bmatrix} 0.0 \\ 1.5i \\ -1.5i \end{bmatrix}$$

3. (a). Let $z = e^{-i2\pi/5}$ and

$$\mathbb{F}_5^{-1} = \frac{1}{5} \begin{bmatrix} 1 & 1 & 1 & 1 & 1 \\ 1 & z^4 & z^3 & z^2 & z \\ 1 & z^3 & z & z^4 & z^2 \\ 1 & z^2 & z^4 & z & z^3 \\ 1 & z & z^2 & z^3 & z^4 \end{bmatrix}$$

Chapter Nine
Section 9.3

Enjoy the image filters.

Bibliography

[1]   Robert L. Borrelli, Courtney S. Coleman, *Differential Equations: a Modeling Perspective,* Wiley, 2004.

[2]   Gonzales and Woods, *Digital Image Processing*, 2nd Edition, Prentice Hall, 2002, http://www.imageprocessingplace.com.

[3]   Center for Research and Scientific Computing, North Carolina State University, http://www.ncsu.edu/crsc/ilfum.htm.

[4]   The MathWorks Inc., http://www.mathworks.com.

[5]   Carl D. Meyer, *Matrix Analysis and Applied Linear Algebra,* SIAM, 2000.

[6]   Cleve Moler, *Numerical Computing with MATLAB*, SIAM, 2002, http://www.mathworks.com/moler.

[7]   R. Kent Nagle, Edward B. Saff and Arthur David Snider, *Fundamentals of Differential Equations and Boundary Value Problems,* Addison Wesley, 2004.

[8]   Gilbert Strang, *Introduction to Applied Mathematics*, Wellesley-Cambridge Press, 1986.

[9]   United Nations, http://www.un.org/esa/population/unpop.htm.

[10] United States Census Bureau, http://www.census.gov/main/www/ cen2000.html.

[11] United States NASA, http://www.nasa.gov/vision/universe/ solarsystem/mer_main.html.

[12] Robert E. White, *Computational Mathematics: Models, Methods and Analysis with MATLAB and MPI*, Chapman and Hall/CRC, 2004.

Index